Springer Proceedings in Mathematics & Statistics

Volume 334

Springer Proceedings in Mathematics & Statistics

This book series features volumes composed of selected contributions from workshops and conferences in all areas of current research in mathematics and statistics, including operation research and optimization. In addition to an overall evaluation of the interest, scientific quality, and timeliness of each proposal at the hands of the publisher, individual contributions are all refereed to the high quality standards of leading journals in the field. Thus, this series provides the research community with well-edited, authoritative reports on developments in the most exciting areas of mathematical and statistical research today.

More information about this series at http://www.springer.com/series/10533

George Jaiani · David Natroshvili
Editors

Applications of Mathematics and Informatics in Natural Sciences and Engineering

AMINSE 2019, Tbilisi, Georgia, September 23–26

Springer

Editors
George Jaiani
I.Vekua Institute of Applied Mathematics &
Faculty of Exact and Natural Sciences
Ivane Javakhishvili Tbilisi State University
Tbilisi, Georgia

David Natroshvili
Georgian Technical University
Tbilisi, Georgia

ISSN 2194-1009 ISSN 2194-1017 (electronic)
Springer Proceedings in Mathematics & Statistics
ISBN 978-3-030-56358-5 ISBN 978-3-030-56356-1 (eBook)
https://doi.org/10.1007/978-3-030-56356-1

Mathematics Subject Classification: 35-XX, 74-XX, 76-XX, 17-XX, 54-XX, 03-XX

This Springer imprint is published by the registered company Springer Nature Switzerland AG
The registered company address is: Gewerbestrasse 11, 6330 Cham, Switzerland

Editorial Preface

The Fourth International Conference on *Applications of Mathematics and Informatics in Natural Sciences and Engineering* (AMINSE 2019) took place in the Ilia Vekua Institute of Applied Mathematics of Ivane Javakhishvili Tbilisi State University (Tbilisi, Georgia) on September 23–26, 2019.

The aim of AMINSE 2019 was to bring together scientists to discuss their research in all the aspects of Mathematics, Informatics, and their Applications in Natural Sciences and Engineering. According to this premise, all the lecturers were invited personally for their active interest in their relative field. There were 80 participants from 10 countries. The main topics of the conference were: Partial Differential Equations, Operator Theory, Numerical Analysis, Mechanics of Deformable Solids, Fluid Mechanics and Computer Science. The program included wo opening lectures, "Recent Developments on Numerical Solutions for Hyperbolic Systems of Conservation Laws" presented by Rolf Jeltsch (Switzerland) and "Progress in Mathematical and Numerical Modelling of Piezoelectric Smart Structures" by Ayech Benjeddou (France). At the conference, 15 plenary lectures, 15 talks, were presented. The Satellite TICMI (Tbilisi International Center of Mathematics and Informatis) Advanced Courses on "Mathematical Models of Piezoelectric Solids and Related Problems" was held parallel to the conference.

This volume includes 14 peer-reviewed papers presented at the conference. The contributions are related to several important directions of Applied Mathematics, Integral Equations, Variational Methods, Continuum Mechanics, Numerical Analysis, Mathematical Modeling in Social Sciences, Financial Mathematics, Theory of Probability and Statistics, Li algebra, and Mathematical Logic.

We are grateful to the organizations and individuals who helped orchestrate the conference. The conference was organized by VIAM, Faculty of Exact and Natural Sciences of TSU, Georgian Mechanical Union, Georgian National Committee of Theoretical and Applied Mechanics, and Tbilisi International Center of Mathematics and Informatics. The conference was sponsored by the Shota Rustaveli National Science Foundation. The scientific organization was entrusted to the international committee consisting of George Jaiani (chair, Georgia), Gia

Avalishvili (Georgia), Nikoloz Avazashvili (Georgia), Ayech Benjeddou (France), Lucian Beznea (Romania), Ramaz Botchorishvili (Georgia), Natalia Chinchaladze (Georgia), Roland Duduchava (Georgia), Maribel Fernandez (UK), ALice Fialowski (Hungary), Temur Jangveladze (Georgia), Rolf Jeltsch (Switzerland), Alexsander Kharazishvili (Georgia), Omar Kikvidze (Georgia), Gela Kipiani (Georgia), Vakhtang Kokilashvili (Georgia), Vakhtang Kvaratskhelia (Georgia), Teimuraz Kutsia (Austria), Mircea Marin (Romania), Bernadette Miara (France), Wolfgang H. Müller (Germany), Elizbar Nadaraya (Georgia), David Natroshvili (Georgia), Jemal Rogava (Georgia), Tamaz Vashakmadze (Georgia). The local arrangements of the conference were in the hands of the committee consisting of Natalia Chinchaladze (Chair) Mariam Beriashvili (Scientific Secretary), Besik Dundua, Bakur Gulua, Mikheil Rukhaia, and Manana Tevdoradze. We would like to thank all of them for their hard and efficient work.

In the present book, the contributions of the participants have been ordered alphabetically by the names of the presenting authors. The responsibility for the contents of the papers lies solely with each author.

The editors wish to express their thanks to Natalia Chinchaladze for spending time, patience, and for her valuable help in editing and layouting the book.

The editors are indebted to Springer-Verlag for their courteous and effective production of these proceedings.

Tbilisi, Georgia
June 2020

George Jaiani
David Natroshvili

Contents

On Variational Methods of Investigation of Mathematical Problems for Thermoelastic Piezoelectric Solids

Gia Avalishvili and Mariam Avalishvili

Abstract In this paper we present the results of investigation of the boundary and initial-boundary value problems corresponding to mathematical models of thermoelastic piezoelectric solids with regard to magnetic field. We consider three-dimensional static and dynamic models of general inhomogeneous anisotropic thermoelastic piezoelectric solids with mixed boundary conditions, when on certain parts of the boundary density of surface force, and normal components of the electric displacement, magnetic induction, and heat flux are given, and on the remaining parts of the boundary mechanical displacement, temperature, electric and magnetic potentials vanish. We obtain variational formulations of the boundary and initial-boundary value problems in suitable function spaces and present the existence, uniqueness and continuous dependence results.

Keywords Thermoelastic piezoelectric solids · Boundary and initial-boundary value problems · Existence and uniqueness of solution · Variational methods Sobolev spaces · Vector-valued distributions

1 Introduction

The important integral parts of modern engineering constructions are smart structures, which involve actuators and sensors, and microprocessors that analyze the responses from the sensors and use actuators to alter construction response. After discovery of piezoelectric effect by the Curie brothers [14], Jacques and Pierre, the

---

G. Avalishvili (✉)
Faculty of Exact and Natural Sciences, I. Javakhishvili Tbilisi State University, 3 I. Tchavtchavadze Ave., 0179 Tbilisi, Georgia
e-mail: gavalish@yahoo.com

M. Avalishvili
School of Science and Technology, University of Georgia,
77a M. Kostava Str., 0175 Tbilisi, Georgia
e-mail: m.avalishvili@ug.edu.ge

G. Jaiani and D. Natroshvili (eds.), *Applications of Mathematics and Informatics in Natural Sciences and Engineering*, Springer Proceedings in Mathematics & Statistics 334, https://doi.org/10.1007/978-3-030-56356-1_1

applications of piezoelectric materials was gradually increasing, and currently piezoelectrics are the most popular smart materials. Therefore, it is important to construct and investigate accurate mathematical models that can predict the coupled response of materials that exhibit not only thermo-elastic, but also electro-magnetic properties.

Mathematical phenomenological theory relating the phenomena of piezoelectricity and pyroelectricity to crystal symmetry first was constructed by Voigt [34]. He determined which of crystal classes can be piezoelectric and rigorously defined the macroscopic relationships among parameters in crystal solids. Later on, Tiersten [31] obtained variational principle for the equations of linear piezoelectricity (with the quasielectrostatic field approximation to Maxwells equations) and studied problems of vibration of piezoelectric plates. The three-dimensional equations of the linear thermopiezoelectricity were considered by Mindlin [23] and two-dimensional equations for plates were derived on the basis of integral energy equation and approximation by polynomials with respect to the variable of plate thickness. Nowacki [26, 27] obtained uniqueness and reciprocity theorems for thermo-piezoelectricity. Dhaliwal and Wang [16] proved a uniqueness theorem for linear three-dimensional thermo-piezoelectricity without restrictions on the coupling constant between temperature and electric field, and positive definiteness assumption imposed on the elasticity tensor, which were used in [27]. Li [21] considered the coupling effects between elastic, electric, magnetic and thermal fields, and generalized the uniqueness result obtained in [16] and reciprocity theorem of Nowacki [26], which further were strengthened by Aouadi [4] and the results were proved without positive definiteness assumption on the thermal conductivity tensor, which was used in [21]. A variational principle for the three-dimensional equations of piezoelectromagnetism and the appropriate boundary conditions for elastic dielectric crystals surrounded by a vacuum or perfect conductor are obtained by Lee [20]. On the basis of the principle of virtual work and Friedrich's transformation variational principles for the discontinuous thermopiezoelectric fields were obtained by Altay and Dökmeci [3]. By introducing the semi-inverse method He [17] obtained a generalized variational principle for the linear magneto-electro-elasticity. On the basis of the Hellinger-Reissner mixed variational principle for three-dimensional model of elastic solids, the modified Hellinger-Reissner mixed variational principle for magnetoelectroelastic solids was obtained by Qing et al. [30]. The analogue of the Reissners mixed variational theorem for thermopiezoelectric multilayered composites was obtained by Benjeddou and Andrianarison [12]. The existence, uniqueness and continuous dependence on given data of a solution of an initial-boundary value problem with the mixed boundary conditions for the mechanical displacement, mechanical stress, electric potential and electric displacement corresponding to the three-dimensional model of an anisotropic inhomogeneous piezoelectric material with quasi-static equations for the electric field were proved in Sobolev spaces by Akamatsu and Nakamura [1]. The well-posedness results in specific function spaces for the three-dimensional model of thermo-piezoelectricity with inhomogeneous material parameters in the cases of homogeneous pure Dirichlet or Neumann type boundary conditions given on the entire boundary were obtained by Mulholland et al. [24]. The well-posedness of the initial value problem, when the electric and magnetic fields, and the mechanical displacement are vanished at the initial time, for the dynamic equations of magneto-electro-elasticity, wherein the Maxwells equations are involved, has been investigated by Yakhno [36]. Applying the potential method and theory of pseudodifferential equations, Natroshvili [25] studied static and pseudo-oscillation problems with basic, mixed and crack-type boundary conditions for homogeneous anisotropic thermo-electro-magneto-elastic solids. The static and dynamic three-dimensional problems for inhomogeneous anisotropic thermo-electro-magneto-elastic solids with general mixed boundary conditions were investigated by Avalishvili et al. [8, 9]. The hierarchies of static and dynamic two-dimensional models for plates with variable thickness and dynamic one-dimensional models for bars with variable cross-section made of thermo-electro-magneto-elastic material were constructed and investigated by Avalishvili and Avalishvili [5–7]. Hierarchical two-dimensional models for cusped prismatic shells were studied by Jaiani [18], and one-dimensional models for cusped bars consisting of piezoelectric material were investigated by Jaiani [19], Chinchaladze [13]. Mathematical models of elastic solids that demonstrate coupling behavior between various physical, in particular, electric, magnetic and thermal, fields were investigated and methods of solutions of the corresponding problems were developed by many researchers (see [2, 10, 11, 28, 29, 32, 33] and the references given therein).

In the present paper, we study the well-posedness of the linear dynamic and static three-dimensional models for piezoelectric thermoelastic body made of anisotropic inhomogeneous material with mixed boundary conditions applying variational approach. We present new existence, uniqueness, and continuous dependence results in suitable Sobolev spaces and the classical spaces of smooth functions.

In Sect. 2, we consider dynamic three-dimensional model for inhomogeneous anisotropic piezoelectric thermoelastic body and the differential formulation of the corresponding initial-boundary value problem, with general mixed boundary conditions, where, on certain parts of the boundary, surface force and components of the electric displacement, magnetic induction, and heat flux along the outward normal vector are given, and, on the remaining parts, the mechanical displacement, electric and magnetic potentials, and temperature vanish. We obtain integral relations that are equivalent to the original differential equations together with the boundary conditions in the space of twice continuously differentiable functions and, on the basis of them, we give the variational formulation of the three-dimensional initial-boundary problem in suitable spaces of vector-valued distributions with respect to the time variable with values in Sobolev spaces. We formulate theorem regarding the existence and uniqueness, and continuous dependence of a solution on given data in suitable function spaces, and energy equality, when the parameters characterizing thermo-elastic and piezo-magnetic properties are Lipschitz continuous or essentially bounded, and the given functions on the boundary of spatial domain and at the initial time satisfy corresponding compatibility conditions.

In Sect. 3, we study static three-dimensional model for inhomogeneous anisotropic piezoelectric thermoelastic body and from the differential formulation of the corresponding boundary value problem, with general mixed boundary conditions, on the basis of suitable integral relations, we obtain variational formulation that is equivalent to the original differential equations together with the boundary conditions in the space of twice continuously differentiable functions. We formulate results regarding the existence and uniqueness, and continuous dependence of a solution on given data in various Sobolev spaces, and applying them we present result regarding the existence and uniqueness of the classical twice continuously differentiable solution.

2 Dynamic Three-Dimensional Problem

In this paper, for each real $s \geq 0$, $0 \leq \check{s} \leq 1$, we denote by $H^s(D)$ and $H^{\check{s}}(\check{\Gamma})$ the Sobolev spaces of real-valued functions based on $H^0(D) = L^2(D)$ and $H^0(\check{\Gamma}) = L^2(\check{\Gamma})$, respectively, where $D \subset \mathbb{R}^n$, $n \in \mathbb{N}$, is a bounded Lipschitz domain and $\check{\Gamma}$ is an element of a Lipschitz dissection of the boundary ∂D [22]. We denote the corresponding spaces of vector-valued functions by $\mathbf{H}^s(D) = [H^s(D)]^3$, $s \geq 0$, $\mathbf{H}^{\check{s}}(\check{\Gamma}) = [H^{\check{s}}(\check{\Gamma})]^3$, $0 \leq \check{s} \leq 1$, $\mathbf{L}^{s_1}(\check{\Gamma}) = [L^{s_1}(\check{\Gamma})]^3$, $s_1 \geq 1$ and by $\mathrm{tr}_{\check{\Gamma}} : H^1(D) \to H^{1/2}(\check{\Gamma})$, $\mathbf{tr}_{\check{\Gamma}} : \mathbf{H}^1(D) \to \mathbf{H}^{1/2}(\check{\Gamma})$ the trace operators. For any measurable set $D \subset \mathbb{R}^n$, $n \in \mathbb{N}$, $(.,.)_{\mathbf{L}^2(D)}$ and $(.,.)_{L^2(D)}$ are the classical scalar products in $\mathbf{L}^2(D)$ and $L^2(D)$, respectively. We denote by $C^{r,1}(\overline{D})$, $r \in \mathbb{N} \cup \{0\}$, the space of function on $\overline{D}$ with Lipschitz-continuous derivatives up to the order r, where $D \subset \mathbb{R}^n$, $n \in \mathbb{N}$, is a bounded Lipschitz domain. $\mathfrak{D}(D)$ denotes the set of infinitely differentiable functions with compact support in D. Along with Lipschitz domains we use the notion of $C^{r,l}$ domain [22], for $r \in \mathbb{N} \cup \{0\}$, $0 \leq l \leq 1$, where the boundary of the domain is locally defined by functions whose derivatives up to the r-th order are Hölder-continuous with exponent l. Note that a Lipschitz domain is a $C^{0,1}$ domain. For bounded $C^{\overline{s},1}$, $\overline{s} \geq 0$, domain $D \subset \mathbb{R}^n$, $n \in \mathbb{N}$, we use the Sobolev space $H^{\check{s}}(\check{\Gamma})$, $0 \leq \check{s} \leq \overline{s} + 1$ [22], of real-valued functions based on $H^0(\check{\Gamma}) = L^2(\check{\Gamma})$, where $\check{\Gamma}$ is an element of a Lipschitz dissection of the boundary ∂D. We denote the corresponding space of vector-valued functions by $\mathbf{H}^{\check{s}}(\check{\Gamma}) = [H^{\check{s}}(\check{\Gamma})]^3$, $0 \leq \check{s} \leq \overline{s} + 1$. For a Banach space X, we denote by $C([0,T]; X)$ the space of continuous vector-functions on $[0,T]$ with values in X. $L^{s_1}(0,T; X)$, $1 \leq s_1 \leq \infty$, is the space of such measurable vector-functions $g : (0,T) \to X$ so that $\|g\|_X \in L^{s_1}(0,T)$ and the generalized derivative of g is denoted by $g' = dg/dt \in \mathfrak{D}'(0,T; X)$ [15]. If $g \in L^1(0,T; X)$ and X is a space of functions of variable $x \in D \subset \mathbb{R}^n$, $n \in \mathbb{N}$, then we identify g with a function $g(x,t)$, and $g(t)$ denotes the function $g(t) : x \to g(x,t)$, for almost all $t \in (0,T)$. We identify the distributional derivative dg/dt with the derivative $\partial g/\partial t$ of g in the space $\mathfrak{D}'(D \times (0,T))$ of distributions on $D \times (0,T)$.

Let us consider a thermoelastic piezoelectric body with initial configuration $\overline{\Omega}$, which consists of a general inhomogeneous anisotropic thermo-electro-magneto-elastic material and it is charachterized by the following consistently spatially dependent parameters:

- the elasticity tensor $c_{ijpq}(x)$, $x \in \Omega$ $(i,j,p,q = 1,2,3)$, which satisfies the following symmetry and positive definiteness conditions:

$$c_{ijpq}(x) = c_{pqij}(x) = c_{jipq}(x), \quad \forall i, j, p, q = 1, 2, 3, \tag{1}$$

$$\sum_{i,j,p,q=1}^{3} c_{ijpq}(x)\xi_{pq}\xi_{ij} \geq \alpha_c \sum_{i,j=1}^{3} (\xi_{ij})^2, \quad \alpha_c = const > 0, \tag{2}$$

for all $\xi_{ij} \in \mathbb{R}$, $\xi_{ij} = \xi_{ji}$, $i, j = 1, 2, 3$;

- the piezoelectric and piezomagnetic coefficients $\varepsilon_{pij}(x)$ and $b_{pij}(x)$, $x \in \Omega$ $(i, j, p = 1, 2, 3)$, which satisfy the following symmetry conditions:

$$\varepsilon_{pij}(x) = \varepsilon_{pji}(x), \quad b_{pij}(x) = b_{pji}(x), \quad i, j, p = 1, 2, 3; \tag{3}$$

- the stress-temperature tensor $\lambda_{ij}(x)$, $x \in \Omega$ $(i, j = 1, 2, 3)$, which satisfy the following symmetry conditions:

$$\lambda_{ij}(x) = \lambda_{ji}(x), \quad i, j = 1, 2, 3; \tag{4}$$

- the mass density $\rho(x)$, $x \in \Omega$;
- the permittivity and permeability tensors $d_{ij}(x)$ and $\zeta_{ij}(x)$, $x \in \Omega$ $(i, j = 1, 2, 3)$, and the coupling coefficients connecting electric and magnetic fields $a_{ij}(x)$, $x \in \Omega$ $(i, j = 1, 2, 3)$, which satisfy the following positive definiteness condition:

$$\sum_{i,j=1}^{3} d_{ij}(x)\xi_j\xi_i + \sum_{i,j=1}^{3} a_{ij}(x)\overline{\xi}_j\xi_i + \sum_{i,j=1}^{3} a_{ij}(x)\xi_j\overline{\xi}_i + \sum_{i,j=1}^{3} \zeta_{ij}(x)\overline{\xi}_j\overline{\xi}_i$$
$$\geq \alpha \sum_{i=1}^{3}((\xi_i)^2 + (\overline{\xi}_i)^2), \quad \alpha = const > 0, \quad \forall \xi_i, \overline{\xi}_i \in \mathbb{R}, \ i = 1, 2, 3; \tag{5}$$

- the piroelectiric and piromagnetic coefficients $\mu_i(x)$ and $m_i(x)$, $x \in \Omega$ $(i = 1, 2, 3)$;
- the thermal conductivity tensor $\eta_{ij}(x)$, $x \in \Omega$ $(i, j = 1, 2, 3)$, which satisfies the following positive definiteness condition:

$$\sum_{i,j=1}^{3} \eta_{ij}(x)\xi_j\xi_i \geq \alpha_\eta \sum_{i=1}^{3}(\xi_i)^2, \quad \alpha_\eta = const > 0, \tag{6}$$

for all $\xi_i \in \mathbb{R}$ $(i = 1, 2, 3)$;

- the thermal capacity $\varkappa(x)$, $x \in \Omega$;
- the temperature $\Theta_0 = const > 0$ of the thermoelastic piezoelectric body in natural state of no deformation and electromagnetic fields, which is considered as a reference temperature.

We consider mixed boundary conditions on the boundary $\Gamma = \partial\Omega$ of the thermoelastic piezoelectric body, such that on certain parts of the boundary the mechanical displacement, electric and magnetic potentials, and temperature vanish, and on the remaining parts the densities of the components of the stress vector, electric displacement and magnetic induction, and heat flux along the unit outward normal vector of the boundary are given. We assume that the body is clamped along a part $\Gamma_0 \subset \Gamma$ of the boundary, the electric potential vanishes along $\Gamma_0^{\varphi} \subset \Gamma$, the magnetic potential vanishes along $\Gamma_0^{\psi} \subset \Gamma$, and the temperature θ vanishes along a part $\Gamma_0^{\theta} \subset \Gamma$ of the boundary. The body is subjected to:

- the applied body force with density $\boldsymbol{f} = (f_i)_{i=1}^3 : \Omega \times (0, T) \to \mathbb{R}^3$;
- the applied surface force, with density $\boldsymbol{g} = (g_i)_{i=1}^3 : \Gamma_1 \times (0, T) \to \mathbb{R}^3$, which is given along the part $\Gamma_1 = \Gamma \backslash \Gamma_0$ of the boundary of Ω, where $\Gamma = \Gamma_0 \cup \Gamma_1$ is a Lipschitz dissection of Γ;
- the electric charges with density $f^{\varphi} : \Omega \times (0, T) \to \mathbb{R}$;
- the component of the electric displacement along the unit outward normal vector of Γ, with density $g^{\varphi} : \Gamma_1^{\varphi} \times (0, T) \to \mathbb{R}$, which is given along the part $\Gamma_1^{\varphi} = \Gamma \backslash \Gamma_0^{\varphi}$ of the boundary Γ, where $\Gamma = \Gamma_0^{\varphi} \cup \Gamma_1^{\varphi}$ is a Lipschitz dissection of Γ;
- the component of the magnetic induction along the unit outward normal vector of Γ, with density $g^{\psi} : \Gamma_1^{\psi} \times (0, T) \to \mathbb{R}$, which is given along the part $\Gamma_1^{\psi} = \Gamma \backslash \Gamma_0^{\psi}$ of the boundary Γ, where $\Gamma = \Gamma_0^{\psi} \cup \Gamma_1^{\psi}$ is a Lipschitz dissection of Γ;
- the heat source with density $f^{\theta} : \Omega \times (0, T) \to \mathbb{R}$;
- the heat flux along the unit outward normal vector of Γ, with density $g^{\theta} : \Gamma_1^{\theta} \times (0, T) \to \mathbb{R}$, which is given along the part $\Gamma_1^{\theta} = \Gamma \backslash \Gamma_0^{\theta}$ of the boundary Γ, where $\Gamma = \Gamma_0^{\theta} \cup \Gamma_1^{\theta}$ is a Lipschitz dissection of Γ.

The dynamic linear three-dimensional model of the stress-strain state of the thermoelastic piezoelectric body $\overline{\Omega}$, with quasi-static equations for electric and magnetic fields, where the rate of the magnetic field is small, i.e. the electric field is curl free, and there is no electric current, i.e. the magnetic field is curl free, is given by the following initial-boundary value problem in differential form [9, 21, 25]:

$$\rho \frac{\partial^2 u_i}{\partial t^2} - \sum_{j=1}^{3} \frac{\partial}{\partial x_j} \left(\sum_{p,q=1}^{3} c_{ijpq} e_{pq}(\boldsymbol{u}) + \sum_{p=1}^{3} \varepsilon_{pij} \frac{\partial \varphi}{\partial x_p} \right.$$

$$\left. + \sum_{p=1}^{3} b_{pij} \frac{\partial \psi}{\partial x_p} - \lambda_{ij} \theta \right) = f_i \quad \text{in } \Omega \times (0, T), i = 1, 2, 3, \tag{7}$$

$$\sum_{i=1}^{3} \frac{\partial}{\partial x_i} \left(\sum_{p,q=1}^{3} \varepsilon_{ipq} e_{pq}(\boldsymbol{u}) - \sum_{j=1}^{3} d_{ij} \frac{\partial \varphi}{\partial x_j} \right.$$

$$\left. - \sum_{j=1}^{3} a_{ij} \frac{\partial \psi}{\partial x_j} + \mu_i \theta \right) = f^{\varphi} \quad \text{in } \Omega \times (0, T), \tag{8}$$

$$\sum_{i=1}^{3}\frac{\partial}{\partial x_i}\left(\sum_{p,q=1}^{3} b_{ipq}e_{pq}(\boldsymbol{u}) - \sum_{j=1}^{3} a_{ij}\frac{\partial \varphi}{\partial x_j} - \sum_{j=1}^{3}\zeta_{ij}\frac{\partial \psi}{\partial x_j} + m_i\theta\right) = 0 \quad \text{in } \Omega \times (0,T), \tag{9}$$

$$\varkappa\frac{\partial \theta}{\partial t} - \sum_{i,j=1}^{3}\frac{\partial}{\partial x_i}\left(\eta_{ij}\frac{\partial \theta}{\partial x_j}\right) + \Theta_0\frac{\partial}{\partial t}\sum_{i,j=1}^{3}\lambda_{ij}e_{ij}(\boldsymbol{u}) - \Theta_0\frac{\partial}{\partial t}\sum_{i=1}^{3}\mu_i\frac{\partial \varphi}{\partial x_i} - \Theta_0\frac{\partial}{\partial t}\sum_{i=1}^{3}m_i\frac{\partial \psi}{\partial x_i} = f^\theta \quad \text{in } \Omega \times (0,T), \tag{10}$$

$$\boldsymbol{u} = 0 \quad \text{on } \Gamma_0 \times (0,T), \qquad \sum_{j=1}^{3}\sigma_{ij}n_j = g_i \quad \text{on } \Gamma_1 \times (0,T),\ i = 1,2,3, \tag{11}$$

$$\varphi = 0 \quad \text{on } \Gamma_0^\varphi \times (0,T), \qquad \sum_{i=1}^{3} D_i n_i = g^\varphi \quad \text{on } \Gamma_1^\varphi \times (0,T), \tag{12}$$

$$\psi = 0 \quad \text{on } \Gamma_0^\psi \times (0,T), \qquad \sum_{i=1}^{3} B_i n_i = g^\psi \quad \text{on } \Gamma_1^\psi \times (0,T), \tag{13}$$

$$\theta = 0 \quad \text{on } \Gamma_0^\theta \times (0,T), \qquad -\sum_{i,j=1}^{3}\eta_{ij}\frac{\partial \theta}{\partial x_j}n_i = g^\theta \quad \text{on } \Gamma_1^\theta \times (0,T), \tag{14}$$

$$\boldsymbol{u}(x,0) = \boldsymbol{u}_0(x), \quad \frac{\partial \boldsymbol{u}}{\partial t}(x,0) = \boldsymbol{u}_1(x), \quad \theta(x,0) = \theta_0(x), \qquad x \in \Omega, \tag{15}$$

where $\boldsymbol{u} = (u_i)_{i=1}^3 : \overline{\Omega} \times [0,T] \to \mathbb{R}^3$ is the mechanical displacement vector-function, $\varphi : \overline{\Omega} \times [0,T] \to \mathbb{R}$ and $\psi : \overline{\Omega} \times [0,T] \to \mathbb{R}$ stand for the electric and magnetic potentials such that the electric and magnetic fields are $\boldsymbol{E} = -(\partial\varphi/\partial x_i)_{i=1}^3$ and $\boldsymbol{H} = -(\partial\psi/\partial x_i)_{i=1}^3$, $\theta : \overline{\Omega} \times [0,T] \to \mathbb{R}$ is the temperature distribution, $\boldsymbol{u}_0 = (u_{0i})_{i=1}^3$ and $\boldsymbol{u}_1 = (u_{1i})_{i=1}^3$ are the initial mechanical displacement and velocity vector-functions, respectively, θ_0 is the initial distribution of temperature; $(\sigma_{ij})_{i,j=1}^3$ is the mechanical stress tensor, which is given by the following linear constitutive equation for a thermo-electro-magneto-elastic solid:

$$\sigma_{ij} = \sum_{p,q=1}^{3} c_{ijpq}e_{pq}(\boldsymbol{u}) + \sum_{p=1}^{3}\varepsilon_{pij}\frac{\partial \varphi}{\partial x_p} + \sum_{p=1}^{3} b_{pij}\frac{\partial \psi}{\partial x_p} - \lambda_{ij}\theta, \quad i,j = 1,2,3, \tag{16}$$

where $e_{ij}(\boldsymbol{v}) = 1/2\left(\partial v_i/\partial x_j + \partial v_j/\partial x_i\right)$, $i,j = 1,2,3$, $\boldsymbol{v} = (v_i)_{i=1}^3$, is the strain tensor; $\boldsymbol{D} = (D_j)_{j=1}^3$ is the electric displacement vector and $\boldsymbol{B} = (B_j)_{j=1}^3$ is the magnetic induction vector, which are given by the following linear constitutive equations:

$$D_i = \sum_{p,q=1}^{3} \varepsilon_{ipq} e_{pq}(\boldsymbol{u}) - \sum_{j=1}^{3} d_{ij} \frac{\partial \varphi}{\partial x_j} - \sum_{j=1}^{3} a_{ij} \frac{\partial \psi}{\partial x_j} + \mu_i \theta, \quad i = 1, 2, 3, \tag{17}$$

$$B_i = \sum_{p,q=1}^{3} b_{ipq} e_{pq}(\boldsymbol{u}) - \sum_{j=1}^{3} a_{ij} \frac{\partial \varphi}{\partial x_j} - \sum_{j=1}^{3} \zeta_{ij} \frac{\partial \psi}{\partial x_j} + m_i \theta, \quad i = 1, 2, 3. \tag{18}$$

If $\boldsymbol{u} = (u_i)_{i=1}^3$, φ, ψ, and θ are twice continuously differentiable, then by multiplying Eqs. (7) by arbitrary continuously differentiable functions $v_i : \overline{\Omega} \to \mathbb{R}$ ($i = 1, 2, 3$), which vanish on Γ_0, Eq. (8) by a continuously differentiable function $\overline{\varphi} : \overline{\Omega} \to \mathbb{R}$, such that $\overline{\varphi} = 0$ on Γ_0^{φ}, Eq. (9) by a continuously differentiable function $\overline{\psi} : \overline{\Omega} \to \mathbb{R}$ vanishing on Γ_0^{ψ}, and Eq. (10) by a continuously differentiable function $\overline{\theta} : \overline{\Omega} \to \mathbb{R}$, such that $\overline{\theta} = 0$ on Γ_0^{θ}, by integrating on Ω and by using Green's formula, and taking into account symmetry condition (1), (3), (4), boundary conditions (11)–(14), and constitutive equations (16)–(18), we obtain the following integral relations:

$$\begin{aligned}
&\int_{\Omega} \rho \sum_{i=1}^{3} \frac{\partial^2 u_i}{\partial t^2} v_i dx + \int_{\Omega} \sum_{i,j,p,q=1}^{3} c_{ijpq} e_{pq}(\boldsymbol{u}) e_{ij}(\boldsymbol{v}) dx \\
&\quad + \int_{\Omega} \sum_{i,j,p=1}^{3} \varepsilon_{pij} \frac{\partial \varphi}{\partial x_p} e_{ij}(\boldsymbol{v}) dx + \int_{\Omega} \sum_{i,j,p=1}^{3} b_{pij} \frac{\partial \psi}{\partial x_p} e_{ij}(\boldsymbol{v}) dx \\
&\quad - \int_{\Omega} \sum_{i,j=1}^{3} \lambda_{ij} \theta e_{ij}(\boldsymbol{v}) dx = \int_{\Omega} \sum_{i=1}^{3} f_i v_i dx + \int_{\Gamma_1} \sum_{i=1}^{3} g_i v_i d\Gamma,
\end{aligned} \tag{19}$$

$$\begin{aligned}
&- \int_{\Omega} \sum_{i,j,p=1}^{3} \varepsilon_{ipq} e_{pq}(\boldsymbol{u}) \frac{\partial \overline{\varphi}}{\partial x_i} dx + \int_{\Omega} \sum_{i,j=1}^{3} d_{ij} \frac{\partial \varphi}{\partial x_j} \frac{\partial \overline{\varphi}}{\partial x_i} dx \\
&\quad + \int_{\Omega} \sum_{i,j=1}^{3} a_{ij} \frac{\partial \psi}{\partial x_j} \frac{\partial \overline{\varphi}}{\partial x_i} dx - \int_{\Omega} \sum_{i=1}^{3} \mu_i \theta \frac{\partial \overline{\varphi}}{\partial x_i} dx = \int_{\Omega} f^{\varphi} \overline{\varphi} dx - \int_{\Gamma_1^{\varphi}} g^{\varphi} \overline{\varphi} d\Gamma,
\end{aligned} \tag{20}$$

$$\begin{aligned}
&- \int_{\Omega} \sum_{i,j,p=1}^{3} b_{ipq} e_{pq}(\boldsymbol{u}) \frac{\partial \overline{\psi}}{\partial x_i} dx + \int_{\Omega} \sum_{i,j=1}^{3} a_{ij} \frac{\partial \varphi}{\partial x_j} \frac{\partial \overline{\psi}}{\partial x_i} dx \\
&\quad + \int_{\Omega} \sum_{i,j=1}^{3} \zeta_{ij} \frac{\partial \psi}{\partial x_j} \frac{\partial \overline{\psi}}{\partial x_i} dx - \int_{\Omega} \sum_{i=1}^{3} m_i \theta \frac{\partial \overline{\psi}}{\partial x_i} dx = - \int_{\Gamma_1^{\psi}} g^{\psi} \overline{\psi} d\Gamma,
\end{aligned} \tag{21}$$

$$\int_{\Omega} \varkappa \frac{\partial \theta}{\partial t} \overline{\theta} dx + \int_{\Omega} \sum_{i,j=1}^{3} \eta_{ij} \frac{\partial \theta}{\partial x_j} \frac{\partial \overline{\theta}}{\partial x_i} dx + \Theta_0 \int_{\Omega} \sum_{i,j=1}^{3} \lambda_{ij} e_{ij} \left(\frac{\partial \boldsymbol{u}}{\partial t} \right) \overline{\theta} dx$$

$$
- \Theta_0 \int_\Omega \sum_{i=1}^3 \mu_i \frac{\partial^2 \varphi}{\partial t \partial x_i} \overline{\theta} dx - \Theta_0 \int_\Omega \sum_{i=1}^3 m_i \frac{\partial^2 \psi}{\partial t \partial x_i} \overline{\theta} dx
$$

$$
= \int_\Omega f^\theta \overline{\theta} dx - \int_{\Gamma_1^\theta} g^\theta \overline{\theta} d\Gamma. \tag{22}
$$

Therefore, if $\boldsymbol{u} = (u_i)_{i=1}^3 : \overline{\Omega} \times [0, T] \to \mathbb{R}^3$, $\varphi : \overline{\Omega} \times [0, T] \to \mathbb{R}$, $\psi : \overline{\Omega} \times [0, T] \to \mathbb{R}$, and $\theta : \overline{\Omega} \times [0, T] \to \mathbb{R}$ are solutions of Eqs. (7)–(10) and satisfy boundary conditions (11)–(14), then $\boldsymbol{u}$, φ, ψ and θ are solutions of Eqs. (19)–(22). Conversely, if $\boldsymbol{u}$, φ, ψ and θ are twice continuously differentiable solutions of Eqs. (19)–(22), then by using Green's formula we obtain:

$$
\int_\Omega \rho \sum_{i=1}^3 \frac{\partial^2 u_i}{\partial t^2} v_i dx + \int_{\Gamma_1} \sum_{i,j=1}^3 \sigma_{ij} n_j v_i d\Gamma
$$

$$
- \int_\Omega \sum_{i,j=1}^3 \frac{\partial}{\partial x_j} \left(\sum_{p,q=1}^3 c_{ijpq} e_{pq}(\boldsymbol{u}) + \sum_{p=1}^3 \varepsilon_{pij} \frac{\partial \varphi}{\partial x_p} \right.
$$

$$
\left. + \sum_{p=1}^3 b_{pij} \frac{\partial \psi}{\partial x_p} - \lambda_{ij} \theta \right) v_i dx = \int_\Omega \sum_{i=1}^3 f_i v_i dx + \int_{\Gamma_1} \sum_{i=1}^3 g_i v_i d\Gamma, \tag{23}
$$

$$
- \int_{\Gamma_1^\varphi} \sum_{i=1}^3 D_i n_i \overline{\varphi} d\Gamma + \int_\Omega \sum_{i=1}^3 \frac{\partial}{\partial x_i} \left(\sum_{p,q=1}^3 \varepsilon_{ipq} e_{pq}(\boldsymbol{u}) \right.
$$

$$
\left. - \sum_{j=1}^3 d_{ij} \frac{\partial \varphi}{\partial x_j} - \sum_{j=1}^3 a_{ij} \frac{\partial \psi}{\partial x_j} + \mu_i \theta \right) \overline{\varphi} dx = \int_\Omega f^\varphi \overline{\varphi} dx - \int_{\Gamma_1^\varphi} g^\varphi \overline{\varphi} d\Gamma, \tag{24}
$$

$$
- \int_{\Gamma_1^\psi} \sum_{i=1}^3 B_i n_i \overline{\psi} d\Gamma + \int_\Omega \sum_{i=1}^3 \frac{\partial}{\partial x_i} \left(\sum_{p,q=1}^3 b_{ipq} e_{pq}(\boldsymbol{u}) \right.
$$

$$
\left. - \sum_{j=1}^3 a_{ij} \frac{\partial \varphi}{\partial x_j} - \sum_{j=1}^3 \zeta_{ij} \frac{\partial \psi}{\partial x_j} + m_i \theta \right) \overline{\psi} dx = - \int_{\Gamma_1^\psi} g^\psi \overline{\psi} d\Gamma, \tag{25}
$$

$$
\int_\Omega \varkappa \frac{\partial \theta}{\partial t} \overline{\theta} dx + \int_{\Gamma_1^\theta} \sum_{i,j=1}^3 \eta_{ij} \frac{\partial \theta}{\partial x_j} n_i \overline{\theta} d\Gamma - \int_\Omega \sum_{i,j=1}^3 \frac{\partial}{\partial x_i} \left(\eta_{ij} \frac{\partial \theta}{\partial x_j} \right) \overline{\theta} dx
$$

$$+\Theta_0\int\limits_\Omega\sum_{i,j=1}^3\lambda_{ij}e_{ij}\left(\frac{\partial\boldsymbol{u}}{\partial t}\right)\overline{\theta}dx-\Theta_0\int\limits_\Omega\sum_{i=1}^3\mu_i\frac{\partial^2\varphi}{\partial t\partial x_i}\overline{\theta}dx$$

$$-\Theta_0\int\limits_\Omega\sum_{i=1}^3 m_i\frac{\partial^2\psi}{\partial t\partial x_i}\overline{\theta}dx=\int\limits_\Omega f^\theta\overline{\theta}dx-\int\limits_{\Gamma_1^\theta}g^\theta\overline{\theta}d\Gamma, \tag{26}$$

where $\boldsymbol{v}=(v_i)_{i=1}^3,\overline{\varphi},\overline{\psi},\overline{\theta}$ are continuously differentiable functions on $\overline{\Omega}$, such that $v_i=0$ on Γ_0 $(i=1,2,3)$, $\overline{\varphi}=0$ on Γ_0^φ, $\overline{\psi}=0$ on Γ_0^ψ, $\overline{\theta}=0$ on Γ_0^θ. By letting $\boldsymbol{v}\in(\mathfrak{D}(\Omega))^3$, $\overline{\varphi}\in\mathfrak{D}(\Omega)$, $\overline{\psi}\in\mathfrak{D}(\Omega)$, $\overline{\theta}\in\mathfrak{D}(\Omega)$ and by taking into account the density of $\mathfrak{D}(\Omega)$ in $L^2(\Omega)$, we obtain, from (23)–(26), that $\boldsymbol{u}$, φ, ψ and θ satisfy Eqs. (7)–(10). Furthermore, if functions $\boldsymbol{v}$, $\overline{\varphi}$, $\overline{\psi}$ and $\overline{\theta}$ are arbitrary continuous functions on the surfaces Γ_1, Γ_1^φ, Γ_1^ψ and Γ_1^θ and vanish on the remaining parts of the boundary Γ, then by applying Eqs. (7)–(10) and density of the sets of continuous functions on Γ_1, Γ_1^φ, Γ_1^ψ and Γ_1^θ vanishing on the boundaries of the corresponding surfaces in spaces $L^2(\Gamma_1)$, $L^2(\Gamma_1^\varphi)$, $L^2(\Gamma_1^\psi)$ and $L^2(\Gamma_1^\theta)$, we infer, from (23)–(26), that $\boldsymbol{u}$, φ, ψ and θ satisfy the boundary conditions (11)–(14).

Hence, the initial-boundary problem (7)–(15) corresponding to the dynamic three-dimensional model of anisotropic inhomogeneous thermoelastic piezoelectric body is equivalent to Eqs. (19)–(22) with initial conditions (15) in the space of twice continuously differentiable functions. Therefore, by identifying the unknown vector-function $\boldsymbol{u}$ and the functions φ, ψ, θ with vector-functions defined on $[0,T]$ with values in suitable spaces of functions defined on Ω, from Eqs. (19)–(22) we obtain the following variational formulation of problem (7)–(15) in the spaces of vector-valued distributions: Find $\boldsymbol{u}\in C([0,T];\mathbf{V}(\Omega))$, $\boldsymbol{u}'\in L^\infty(0,T;\mathbf{V}(\Omega))$, $\boldsymbol{u}''\in L^\infty(0,T;\mathbf{L}^2(\Omega))$, $\varphi\in C([0,T];V^\varphi(\Omega))$, $\varphi'\in L^\infty(0,T;V^\varphi(\Omega))$, $\psi\in C([0,T];V^\psi(\Omega))$, $\psi'\in L^\infty(0,T;V^\psi(\Omega))$, $\theta\in C([0,T];V^\theta(\Omega))$, $\theta'\in L^\infty(0,T;L^2(\Omega))\cap L^2(0,T;V^\theta(\Omega))$, which satisfy the following equations in the sense of distributions on $(0,T)$,

$$(\rho\boldsymbol{u}'',\boldsymbol{v})_{\mathbf{L}^2(\Omega)}+c(\boldsymbol{u},\boldsymbol{v})+\varepsilon(\varphi,\boldsymbol{v})+b(\psi,\boldsymbol{v})-\lambda(\theta,\boldsymbol{v})=L^{\boldsymbol{u}}(\boldsymbol{v}),\quad\forall\boldsymbol{v}\in\mathbf{V}(\Omega), \tag{27}$$

$$-\varepsilon(\overline{\varphi},\boldsymbol{u})+d(\varphi,\overline{\varphi})+a(\psi,\overline{\varphi})-\mu(\theta,\overline{\varphi})=L^\varphi(\overline{\varphi}),\quad\forall\overline{\varphi}\in V^\varphi(\Omega), \tag{28}$$

$$-b(\overline{\psi},\boldsymbol{u})+a(\varphi,\overline{\psi})+\zeta(\psi,\overline{\psi})-m(\theta,\overline{\psi})=L^\psi(\overline{\psi}),\quad\forall\overline{\psi}\in V^\psi(\Omega), \tag{29}$$

$$(\varkappa\theta',\overline{\theta})_{L^2(\Omega)}+\eta(\theta,\overline{\theta})+\Theta_0\lambda(\overline{\theta},\boldsymbol{u}')$$
$$-\Theta_0\mu(\overline{\theta},\varphi')-\Theta_0 m(\overline{\theta},\psi')=L^\theta(\overline{\theta}),\quad\forall\overline{\theta}\in V^\theta(\Omega), \tag{30}$$

together with the initial conditions

$$\boldsymbol{u}(0)=\boldsymbol{u}_0,\quad\boldsymbol{u}'(0)=\boldsymbol{u}_1,\quad\theta(0)=\theta_0, \tag{31}$$

where $\mathbf{V}(\Omega)=\{\boldsymbol{v}\in\mathbf{H}^1(\Omega);\,\mathbf{tr}_\Gamma(\boldsymbol{v})=\boldsymbol{0}$ on $\Gamma_0\}$, $V^\varphi(\Omega)=\{\overline{\varphi}\in H^1(\Omega);\,\mathrm{tr}_\Gamma(\overline{\varphi})=0$ on $\Gamma_0^\varphi\}$, $V^\psi(\Omega)=\{\overline{\psi}\in H^1(\Omega);\,\mathrm{tr}_\Gamma(\overline{\psi})=0$ on $\Gamma_0^\psi\}$, $V^\theta(\Omega)=\{\overline{\theta}\in H^1(\Omega)$;

$\mathrm{tr}_{\Gamma}(\overline{\theta}) = 0$ on $\Gamma_0^{\theta}\}$, and

$$c(\boldsymbol{u}, \boldsymbol{v}) = \int_{\Omega} \sum_{i,j,p,q=1}^{3} c_{ijpq} e_{pq}(\boldsymbol{u}) e_{ij}(\boldsymbol{v}) dx, \quad \varepsilon(\varphi, \boldsymbol{v}) = \int_{\Omega} \sum_{i,j,p=1}^{3} \varepsilon_{pij} \frac{\partial \varphi}{\partial x_p} e_{ij}(\boldsymbol{v}) dx,$$

$$b(\psi, \boldsymbol{v}) = \int_{\Omega} \sum_{i,j,p=1}^{3} b_{pij} \frac{\partial \psi}{\partial x_p} e_{ij}(\boldsymbol{v}) dx, \quad \lambda(\theta, \boldsymbol{v}) = \int_{\Omega} \sum_{i,j=1}^{3} \lambda_{ij} \theta e_{ij}(\boldsymbol{v}) dx,$$

$$d(\varphi, \overline{\varphi}) = \int_{\Omega} \sum_{i,j=1}^{3} d_{ij} \frac{\partial \varphi}{\partial x_j} \frac{\partial \overline{\varphi}}{\partial x_i} dx, \quad a(\psi, \overline{\varphi}) = \int_{\Omega} \sum_{i,j=1}^{3} a_{ij} \frac{\partial \psi}{\partial x_j} \frac{\partial \overline{\varphi}}{\partial x_i} dx,$$

$$\mu(\theta, \overline{\varphi}) = \int_{\Omega} \sum_{i=1}^{3} \mu_i \theta \frac{\partial \overline{\varphi}}{\partial x_i} dx, \quad \zeta(\psi, \overline{\psi}) = \int_{\Omega} \sum_{i,j=1}^{3} \zeta_{ij} \frac{\partial \psi}{\partial x_j} \frac{\partial \overline{\psi}}{\partial x_i} dx,$$

$$m(\theta, \overline{\psi}) = \int_{\Omega} \sum_{i=1}^{3} m_i \theta \frac{\partial \overline{\psi}}{\partial x_i} dx, \quad \eta(\theta, \overline{\theta}) = \int_{\Omega} \sum_{i,j=1}^{3} \eta_{ij} \frac{\partial \theta}{\partial x_j} \frac{\partial \overline{\theta}}{\partial x_i} dx,$$

$$L^{\boldsymbol{u}}(\boldsymbol{v}) = \int_{\Omega} \sum_{i=1}^{3} f_i v_i dx + \int_{\Gamma_1} \sum_{i=1}^{3} g_i \mathrm{tr}_{\Gamma_1}(v_i) d\Gamma, \quad L^{\psi}(\overline{\psi}) = -\int_{\Gamma_1^{\psi}} g^{\psi} \mathrm{tr}_{\Gamma_1^{\psi}}(\overline{\psi}) d\Gamma,$$

$$L^{\theta}(\overline{\theta}) = \int_{\Omega} f^{\theta} \overline{\theta} dx - \int_{\Gamma_1^{\theta}} g^{\theta} \mathrm{tr}_{\Gamma_1^{\theta}}(\overline{\theta}) d\Gamma, \quad L^{\varphi}(\overline{\varphi}) = \int_{\Omega} f^{\varphi} \overline{\varphi} dx - \int_{\Gamma_1^{\varphi}} g^{\varphi} \mathrm{tr}_{\Gamma_1^{\varphi}}(\overline{\varphi}) d\Gamma.$$

Note that if $\varepsilon_{pij}, b_{pij}, d_{ij}, a_{ij}, \zeta_{ij}, \mu_i, m_i \in C^{0,1}(\overline{\Omega})$, then, from Rademacher's theorem [35], we have that the functions $\varepsilon_{pij}, b_{pij}, d_{ij}, a_{ij}, \zeta_{ij}, \mu_i, m_i$ are differentiable almost everywhere in Ω and their derivatives belong to $L^{\infty}(\Omega)$. If $\boldsymbol{u}_0 \in \mathbf{H}^2(\Omega)$, $f^{\varphi}(0) \in L^2(\Omega)$, $g^{\varphi}(0) \in H^{1/2}(\Gamma_1^{\varphi})$, $g^{\psi}(0) \in H^{1/2}(\Gamma_1^{\psi})$, $\theta_0 \in H^1(\Omega)$, then by applying Green's formula, Eqs. (28), (29) can be written as follows:

$$\begin{aligned} d(\varphi_0, \overline{\varphi}) + a(\psi_0, \overline{\varphi}) = & \sum_{i,j,p=1}^{3} \int_{\Gamma_1^{\varphi}} \mathrm{tr}_{\Gamma_1^{\varphi}}(\overline{\varphi}) \mathrm{tr}_{\Gamma_1^{\varphi}}(\varepsilon_{pij} e_{ij}(\boldsymbol{u}_0)) n_p d\Gamma \\ & - \sum_{i,j,p=1}^{3} \int_{\Omega} \overline{\varphi} \frac{\partial(\varepsilon_{pij} e_{ij}(\boldsymbol{u}_0))}{\partial x_p} dx + \sum_{i=1}^{3} \int_{\Gamma_1^{\varphi}} \mathrm{tr}_{\Gamma_1^{\varphi}}(\overline{\varphi}) \mathrm{tr}_{\Gamma_1^{\varphi}}(\mu_i \theta_0) n_i d\Gamma \\ & - \sum_{i=1}^{3} \int_{\Omega} \overline{\varphi} \frac{\partial(\mu_i \theta_0)}{\partial x_i} dx + (f^{\varphi}(0), \overline{\varphi})_{L^2(\Omega)} - (g^{\varphi}(0), \mathrm{tr}_{\Gamma_1^{\varphi}}(\overline{\varphi}))_{L^2(\Gamma_1^{\varphi})}, \end{aligned} \tag{32}$$

$$a(\varphi_0, \overline{\psi}) + \zeta(\psi_0, \overline{\psi}) = \sum_{i,j,p=1}^{3} \int_{\Gamma_1^\psi} \mathrm{tr}_{\Gamma_1^\psi}(\overline{\psi})\mathrm{tr}_{\Gamma_1^\psi}(b_{pij}e_{ij}(\boldsymbol{u}_0))n_p d\Gamma$$

$$- \sum_{i,j,p=1}^{3} \int_{\Omega} \overline{\psi} \frac{\partial(b_{pij}e_{ij}(\boldsymbol{u}_0))}{\partial x_p} dx + \sum_{i=1}^{3} \int_{\Gamma_1^\psi} \mathrm{tr}_{\Gamma_1^\psi}(\overline{\psi})\mathrm{tr}_{\Gamma_1^\psi}(m_i\theta_0)n_i d\Gamma$$

$$- \sum_{i=1}^{3} \int_{\Omega} \overline{\psi} \frac{\partial(m_i\theta_0)}{\partial x_i} dx - (g^\psi(0), \mathrm{tr}_{\Gamma_1^\psi}(\overline{\psi}))_{L^2(\Gamma_1^\psi)}, \tag{33}$$

where $\boldsymbol{n} = (n_i)_{i=1}^3$ is the unit outward normal vector of the boundary Γ, and the given functions in the right-hand parts of Eqs. (32), (33) have the following properties:

$$\sum_{i,j,p=1}^{3} \mathrm{tr}_{\Gamma_1^\varphi}(\varepsilon_{pij}e_{ij}(\boldsymbol{u}_0))n_p + \sum_{i=1}^{3} \mathrm{tr}_{\Gamma_1^\varphi}(\mu_i\theta_0)n_i - g^\varphi(0) \in H^{1/2}(\Gamma_1^\varphi),$$

$$- \sum_{i,j,p=1}^{3} \frac{\partial(\varepsilon_{pij}e_{ij}(\boldsymbol{u}_0))}{\partial x_p} - \sum_{i=1}^{3} \frac{\partial(\mu_i\theta_0)}{\partial x_i} + f^\varphi(0) \in L^2(\Omega),$$

$$\sum_{i,j,p=1}^{3} \mathrm{tr}_{\Gamma_1^\psi}(b_{pij}e_{ij}(\boldsymbol{u}_0))n_p + \sum_{i=1}^{3} \mathrm{tr}_{\Gamma_1^\psi}(m_i\theta_0)n_i - g^\psi(0) \in H^{1/2}(\Gamma_1^\psi),$$

$$- \sum_{i,j,p=1}^{3} \frac{\partial(b_{pij}e_{ij}(\boldsymbol{u}_0))}{\partial x_p} - \sum_{i=1}^{3} \frac{\partial(m_i\theta_0)}{\partial x_i} \in L^2(\Omega).$$

It follows from the positive definiteness condition (6) that (32), (33) constitute a boundary value problem for a strongly elliptic system of the second-order partial differential equations [22] with respect to $\varphi_0 \in V^\varphi(\Omega)$ and $\psi_0 \in V^\psi(\Omega)$, which possesses a unique solution when $\Gamma_0^\varphi \neq \varnothing$ and $\Gamma_0^\psi \neq \varnothing$, and if Ω is a bounded $C^{1,1}$ domain and $\overline{\Gamma_0^\varphi} \cap \overline{\Gamma_1^\varphi} = \varnothing$, $\overline{\Gamma_0^\psi} \cap \overline{\Gamma_1^\psi} = \varnothing$, then, by applying the regularity theorem [22], we infer that the solutions φ_0 and ψ_0 of (32), (33) belong to $H^2(\Omega)$.

For problem (27)–(31), which is equivalent to the initial-boundary value problem (7)–(15) in the space of classical twice continuously differentiable functions, the following existence, uniqueness and continuous dependence theorem is valid.

Theorem 1 *Suppose that $\Omega \subset \mathbb{R}^3$ is a bounded $C^{1,1}$ domain, $\Gamma_0^\varphi \neq \varnothing$, $\Gamma_0^\psi \neq \varnothing$ and $\overline{\Gamma_0^\varphi} \cap \overline{\Gamma_1^\varphi} = \varnothing$, $\overline{\Gamma_0^\psi} \cap \overline{\Gamma_1^\psi} = \varnothing$, the parameters characterizing thermal, electro-magnetic and elastic properties of the body $\rho, \varkappa \in L^\infty(\Omega)$, $c_{ijpq}, \varepsilon_{pij}, b_{pij}, d_{ij}, a_{ij}, \zeta_{ij}, \lambda_{ij}, \mu_i, m_i, \eta_{ij} \in C^{0,1}(\overline{\Omega})$ $(i, j, p, q = 1, 2, 3)$, for all $x \in \Omega$ satisfy the symmetry conditions* (1), (3), (4) *and*

$$d_{ij}(x) = d_{ji}(x), \quad a_{ij}(x) = a_{ji}(x), \quad \zeta_{ij}(x) = \zeta_{ji}(x), \quad i, j = 1, 2, 3,$$

and positive definiteness conditions (2), (7), *and for almost all* $x \in \Omega$,

$$\rho(x) > \alpha_\rho = const > 0, \quad \varkappa(x) > \alpha_\varkappa = const > 0,$$

$$\sum_{i,j=1}^{3} d_{ij}(x)\xi_j\xi_i + 2\sum_{i,j=1}^{3} a_{ij}(x)\xi_j\overline{\xi}_i + \sum_{i,j=1}^{3} \zeta_{ij}(x)\overline{\xi}_j\overline{\xi}_i + \frac{1}{\Theta_0}\varkappa\xi\xi - 2\sum_{i=1}^{3}\mu_i(x)\xi\xi_i$$
$$-2\sum_{i=1}^{3} m_i(x)\xi\overline{\xi}_i \geq \widetilde{\alpha}\sum_{i=1}^{3}\left((\xi_i)^2 + (\overline{\xi}_i)^2 + \xi^2\right), \quad \forall x \in \Omega, \ \xi, \xi_i, \overline{\xi}_i \in \mathbb{R}, \ i = 1, 2, 3,$$

where $\widetilde{\alpha} = const > 0$. *If* $\boldsymbol{f}, \boldsymbol{f}' \in L^2(0, T; \mathbf{L}^2(\Omega))$, $\boldsymbol{g}, \boldsymbol{g}', \boldsymbol{g}'' \in L^2(0, T; \mathbf{L}^{4/3}(\Gamma_1))$, $f^\varphi, (f^\varphi)', (f^\varphi)'' \in L^2(0, T; L^{6/5}(\Omega))$, $f^\varphi(0) \in L^2(\Omega)$, $g^\varphi, (g^\varphi)', (g^\varphi)'' \in L^2(0, T; L^{4/3}(\Gamma_1^\varphi))$, $g^\varphi(0) \in H^{1/2}(\Gamma_1^\varphi)$, $g^\psi, (g^\psi)', (g^\psi)'' \in L^2(0, T; L^{4/3}(\Gamma_1^\psi))$, $g^\psi(0) \in H^{1/2}(\Gamma_1^\psi)$, $f^\theta, (f^\theta)' \in L^2(0, T; L^2(\Omega))$, $g^\theta, (g^\theta)' \in L^2(0, T; L^{4/3}(\Gamma_1^\theta))$ *and the initial data* $\boldsymbol{u}_0 \in \mathbf{V}(\Omega) \cap \mathbf{H}^2(\Omega)$, $\boldsymbol{u}_1 \in \mathbf{V}(\Omega)$, $\theta_0 \in V^\theta(\Omega) \cap H^2(\Omega)$ *satisfy the following compatibility conditions:*

$$g^\theta(0) = -\sum_{i,j=1}^{3} \mathrm{tr}_{\Gamma_1^\theta}\left(\eta_{ij}\frac{\partial\theta_0}{\partial x_j}\right)n_i^\theta,$$

$$g_i(0) = \sum_{j=1}^{3}\mathrm{tr}_{\Gamma_1}\left(\sum_{p,q=1}^{3} c_{ijpq}e_{pq}(\boldsymbol{u}_0) + \sum_{p=1}^{3}\varepsilon_{pij}\frac{\partial\varphi_0}{\partial x_p} + \sum_{p=1}^{3} b_{pij}\frac{\partial\psi_0}{\partial x_p} - \lambda_{ij}\theta_0\right)n_j,$$

where $i = 1, 2, 3$, $\boldsymbol{n}^\theta = (n_i^\theta)_{i=1}^3$ *and* $\boldsymbol{n} = (n_i)_{i=1}^3$ *are the unit outward normal vectors to* Γ_1^θ *and* Γ_1, *respectively, then problem* (27)–(31) *possesses a unique solution, which continuously depends on the given data, i.e., the mapping*

$$(\boldsymbol{u}_0, \boldsymbol{u}_1, \theta_0, \boldsymbol{f}, \boldsymbol{g}, \boldsymbol{g}', f^\varphi, (f^\varphi)', g^\varphi, (g^\varphi)', g^\psi, (g^\psi)', f^\theta, g^\theta) \to (\boldsymbol{u}, \boldsymbol{u}', \varphi, \psi, \theta)$$

is linear and continuous from space

$$\mathbf{V}(\Omega) \times \mathbf{L}^2(\Omega) \times L^2(\Omega) \times L^2(0, T; \mathbf{L}^2(\Omega)) \times L^2(0, T; \mathbf{L}^{4/3}(\Gamma_1)) \times L^2(0, T; \mathbf{L}^{4/3}(\Gamma_1))$$
$$\times L^2(0, T; L^{6/5}(\Omega)) \times L^2(0, T; L^{6/5}(\Omega)) \times L^2(0, T; L^{4/3}(\Gamma_1^\varphi)) \times L^2(0, T; L^{4/3}(\Gamma_1^\varphi))$$
$$\times L^2(0, T; L^{4/3}(\Gamma_1^\psi)) \times L^2(0, T; L^{4/3}(\Gamma_1^\psi)) \times L^2(0, T; L^2(\Omega)) \times L^2(0, T; L^{4/3}(\Gamma_1^\theta))$$

to space

$$C([0, T]; \mathbf{V}(\Omega)) \times C([0, T]; \mathbf{L}^2(\Omega)) \times C([0, T]; V^\varphi(\Omega))$$
$$\times C([0, T]; V^\psi(\Omega)) \times C([0, T]; L^2(\Omega)),$$

and the following energy equality is valid

$$E(t) = E(0) + L(t), \quad \forall t \in [0, T],$$

where

$$E(t) = (\rho \boldsymbol{u}'(t), \boldsymbol{u}'(t))_{\mathbf{L}^2(\Omega)} + c(\boldsymbol{u}(t), \boldsymbol{u}(t)) + \frac{1}{\Theta_0}(\varkappa\theta(t), \theta(t))_{L^2(\Omega)}$$
$$+ \frac{2}{\Theta_0}\int_0^t \eta(\theta, \theta)d\tau + d(\varphi(t), \varphi(t)) + 2a(\varphi(t), \psi(t))$$
$$+ \zeta(\psi(t), \psi(t)) - 2\mu(\theta(t), \varphi(t)) - 2m(\theta(t), \psi(t)),$$
$$L(t) = 2\int_0^t (\boldsymbol{f}(\tau), \boldsymbol{u}'(\tau))_{\mathbf{L}^2(\Omega)}d\tau + 2(\boldsymbol{g}(t), \mathbf{tr}_{\Gamma_1}(\boldsymbol{u}(t)))_{\mathbf{L}^2(\Gamma_1)}$$
$$- 2(\boldsymbol{g}(0), \mathbf{tr}_{\Gamma_1}(\boldsymbol{u}(0)))_{\mathbf{L}^2(\Gamma_1)} - 2\int_0^t (\boldsymbol{g}'(\tau), \mathbf{tr}_{\Gamma_1}(\boldsymbol{u}(\tau)))_{\mathbf{L}^2(\Gamma_1)}d\tau$$
$$+ 2\int_0^t ((f^\varphi)'(\tau), \varphi(\tau))_{L^2(\Omega)}d\tau - 2\int_0^t ((g^\varphi)'(\tau), \mathrm{tr}_{\Gamma_1^\varphi}(\varphi(\tau)))_{L^2(\Gamma_1^\varphi)}d\tau$$
$$- 2\int_0^t ((g^\psi)'(\tau), \mathrm{tr}_{\Gamma_1^\psi}(\psi(\tau)))_{L^2(\Gamma_1^\psi)}d\tau + \frac{2}{\Theta_0}\int_0^t (f^\theta(\tau), \theta(\tau))_{L^2(\Omega)}d\tau$$
$$- \frac{2}{\Theta_0}\int_0^t (g^\theta(\tau), \mathrm{tr}_{\Gamma_1^\theta}(\theta(\tau)))_{L^2(\Gamma_1^\theta)}d\tau, \quad \forall t \in [0, T].$$

3 Static Three-Dimensional Problem

The linear three-dimensional model [8, 25] of the static equilibrium of the thermoelastic piezoelectric body $\overline{\Omega}$ in differential form is given by the partial differential equations (7)–(10) together with the boundary conditions (11)–(14), where all the unknown and the given functions do not depend on time variable t, the corresponding governing equations are given in Ω and the boundary conditions are prescribed on the corresponding parts of the boundary Γ. Hence, in the static model, instead of Eqs. (7) and (10) we have:

$$
-\sum_{j=1}^{3}\frac{\partial}{\partial x_j}\left(\sum_{p,q=1}^{3} c_{ijpq}e_{pq}(\boldsymbol{u})+\sum_{p=1}^{3}\varepsilon_{pij}\frac{\partial\varphi}{\partial x_p}\right.
$$

$$
\left.+\sum_{p=1}^{3} b_{pij}\frac{\partial\psi}{\partial x_p}-\lambda_{ij}\theta\right)=f_i \quad \text{in } \Omega,\quad i=1,2,3, \tag{34}
$$

$$
-\sum_{i,j=1}^{3}\frac{\partial}{\partial x_i}\left(\eta_{ij}\frac{\partial\theta}{\partial x_j}\right)=f^{\theta} \quad \text{in } \Omega. \tag{35}
$$

By multiplying Eqs. (34) by arbitrary continuously differentiable functions $v_i : \overline{\Omega} \to \mathbb{R}$ $(i = 1, 2, 3)$, which vanish on Γ_0 and Eq. (35) by a continuously differentiable function $\overline{\theta} : \overline{\Omega} \to \mathbb{R}$, such that $\overline{\theta} = 0$ on Γ_0^{θ}, by integrating on Ω, by using Green's formula, and taking into account symmetry conditions (1), (4), boundary conditions (11), (14), and constitutive equations (16), instead of Eqs. (19) and (22), we obtain the following equations:

$$
\int_{\Omega}\sum_{i,j,p,q=1}^{3} c_{ijpq}e_{pq}(\boldsymbol{u})e_{ij}(\boldsymbol{v})dx+\int_{\Omega}\sum_{i,j,p=1}^{3}\varepsilon_{pij}\frac{\partial\varphi}{\partial x_p}e_{ij}(\boldsymbol{v})dx
$$

$$
+\int_{\Omega}\sum_{i,j,p=1}^{3} b_{pij}\frac{\partial\psi}{\partial x_p}e_{ij}(\boldsymbol{v})dx-\int_{\Omega}\sum_{i,j=1}^{3}\lambda_{ij}\theta e_{ij}(\boldsymbol{v})dx
$$

$$
=\int_{\Omega}\sum_{i=1}^{3} f_i v_i dx+\int_{\Gamma_1}\sum_{i=1}^{3} g_i v_i d\Gamma, \tag{36}
$$

$$
\int_{\Omega}\sum_{i,j=1}^{3}\eta_{ij}\frac{\partial\theta}{\partial x_j}\frac{\partial\overline{\theta}}{\partial x_i}dx=\int_{\Omega} f^{\theta}\overline{\theta}dx-\int_{\Gamma_1^{\theta}} g^{\theta}\overline{\theta}d\Gamma. \tag{37}
$$

Conversely, if $\boldsymbol{u} = (u_i)_{i=1}^3 : \overline{\Omega} \to \mathbb{R}^3, \varphi : \overline{\Omega} \to \mathbb{R}, \psi : \overline{\Omega} \to \mathbb{R}$, and $\theta : \overline{\Omega} \to \mathbb{R}$ are twice continuously differentiable solutions of Eqs. (36), (37), then by using Green's formula we infer, as for the dynamic problem, that $\boldsymbol{u}$, φ, ψ and θ are solutions of Eqs. (34), (35) satisfying the boundary conditions (11), (14).

Therefore, the boundary value problem (8), (9), (34), (35), (11)–(14), corresponding to the static three-dimensional model of the thermoelastic piezoelectric body $\overline{\Omega}$, is equivalent to Eqs. (20), (21), (36), (37) in the space of twice continuously differentiable functions, and on the basis of them we obtain the following variational formulation of the boundary value problem (8), (9), (34), (35), (11)–(14): Find $\boldsymbol{u} \in \mathbf{V}(\Omega)$, $\varphi \in V^{\varphi}(\Omega)$, $\psi \in V^{\psi}(\Omega)$, $\theta \in V^{\theta}(\Omega)$ such that

$$c(\boldsymbol{u},\boldsymbol{v})+\varepsilon(\varphi,\boldsymbol{v})+b(\psi,\boldsymbol{v})-\lambda(\theta,\boldsymbol{v})=L^{u}(\boldsymbol{v}),\quad \forall \boldsymbol{v}\in \mathbf{V}(\Omega), \tag{38}$$

$$-\varepsilon(\overline{\varphi},\boldsymbol{u})+d(\varphi,\overline{\varphi})+a(\psi,\overline{\varphi})-\mu(\theta,\overline{\varphi})=L^{\varphi}(\overline{\varphi}),\quad \forall \overline{\varphi}\in V^{\varphi}(\Omega), \tag{39}$$

$$-b(\overline{\psi},\boldsymbol{u})+a(\varphi,\overline{\psi})+\zeta(\psi,\overline{\psi})-m(\theta,\overline{\psi})=L^{\psi}(\overline{\psi}),\quad \forall \overline{\psi}\in V^{\psi}(\Omega), \tag{40}$$

$$\eta(\theta,\overline{\theta})=L^{\theta}(\overline{\theta}),\quad \forall \overline{\theta}\in V^{\theta}(\Omega). \tag{41}$$

For problem (38)–(41) the following theorem regarding the existence, uniqueness, regularity and continuous dependence on the given data of a solution of the boundary value problem in suitable function spaces is valid.

Theorem 2 *Suppose that $\Omega \subset \mathbb{R}^3$ is a bounded Lipschitz domain, $\Gamma_0 \neq \varnothing$, $\Gamma_0^{\varphi} \neq \varnothing$, $\Gamma_0^{\psi} \neq \varnothing$, $\Gamma_0^{\theta} \neq \varnothing$, the parameters c_{ijpq}, ε_{pij}, b_{pij}, d_{ij}, a_{ij}, ζ_{ij}, λ_{ij}, μ_i, m_i, $\eta_{ij} \in L^{\infty}(\Omega)$, $i,j,p,q = 1,2,3$, for almost all $x \in \Omega$ satisfy the symmetry conditions* (1), (3), (4), *and positive definiteness conditions* (2), (5), (6). *If $\boldsymbol{f} \in \mathbf{L}^{6/5}(\Omega)$, $\boldsymbol{g} \in \mathbf{L}^{4/3}(\Gamma_1)$, $f^{\varphi} \in L^{6/5}(\Omega)$, $g^{\varphi} \in L^{4/3}(\Gamma_1^{\varphi})$, $g^{\psi} \in L^{4/3}(\Gamma_1^{\psi})$, $f^{\theta} \in L^{6/5}(\Omega)$, $g^{\theta} \in L^{4/3}(\Gamma_1^{\theta})$, then problem* (38)–(41) *possesses a unique solution $(\boldsymbol{u},\varphi,\psi,\theta) \in \mathbf{V}(\Omega) \times V^{\varphi}(\Omega) \times V^{\psi}(\Omega) \times V^{\theta}(\Omega)$, which continuously depends on the given data, i.e., the following estimate is valid:*

$$\begin{aligned}||\boldsymbol{u}||_{\mathbf{H}^1(\Omega)} + ||\varphi||_{H^1(\Omega)} + ||\psi||_{H^1(\Omega)} + ||\theta||_{H^1(\Omega)} \leq \hat{\alpha}\Big(&||\boldsymbol{f}||_{\mathbf{L}^{6/5}(\Omega)} + ||\boldsymbol{g}||_{\mathbf{L}^{4/3}(\Gamma_1)} \\ &+ ||f^{\varphi}||_{L^{6/5}(\Omega)} + ||g^{\varphi}||_{L^{4/3}(\Gamma_1^{\varphi})} + ||g^{\psi}||_{L^{4/3}(\Gamma_1^{\psi})} \\ &+ ||f^{\theta}||_{L^{6/5}(\Omega)} + ||g^{\theta}||_{L^{4/3}(\Gamma_1^{\theta})}\Big), \quad \hat{\alpha} = const > 0.\end{aligned}$$

Furthermore, if $\Omega \subset \mathbb{R}^3$ is a $C^{r+1,1}$ $(r \in \mathbb{N} \cup \{0\})$ domain, $\overline{\Gamma_0} \cap \overline{\Gamma_1} = \varnothing$, $\overline{\Gamma_0^{\varphi}} \cap \overline{\Gamma_1^{\varphi}} = \varnothing$, $\overline{\Gamma_0^{\psi}} \cap \overline{\Gamma_1^{\psi}} = \varnothing$, $\overline{\Gamma_0^{\theta}} \cap \overline{\Gamma_1^{\theta}} = \varnothing$, c_{ijpq}, ε_{pij}, b_{pij}, d_{ij}, a_{ij}, ζ_{ij}, λ_{ij}, μ_i, m_i, $\eta_{ij} \in C^{r,1}(\overline{\Omega})$, $i,j,p,q = 1,2,3$, $\boldsymbol{f} \in \mathbf{H}^r(\Omega)$, $\boldsymbol{g} \in \mathbf{H}^{r+1/2}(\Gamma_1)$, $f^{\varphi} \in H^r(\Omega)$, $g^{\varphi} \in H^{r+1/2}(\Gamma_1^{\varphi})$, $g^{\psi} \in H^{r+1/2}(\Gamma_1^{\psi})$, $f^{\theta} \in H^r(\Omega)$, $g^{\theta} \in H^{r+1/2}(\Gamma_1^{\theta})$, then solution $(\boldsymbol{u},\varphi,\psi,\theta)$ of problem (38)–(41) *has additional regularity $\boldsymbol{u} \in \mathbf{V}(\Omega) \cap \mathbf{H}^{r+2}(\Omega)$, $\varphi \in V^{\varphi}(\Omega) \cap H^{r+2}(\Omega)$, $\psi \in V^{\psi}(\Omega) \cap H^{r+2}(\Omega)$, $\theta \in V^{\theta}(\Omega) \cap H^{r+2}(\Omega)$, and the mapping*

$$(\boldsymbol{f},\boldsymbol{g},f^{\varphi},g^{\varphi},g^{\psi},f^{\theta},g^{\theta}) \to (\boldsymbol{u},\varphi,\psi,\theta)$$

is linear and continuous from space

$$\mathbf{H}^r(\Omega) \times \mathbf{H}^{r+1/2}(\Gamma_1) \times H^r(\Omega) \times H^{r+1/2}(\Gamma_1^{\varphi}) \times H^{r+1/2}(\Gamma_1^{\psi}) \times H^r(\Omega) \times H^{r+1/2}(\Gamma_1^{\theta})$$

to space $\mathbf{H}^{r+2}(\Omega) \times H^{r+2}(\Omega) \times H^{r+2}(\Omega) \times H^{r+2}(\Omega)$.

Corollary 1 *If $\Omega \subset \mathbb{R}^3$ is a $C^{3,1}$ domain, $\Gamma_0 \neq \varnothing$, $\Gamma_0^{\varphi} \neq \varnothing$, $\Gamma_0^{\psi} \neq \varnothing$, $\Gamma_0^{\theta} \neq \varnothing$, and $\overline{\Gamma_0} \cap \overline{\Gamma_1} = \varnothing$, $\overline{\Gamma_0^{\varphi}} \cap \overline{\Gamma_1^{\varphi}} = \varnothing$, $\overline{\Gamma_0^{\psi}} \cap \overline{\Gamma_1^{\psi}} = \varnothing$, $\overline{\Gamma_0^{\theta}} \cap \overline{\Gamma_1^{\theta}} = \varnothing$, the parameters c_{ijpq}, ε_{pij}, b_{pij}, d_{ij}, a_{ij}, ζ_{ij}, λ_{ij}, μ_i, m_i, $\eta_{ij} \in C^{2,1}(\overline{\Omega})$, $i,j,p,q = 1,2,3$, satisfy the symmetry conditions* (1), (3), (4), *and positive definiteness conditions* (2), (5), (6), *and $\boldsymbol{f} \in \mathbf{H}^2(\Omega)$, $\boldsymbol{g} \in \mathbf{H}^{5/2}(\Gamma_1)$, $f^{\varphi} \in H^2(\Omega)$, $g^{\varphi} \in H^{5/2}(\Gamma_1^{\varphi})$, $g^{\psi} \in H^{5/2}(\Gamma_1^{\psi})$,*

$f^\theta \in H^2(\Omega)$, $g^\theta \in H^{5/2}(\Gamma_1^\theta)$, *then the boundary value problem* (8), (9), (34), (35), (11)–(14) possesses a unique classical solution $(\boldsymbol{u}, \varphi, \psi, \theta)$, which is twice continuously differentiable on $\overline{\Omega}$, satisfies equations (8), (9), (34), (35) in Ω, and boundary conditions (11)–(14) on the corresponding parts of the boundary Γ.

4 Conclusions

We studied initial-boundary and boundary value problems with general mixed boundary conditions for mechanical displacement, electric and magnetic potentials, and temperature corresponding to the linear dynamic and static three-dimensional models of inhomogeneous anisotropic thermoelastic piezoelectric bodies with regard to magnetic field. We obtained the variational formulations of the three-dimensional problems in corresponding spaces of vector-valued distributions with respect to the time variable or Sobolev spaces that are equivalent to the original differential formulations of the initial-boundary and boundary value problems in the spaces of twice continuously differentiable functions. We formulated new results regarding the existence, uniqueness and continuous dependence on the given data of solutions of the three-dimensional initial-boundary and boundary value problems in suitable function spaces.

Acknowledgement This work was supported by Shota Rustaveli National Science Foundation (SRNSF) [Grant number 217596, Construction and investigation of hierarchical models for thermoelastic piezoelectric structures].

References

1. Akamatsu, M., Nakamura, G.: Well-posedness of initial-boundary value problems for piezoelectric equations. Appl. Anal. **81**, 129–141 (2002)
2. Allam, M.N.M., Tantawy, R., Zenkour, A.M.: Magneto-thermo-elastic response of exponentially graded piezoelectric hollow spheres. Adv. Comput. Design **3**(3), 303–318 (2018)
3. Altay, G.A., Dökmeci, M.C.: Fundamental variational equations of discontinuous thermopiezoelectric fields. Int. J. Eng. Sci. **34**(7), 769–782 (1996)
4. Aouadi, M.: On the coupled theory of thermo-magnetoelectroelasticity. Quart. J. Mech. Appl. Math. **60**(4), 443–456 (2007)
5. Avalishvili, G., Avalishvili, M.: On static hierarchical two-dimensional models of thermoelastic piezoelectric plates with variable thickness. WSEAS Trans. Appl. Theor. Mech. **13**, 76–84 (2018)
6. Avalishvili, G., Avalishvili, M.: On approximation of three-dimensional model of thermoelastic piezoelectric plates by two-dimensional problems. Bull. Georgian Natl. Acad. Sci. **12**(4), 23–32 (2018)
7. Avalishvili, G., Avalishvili, M.: On the investigation of one-dimensional models of thermo-electro-magneto-elastic bars. Bull. Georgian Natl. Acad. Sci. **14**(1), 7–17 (2020)
8. Avalishvili, G., Avalishvili, M., Müller, W.H.: Investigation of the three-dimensional boundary value problem for thermoelastic piezoelectric solids. Bull. Tbil. Int. Cent. Math. Inform. **21**(2), 65–79 (2017)

9. Avalishvili, G., Avalishvili, M., Müller, W.H.: On investigation of dynamical three-dimensional model of thermoelastic piezoelectric solids. Bull. Georgian Natl. Acad. Sci. **11**(4), 13–21 (2017)
10. Bardzokas, D.I., Filshtinsky, M.L., Filshtinsky, L.A.: Mathematical Methods in Electro-Magneto-Elasticity. Springer, Berlin (2007)
11. Benjeddou, A.: Field-dependent nonlinear piezoelectricity: a focused review. Int. J. Smart Nano Mater. **9**(1), 68–84 (2018)
12. Benjeddou, A., Andrianarison, O.: A thermopiezoelectric mixed variational theorem for smart multilayered composites. Comput. Struct. **83**, 1266–1276 (2005)
13. Chinchaladze, N.: On a vibration problem of the transversely isotropic bars. Bull. Tbil. Int. Cent. Math. Inform. **23**(2), 87-96 (2019)
14. Curie, J., Curie, P.: Développement, par pression, de lélectricité polaire dans les cristaux hémièdre à faces inclinées. Comptes Rend. Acad. Sci. **91**, 294–295 (1880)
15. Dautray, R., Lions, J.-L.: Mathematical Analysis and Numerical Methods for Science and Technology, Vol. 5: Evolution Problems I. Springer, Berlin (2000)
16. Dhaliwal, R.S., Wang, J.: A uniqueness theorem for linear theory of thermopiezoelectricity. Z. Angew. Math. Mech. **74**, 558–560 (1994)
17. He, J.-H.: Variational theory for linear magneto-electro-elasticity. Int. J. Nonl. Sci. Num. Simul. **2**, 309–316 (2001)
18. Jaiani, G.: Piezoelectric viscoelastic Kelvin-Voigt cusped prismatic shells. Lect. Notes Tbil. Int. Cent. Math. Inform. **19** (2018)
19. Jaiani, G.: On BVPs for piezoelectric transversely isotropic cusped bars. Bull. Tbil. Int. Cent. Math. Inform. **23**(1), 35–66 (2019)
20. Lee, P.C.Y.: A variational principle for the equations of piezoelectromagnetism in elastic dielectric crystals. J. Appl. Phys. **69**(11), 7470–7473 (1991)
21. Li, J.Y.: Uniqueness and reciprocity theorems for linear thermo-electro-magnetoelasticity. Quart. J. Mech. Appl. Math. **56**(1), 35–43 (2003)
22. McLean, W.: Strongly Elliptic Systems and Boundary Integral Equations. Cambridge University Press, Cambridge (2000)
23. Mindlin, R.D.: Equations of high frequency vibrations of thermopiezoelectric crystal plates. Int. J. Solids Struct. **10**, 625–637 (1974)
24. Mulholland, A.J., Picard, R., Trostorff, S., Waurick, M.: On well-posedness for some thermo-piezoelectric coupling models. Math. Meth. Appl. Sci. **39**(15), 4375–4384 (2016)
25. Natroshvili, D.: Mathematical problems of thermo-electro-magneto-elasticity. Lect. Notes Tbil. Int. Cent. Math. Inform. **12** (2011)
26. Nowacki, W.: A reciprocity theorem for coupled mechanical and thermoelectric fields in piezoelectric crystals. Proc. Vibr. Prob. **1**, 3–11 (1965)
27. Nowacki, W.: Some general theorems of thermopiezoelectricity. J. Thermal Stresses **1**, 171–182 (1978)
28. Qin, Q.-H.: Advanced Mechanics of Piezoelectricity. Springer, New York (2013)
29. Qin, Q.-H., Yang, Q.-S.: Macro-Micro Theory on Multifield Coupling Behavior of Heterogeneous Materials. Springer, Berlin (2008)
30. Qing, G., Qiu, J., Liu, Y.: Modified H-R mixed variational principle for magnetoelectroelastic bodies and state-vector equation. Appl. Math. Mech. **26**(6), 722–728 (2005)
31. Tiersten, H.F.: Linear Piezoelectric Plate Vibrations. Plenum, New York (1964)
32. Tiwari, R., Mukhopadhyay, S.: On electromagnetothermoelastic plane waves under Green-Naghdi theory of thermoelasticity-II. J. Thermal Stresses **40**(8), 1040–1062 (2017)
33. Vashishth, A.K., Sukhija, H.: Reflection and transmission of plane waves from fluid-piezothermoelastic solid interface. Appl. Math. Mech. **36**(1), 11–36 (2015)
34. Voigt, W.: Allgemeine Theorie der piëzo- und pyroelectrischen Erscheinungen an Krystallen. Abhandlungen der Königlichen Gesellschaft der Wissenschaften zu Göttingen **36**, 1–99 (1890)
35. Whitney, H.: Geometric Integration Theory. Princeton University Press, Princeton (1957)
36. Yakhno, V.: The well-posedness of dynamical equations of magneto-electro-elasticity. Mediterr. J. Math. **15**(21) (2018). https://doi.org/10.1007/s00009-018-1065-4

On Nonparametric Kernel-Type Estimate of the Bernoulli Regression Function

Petre K. Babilua and Elizbar A. Nadaraya

Abstract In the paper, the limit distribution is established for an integral mean-square deviation of a nonparametric generalized kernel-type estimate of the Bernoulli regression function. A test criterion is constructed for the hypothesis on the Bernoulli regression function. The question of consistency is considered, and for some close alternatives the asymptotics of test power behavior is investigated.

Keywords Bernoulli regression function · Limiting distribution · Consistency · Test power

1 Introduction

Let a random value Y have two values 1 and 0 with probabilities p ("success") and $1-p$ ("failure"). Assume that the success probability p is a function of an independent variable $x \in [0,1]$, i.e. $p = p(x) = \mathbf{P}\{Y=1 \mid x\}$ [2, 3, 9, 10]. Let x_i, $i=1,\dots,n$, be the partition points of the interval $[0,1]$:

$$x_i = \frac{2i-1}{2n}, \quad i=1,\dots,n.$$

Let, further, Y_{ij}, $j=1,\dots,m_i$, $m_i \geq 1$, $i=1,\dots,n$, be mutually independent Bernoulli random variables with $\mathbf{P}\{Y_{ij}=1 \mid x_i\} = p(x_i)$, $\mathbf{P}\{Y_{ij}=0 \mid x_i\} = 1-p(x_i)$, $j=1,\dots,m_i$, $i=1,\dots,n$ [9, 10]. The problem consists in estimating the function $p(x)$, $x \in [0,1]$, by the group sampling Y_{ij}, $j=1,\dots,m_i$, $i=1,\dots,n$. Such problems arise, for example, in biology [9, 10], medicine [3], and so on.

P. K. Babilua (✉) · E. A. Nadaraya
Department of Mathemetics, Faculty of Exact and Natural Sciences,
Ivane Javakhishvili Tbilisi State University, 13 University Str., Tbilisi 0186, Georgia
e-mail: petre.babilua@tsu.ge

E. A. Nadaraya
e-mail: elizbar.nadaraya@tsu.ge

G. Jaiani and D. Natroshvili (eds.), *Applications of Mathematics and Informatics in Natural Sciences and Engineering*, Springer Proceedings in Mathematics & Statistics 334, https://doi.org/10.1007/978-3-030-56356-1_2

As an estimate for $p(x)$ we consider the following statistic [6, 11]

$$\widehat{p}_n(x) = p_{1n}(x)p_{2n}^{-1}(x),$$

$$p_{\nu n}(x) = \frac{1}{nb_n}\sum_{i=1}^{n} K\Big(\frac{x-x_i}{b_n}\Big)\overline{Y}_i^{2-\nu}, \quad \nu = 1,2, \quad \overline{Y}_i = \frac{1}{m_i}\sum_{j=1}^{m_i} Y_{ij}, \quad i = 1,\dots,n,$$

where $K(x)$ is some distribution density that satisfies the requirements formulated below, and $b_n \to 0$ is a sequence of positive integers.

2 Assumptions and Notation

Assume that the kernel $K(x) \ge 0$ is chosen such that it is a function with finite variation and satisfies the conditions: $K(x) = K(-x)$, $K(x) = 0$ for $|x| \ge \tau > 0$, $\int K(x)\,dx = 1$. The class of such functions is denoted by $H(\tau)$.

Denote by $C^{(i)}$ the set of functions $p(x)$, $0 \le p(x) \le 1$, $x \in [0,1]$, having bounded derivatives of up to i-th order, $i = 1, 2$.

Let us also introduce the following notation:

$$U_n = nb_n \int\limits_{\Omega_n(\tau)} \big[p_{1n}(x) - \mathbf{E}\, p_{1n}(x)\big]^2 dx, \quad \Omega_n(\tau) = [\tau b_n, 1-\tau b_n],$$

$$T_n = nb_n N_n \int\limits_{\Omega_n(\tau)} \big[\widehat{p}_n(x) - p(x)\big]^2 p_{2n}^2(x)\,dx, \quad N_n = \max_{1\le k\le n} m_k,$$

$$Q_{ij} = \psi_n(x_i, x_j), \quad \psi_n(u,v) = \int\limits_{\Omega_n(\tau)} K\Big(\frac{x-u}{b_n}\Big)K\Big(\frac{x-v}{b_n}\Big)\,dx,$$

$$B_n^2 = 4(nb_n)^{-2}\sum_{k=2}^{n}\frac{p_k(1-p_k)}{m_k}\sum_{i=1}^{k-1}\frac{p_i(1-p_i)}{m_i}\,Q_{ik}^2, \quad p_i = p(x_i), \quad i = 1,\dots,n,$$

$$\eta_{ij}^{(n)} = \frac{2\varepsilon_i\varepsilon_j Q_{ij}}{nb_n B_n}, \quad \varepsilon_i = \overline{Y}_i - p(x_i),$$

$$\xi_k^{(n)} = \sum_{i=1}^{k-1}\eta_{ik}^{(n)}, \quad k = \overline{2,n}, \quad \xi_1^{(n)} = 0, \quad \xi_k^{(n)} = 0, \quad k > n,$$

$$\mathscr{F}_k^{(n)} = \sigma(\omega : \varepsilon_1, \dots, \varepsilon_k),$$

where $\mathscr{F}_k^{(n)}$ is a σ-algebra generated by random variables $\varepsilon_1,\dots,\varepsilon_k$, $\mathscr{F}_0^{(n)} = (\varnothing, \Omega)$ (in the sequel, for the sake of simplicity, we will write ξ_k and η_{ij}) instead of $\xi_k^{(n)}$ and $\eta_{ij}^{(n)}$.

Lemma 1 *A stochastic sequence $(\xi_k, \mathscr{F}_k)_{k\ge1}$ is a martingale difference.*

Lemma 2 *Let $K(x) \in H(\tau)$ and $p(x)$, $0 \le x \le 1$, be also a function of bounded variation. If $nb_n \to \infty$, then*

$$\frac{1}{nb_n}\sum_{i=1}^{n} K^{\nu_1}\Big(\frac{x-x_i}{b_n}\Big)K^{\nu_2}\Big(\frac{y-x_i}{b_n}\Big)p^{\nu_3}(x_i)$$

$$= \frac{1}{b_n}\int_0^1 K^{\nu_1}\Big(\frac{x-u}{b_n}\Big)K^{\nu_2}\Big(\frac{y-u}{b_n}\Big)p^{\nu_3}(u)\,du + O\Big(\frac{1}{nb_n}\Big) \quad (1)$$

uniformly with respect to $x, y \in [0, 1]$, $\nu_i \in N \cup \{0\}$, $i = 1, 2, 3$.

Proof Relation (1) is proved analogously to Lemma 1 in [8, p. 1643].

Lemma 3 *Let $K(x) \in H(\tau)$ and $p(x) \in C^{(1)}$. If $nb_n^2 \to \infty$, then*

$$b_n^{-1}N_n^2B_n^2 \ge b_n^{-1}\sigma_n^2 \ \ \text{and} \ \ b_n^{-1}\sigma_n^2 \to \sigma^2(p) \ \ \text{as} \ \ n \to \infty, \quad (2)$$

where

$$\sigma_n^2 = 4(nb_n)^{-2}\sum_{k=2}^{n} p_k(1-p_k)\sum_{i=1}^{k-1} p_i(1-p_i)Q_{ik}^2,$$

$$\sigma^2(p) = 2\int_0^1 p^2(x)(1-p(x))^2\,dx \int_{|x|\le 2\tau} K_0^2(x)\,dx,$$

$$N_n = \max_{1\le k\le n} m_k, \quad p_i = p(x_i), \quad i = 1,\dots,n, \quad K_0 = K * K.$$

In particular, if $m_k = N_n$, $k = 1,\dots,n$, then

$$\lim_{n\to\infty} \frac{N_n^2}{b_n} B_n^2 = \lim_{n\to\infty} b_n^{-1}\sigma_n^2 = \sigma^2(p).$$

Proof Clearly, $N_n^2 b_n^{-1} B_n^2 \ge b_n^{-1}\sigma_n^2$ and we have

$$\sigma_n^2 = 2(nb_n)^{-2}\Big\{\sum_{k,i=1}^{n} p_k(1-p_k)p_i(1-p_i)Q_{ik}^2 - \sum_{i=1}^{n} p_i^2(1-p_i)^2Q_{ii}^2\Big\}$$
$$= d_1(n) + d_2(n), \quad p_i = p(x_i), \quad i = 1,\dots,n.$$

It is easy to verify that

$$b_n^{-1}|d_2(n)| = 2n^{-2}b_n^{-3}\sum_{i=1}^{n} p_i^2(1-p_i)^2\Big(\int_{\Omega_n(\tau)} K^2\Big(\frac{x-x_i}{b_n}\Big)dx\Big)^2 \le c_1\frac{1}{nb_n}. \quad (3)$$

Further, using the definition of Q_{ki}, we obtain

$$d_1(n)=\frac{2}{(nb_n)^2}\int\limits_{\Omega_n(\tau)}\int\limits_{\Omega_n(\tau)}\Big(\sum_{i=1}^{n}p(x_i)(1-p(x_i))K\Big(\frac{x-x_i}{b_n}\Big)K\Big(\frac{y-x_i}{b_n}\Big)\Big)^2\,dx\,dy.$$

Hence, by Lemma 2, we have

$$\begin{aligned}d_1(n)&=2\int\limits_{\Omega_n(\tau)}\int\limits_{\Omega_n(\tau)}\Big\{\frac{1}{b_n}\int_0^1 p(u)(1-p(u))\\&\qquad\times K\Big(\frac{x-u}{b_n}\Big)K\Big(\frac{y-u}{b_n}\Big)\,du\Big\}^2\,dx\,dy+O\Big(\frac{1}{nb_n}\Big)\\&=2\int\limits_{\Omega_n(\tau)}\int\limits_{\Omega_n(\tau)}\Big\{\int_{\frac{x-1}{b_n}}^{\frac{x}{b_n}} p(x-ub_n)(1-p(x-ub_n))\\&\qquad\times K(u)K\Big(\frac{y-x}{b_n}-u\Big)\,du\Big\}^2\,dx\,dy+O\Big(\frac{1}{nb_n}\Big).\end{aligned}\tag{4}$$

Since $p(x)\in C^{(1)}$ and $\big[\frac{x-1}{b_n},\frac{x_n}{b_n}\big]\supseteq[-\tau,\tau]$ for all $x\in\Omega_n(\tau)$, we find from (4) that

$$d_1(n)=2\int\limits_{\Omega_n(\tau)}\int\limits_{\Omega_n(\tau)}p^2(x)(1-p(x))^2K_0^2\Big(\frac{x-y}{b_n}\Big)\,dx\,dy+O(b_n^2)+O\Big(\frac{1}{nb_n}\Big).$$

It can be easily established that

$$b_n^{-1}d_1(n)=2\int_0^1 p^2(x)(1-p(x))^2\Big(\int_{\frac{x-1}{b_n}+\tau_n}^{\frac{x}{b_n}-\tau}K_0^2(u)\,du\Big)\,dx+O(b_n)+O\Big(\frac{1}{nb_n^2}\Big).$$

Therefore

$$b_n^{-1}d_1(n)\longrightarrow 2\int_0^1 p^2(x)(1-p(x))^2\int\limits_{|x|\le 2\tau}K_0^2(x)\,dx.\tag{5}$$

From (3) and (5) assertion (2) follows.

3 Asymptotic Normality of Statistics U_n and T_n

The following assertion holds true.

Theorem 1 *Let $K(x) \in H(\tau)$ and $p(x) \in C^{(1)}$. If $\frac{N_n^4}{nb_n^2} \to 0$ and $N_n^4 b_n \to 0$ as $n \to \infty$, then*

$$\frac{U_n - \mathbf{E}\, U_n}{B_n} \xrightarrow{d} N(0,1),$$

where $\xrightarrow{d}$ denotes convergence in distribution , and $N(0,1)$ is a random variable having a standard normal distribution $\Phi(x)$.

Proof We have

$$\frac{U_n - \mathbf{E}\, U_n}{B_n} = H_n^{(1)} + H_n^{(2)},$$

where

$$H_n^{(1)} = \sum_{k=1}^{n} \xi_k, \quad H_n^{(2)} = \frac{\sum\limits_{i=1}^{n} (\varepsilon_i^2 - \mathbf{E}\, \varepsilon_i^2) Q_{ii}}{nb_n B_n}.$$

Let us show that $H_n^{(2)}$ converges to zero in probability. Indeed,

$$\operatorname{var} H_n^{(2)} \le \frac{1}{(nb_n B_n)^2} \sum_{i=1}^{n} \mathbf{E}\, \varepsilon_i^4 Q_{ii}^2.$$

Since $Q_{ij} \le c_2 b_n$ and $\mathbf{E}\, \varepsilon_i^4 \le c_3 m_i^{-2}$ and $\min\limits_{1 \le i \le n} m_i \ge 1$, by Lemma 3 we have

$$\operatorname{var} H_n^{(2)} \le c_4 \frac{N_n^2}{nb_n^2} \longrightarrow 0.$$

Therefore $H_n^{(2)} \xrightarrow{\mathbf{P}} 0$ (here and below the letter **P** over the arrow denotes convergence in probability).

Now let us show that $H_n^{(1)} \xrightarrow{d} N(0,1)$. For this, we need to verify the validity of Corollaries 2 and 6 of Theorem 2 in [5]. We have to show the fulfillment of the conditions contained in these corollaries and guaranteeing the asymptotic normality of the square integrable martingale difference, which, according to Lemma 1, is our sequence $\{\xi_k, \mathscr{F}_k\}_{k \ge 1}$. Direct calculations show that $\sum\limits_{k=1}^{n} \mathbf{E}\, \xi_k^2 = 1$. Asymptotic normality takes place if for $n \to \infty$

$$\sum_{k=1}^{n} \mathbf{E}\left[\xi_k^2 I\big(|\xi_k| \ge \varepsilon\big) \mid \mathscr{F}_{k-1}\right] \xrightarrow{\mathbf{P}} 0 \tag{6}$$

and

$$\sum_{k=1}^{n} \xi_k^2 \xrightarrow{\mathbf{P}} 1. \tag{7}$$

In [5], it is proved that if (7) and the condition $\sup_{1\le k\le n} |\xi_k| \xrightarrow{\mathbf{P}} 0$ are fulfilled, then condition (6) is fulfilled too. Since for $\varepsilon > 0$

$$\mathbf{P}\Big\{\sup_{1\le k\le n} |\xi_k| \ge \varepsilon\Big\} \le \varepsilon^{-4}\sum_{k=1}^{n} \mathbf{E}\,\xi_k^4,$$

according to relation (8) given below, to prove $H_n^{(1)} \xrightarrow{d} N(0, 1)$. It remains only to verify (7). For this, it suffices to ascertain that

$$\mathbf{E}\Big(\sum_{k=1}^{n} \xi_k^2 - 1\Big)^2 \longrightarrow 0 \text{ as } n \to \infty,$$

i.e. that since $\sum_{k=1}^{n} \mathbf{E}\,\xi_k^2 = 1$,

$$\mathbf{E}\Big(\sum_{k=1}^{n} \xi_k^2\Big)^2 = \sum_{k=1}^{n} \mathbf{E}\,\xi_k^4 + 2\sum_{1\le k_1<k_2\le n} \mathbf{E}\,\xi_{k_1}^2\xi_{k_2}^2 \longrightarrow 1.$$

In the first place we establish that $\sum_{k=1}^{n} \mathbf{E}\,\xi_k^4 \longrightarrow 0$ as $n \to \infty$. Taking the definitions of ξ_k and η_k into account, we write

$$\sum_{k=1}^{n} \mathbf{E}\,\xi_k^4 = I_n^{(1)} + I_n^{(2)},$$

where

$$I_n^{(1)} = \frac{16}{(nb_n)^4 B_n^4}\sum_{k=2}^{n} \mathbf{E}\,\varepsilon_k^4 \sum_{j=1}^{k-1} \mathbf{E}\,\varepsilon_j^4 Q_{jk}^4,$$

$$I_n^{(2)} = \frac{48}{(nb_n)^4 B_n^4}\sum_{k=2}^{n}\sum_{i\ne j} \mathbf{E}\,\varepsilon_j^2 \mathbf{E}\,\varepsilon_i^2 Q_{jk}^2 Q_{ik}^2.$$

Since $Q_{ij} \le c_4 b_n$, $\mathbf{E}\,\varepsilon_j^4 \le c_5 m_j^{-2}$, $\mathbf{E}\,\varepsilon_j^2 \le m_j^{-1}$ and $b_n^{-1}\sigma_n^2 \to \sigma^2(p)$, we have

$$I_n^{(1)} = O\Big(\frac{N_n^4}{(nb_n)^2}\Big), \quad I_n^{(2)} = O\Big(\frac{N_n^4}{nb_n^2}\Big).$$

Therefore

$$\sum_{k=1}^{n} \mathbf{E}\,\xi_k^4 \longrightarrow 0 \text{ as } n \to \infty. \tag{8}$$

Now, it will be shown that

$$2 \sum_{1 \le k_1 < k_2 \le n} \mathbf{E}\,\xi_{k_1}^2 \xi_{k_2}^2 \longrightarrow 1 \text{ as } n \to \infty.$$

From the definition of ξ_i it follows that

$$\xi_{k_1}^2 \xi_{k_2}^2 = B_{k_1 k_2}^{(1)} + B_{k_1 k_2}^{(2)} + B_{k_1 k_2}^{(3)} + B_{k_1 k_2}^{(4)},$$

where

$$\begin{aligned} B_{k_1 k_2}^{(1)} &= \sigma_2(k_1)\sigma_2(k_2), \quad B_{k_1 k_2}^{(2)} = \sigma_2(k_1)\sigma_1(k_2), \\ B_{k_1 k_2}^{(3)} &= \sigma_1(k_1)\sigma_2(k_2), \quad B_{k_1 k_2}^{(4)} = \sigma_1(k_1)\sigma_1(k_2), \\ \sigma_1(k) &= \sum_{i \ne j}^{k-1} \eta_{ik}\eta_{jk}, \quad \sigma_2(k) = \sum_{i=1}^{k-1} \eta_{ik}^2. \end{aligned}$$

Consequently,

$$2 \sum_{1 \le k_1 < k_2 \le n} \mathbf{E}\,\xi_{k_1}^2 \xi_{k_2}^2 = \sum_{i=1}^{4} A_n^{(i)},$$

where

$$A_n^{(i)} = 2 \sum_{1 \le k_1 < k_2 \le n} \mathbf{E}\,B_{k_1 k_2}^{(i)}, \quad i = 1, 2, 3, 4.$$

Let us consider $A_n^{(3)}$. Using the definition of η_{ij}, it can be easily shown that $\mathbf{E}\,B_{k_1 k_2}^{(3)}=0$ and therefore

$$A_n^{(3)} = 0. \tag{9}$$

Let us estimate $A_n^{(2)}$. We have

$$\begin{aligned} |\mathbf{E}\,B_{k_1 k_2}^{(2)}| &= \frac{16}{n^4 b_n^4 B_n^4} \Big| \sum_{i=1}^{k_1-1} \mathbf{E}\,\varepsilon_i^3 \mathbf{E}\,\varepsilon_{k_1}^3 \mathbf{E}\,\varepsilon_{k_2}^2 Q_{ik_1}^2 Q_{ik_2} Q_{k_1 k_2} \Big| \\ &= \frac{16}{n^4 b_n^4 B_n^4} \Big| \sum_{i=1}^{k_1-1} (1-p_i)p_i(1-2p_i)(1-p_{k_1}) \end{aligned}$$

$$\times p_{k_1}(1-2p_{k_1})(1-p_{k_2})p_{k_2}(1-2p_{k_2})Q^2_{ik_1}Q_{ik_2}Q_{k_1k_2}\Big|$$

$$\le \frac{c_7(k_1-1)}{(nB_n)^4}.$$

Since $\sum\limits_{1\le k_1<k_2\le n}(k_1-1)=O(n^3)$ and $b_n^{-1}\sigma_n^2\to\sigma^2(p)>0$, we obtain

$$|A_n^{(2)}|\le \frac{c_8 n^3}{n^4B_n^4}=c_8\,\frac{N_n^4}{nb_n^2(\frac{N_n^2B_n^2}{b_n})^2}=O\Big(\frac{N_n^4}{nb_n^2}\Big). \tag{10}$$

Now we will establish that $A_n^{(1)}\to 1$ as $n\to\infty$. It is obvious that

$$A_n^{(1)}=2\sum_{1\le k_1<k_2\le n}\mathbf{E}\,B^{(1)}_{k_1k_2}=D_n^{(1)}+D_n^{(2)},$$

where

$$D_n^{(1)}=2\sum_{1\le k_1<k_2\le n}\Big(\sum_{i=1}^{k_1-1}\mathbf{E}\,\eta^2_{ik_1}\Big)\Big(\sum_{j=1}^{k_2-1}\mathbf{E}\,\eta^2_{jk_2}\Big),$$

$$D_n^{(2)}=2\Big(\sum_{k_1<k_2}\mathbf{E}\,B^{(1)}_{k_1k_2}-\sum_{k_1<k_2}\Big(\sum_{i=1}^{k_1-1}\mathbf{E}\,\eta^2_{ik_1}\Big)\Big(\sum_{j=1}^{k_2-1}\mathbf{E}\,\eta^2_{jk_2}\Big)\Big).$$

From the definition of B_n^2 it follows that

$$D_n^{(1)}=1-\sum_{k=2}^{n}\Big(\sum_{i=1}^{k-1}\mathbf{E}\,\eta^2_{ik}\Big)^2.$$

But

$$\sum_{k=2}^{n}\Big(\sum_{i=1}^{k-1}\mathbf{E}\,\eta^2_{ik}\Big)^2\le c_9\,\frac{b_n^4n^3}{(nb_n)^4B_n^4}=O\Big(\frac{N_n^4}{nb_n^2}\Big).$$

Thus

$$D_n^{(1)}\to 1\ \text{ as }\ n\to\infty. \tag{11}$$

We will further show that $D_n^{(2)}\to 0$. It is easy to see that

$$D_n^{(2)}=2\sum_{k_1<k_2}\Big[\sum_{i=1}^{k_1-1}\operatorname{cov}(\eta^2_{ik_1},\eta^2_{ik_2})+\sum_{i=1}^{k_1-1}\operatorname{cov}(\eta^2_{ik_1},\eta^2_{k_1k_2})\Big].$$

But

$$\mathbf{E}\,\eta_{ik_1}^2\eta_{ik_2}^2 \le c_{10}\,\frac{Q_{ik_1}^2 Q_{ik_2}^2}{(nb_n)^4 B_n^4} \le c_{11}\,\frac{1}{n^4 B_n^4}\,.$$

Analogously,

$$\mathbf{E}\,\eta_{ij}^2 = O\Big(\frac{1}{n^2 B_n^2}\Big).$$

Therefore

$$\mathrm{cov}\,(\eta_{ik_1}^2, \eta_{ik_2}^2) = O\Big(\frac{1}{n^4 B_n^4}\Big). \tag{12}$$

Further, since $\sum\limits_{1\le k_1<k_2\le n}(k_1-1) = O(n^3)$, from (12) we obtain

$$D_n^{(2)} = O\Big(\frac{1}{nB_n^4}\Big) = O\Big(\frac{N_n^4}{nb_n^2}\Big). \tag{13}$$

Thus, by (11) and (13),

$$A_n^{(1)} = 1 + O\Big(\frac{N_n^4}{nb_n^2}\Big). \tag{14}$$

Finally, we will prove that $A_n^{(4)} \to 0$ as $n\to\infty$. Using the definition of η_{ij} and the relations $Q_{ij}\ge 0$ and $\mathbf{E}\,(\overline{Y}_i - p(x_i))^2 \le c_{12}m_i^{-1}$, we have

$$\begin{aligned}
|\mathbf{E}\,B_n^{(4)}| &= 4\Big|\sum_{t<s}^{k_1-1}\mathbf{E}\,\eta_{sk_1}\eta_{tk_1}\eta_{sk_2}\eta_{tk_2}\Big| \\
&= 4\,\frac{1}{n^4 b_n^4 B_n^4}\,\Big|\sum_{t<s}^{k_1-1}\mathbf{E}\,\varepsilon_s^2\varepsilon_t^2\varepsilon_{k_1}^2\varepsilon_{k_2}^2\,Q_{sk_1}Q_{tk_1}Q_{sk_2}Q_{tk_2}\Big| \\
&\le \frac{c_{13}}{n^4 b_n^4 B_n^4}\sum_{t<s}^{k_1-1} Q_{sk_1}Q_{tk_1}Q_{sk_2}Q_{tk_2}.
\end{aligned}$$

Therefore

$$|A_n^{(4)}| \le \frac{c_{14}}{n^2 b_n^4 B_n^4}\sum_{k_1<k_2} A_{k_1k_2},$$

where

$$A_{k_1k_2} = \sum_{1\le t<s\le k_1-1} Q_{sk_1}Q_{tk_1}Q_{sk_2}Q_{tk_2}.$$

But

$$\sum_{k_1<k_2} A_{k_1k_2} \le \frac{1}{n^2}\sum_{k_1,k_2}^{n}\sum_{t=1}^{n}\sum_{s=1}^{n} Q_{sk_1}Q_{tk_1}Q_{tk_2}Q_{sk_2} = \sum_{k_1,k_2}^{n}\Big(\frac{1}{n}\sum_{t=1}^{n} Q_{tk_1}Q_{tk_2}\Big)^2.$$

Thus

$$\begin{aligned}
|A_n^{(4)}| &\le c_{15}\,\frac{1}{n^2b_n^4B_n^4}\sum_{k_1,k_2}^{n}\Big(\frac{1}{n}\sum_{t=1}^{n} Q_{tk_1}Q_{tk_2}\Big)^2\\
&= c_{15}\,\frac{1}{n^2b_n^4B_n^4}\sum_{k_1,k_2}^{n}\Big(\sum_{i=1}^{n}\int\limits_{\Omega_n(\tau)} K\Big(\frac{x-x_i}{b_n}\Big)K\Big(\frac{x-x_{k_1}}{b_n}\Big)\,dx\int\limits_{\Omega_n(\tau)} K\Big(\frac{y-x_i}{b_n}\Big)K\Big(\frac{y-x_{k_2}}{b_n}\Big)\,dy\Big)^2\\
&= c_{15}\frac{1}{n^2b_n^4B_n^4}\sum_{k_1,k_2}^{n}\Big[\iint\limits_{\Omega_n(\tau)\Omega_n(\tau)} K\Big(\frac{x-x_{k_1}}{b_n}\Big)K\Big(\frac{y-x_{k_2}}{b_n}\Big)\,dx\,dy\\
&\qquad\times\Big(\frac{1}{n}\sum_{i=1}^{n}K\Big(\frac{x-x_i}{b_n}\Big)K\Big(\frac{y-x_i}{b_n}\Big)\,dx\,dy\Big)\Big]^2.
\end{aligned} \tag{15}$$

Next, applying Lemma 2, from (15) we conclude that

$$\begin{aligned}
|A_n^{(4)}| &\le c_{15}\,\frac{1}{n^2b_n^4B_n^4}\sum_{k_1,k_2}^{n}\Big\{\iint\limits_{\Omega_n(\tau)\Omega_n(\tau)} K\Big(\frac{x-x_{k_1}}{b_n}\Big)K\Big(\frac{y-x_{k_2}}{b_n}\Big)\,dx\,dy\\
&\qquad\times\Big[\int_0^1 K\Big(\frac{x-u}{b_n}\Big)K\Big(\frac{y-u}{b_n}\Big)\,du + O\Big(\frac{1}{n}\Big)\Big]\Big\}^2\\
&= c_{15}\,\frac{1}{b_n^4B_n^4}\sum_{k_1,k_2}^{n}\Big\{\frac{1}{n}\int_0^1\iint\limits_{\Omega_n(\tau)\Omega_n(\tau)} K\Big(\frac{x-x_{k_1}}{b_n}\Big)K\Big(\frac{y-x_{k_2}}{b_n}\Big)\\
&\qquad\times K\Big(\frac{x-u}{b_n}\Big)K\Big(\frac{y-u}{b_n}\Big)\,du\,dx\,dy\Big\}^2 + O\Big(\frac{N_n^4}{nb_n^2}\Big).
\end{aligned} \tag{16}$$

Analogously, again applying Lemma 2 in (16), it can be shown that

$$\begin{aligned}
|A_n^{(4)}| &\le \frac{c_{15}}{b_n^4B_n^4}\int_0^1\int_0^1\int_0^1\int_0^1 \psi_n(u_1,v_2)\psi_n(u_1,v_1)\\
&\qquad\times\psi_n(u_2,v_1)\psi_n(u_2,v_2)\,du_1\,du_2\,dv_1\,dv_2 + O\Big(\frac{N_n^4}{nb_n^2}\Big),
\end{aligned} \tag{17}$$

where

$$\psi_n(x,y) = \int\limits_{\Omega_n} K\Big(\frac{u-x}{b_n}\Big)K\Big(\frac{u-y}{b_n}\Big)\,du.$$

Let us now estimate the integral I_n contained in (17). We have

$$I_n = \int_0^1\int_0^1\int_0^1 \psi_n(u_2, v_1)\psi_n(u_2, v_2)\, dv_1\, dv_2\, du_2 \int_0^1 \psi_n(u_1, v_2)\psi_n(u_1, v_1)\, du_1.$$

But since $\left[\frac{x-1}{b_n}, \frac{x}{b_n}\right] \supseteq [-\tau, \tau]$ for all $x \in \Omega_n(\tau)$, we have

$$\begin{aligned}
&\int_0^1 \psi_n(u_1, v_2)\psi_n(u_1, v_1)\, du_1 \\
&\quad = b_n \int_{\Omega_n(\tau)}\int_{\Omega_n(\tau)} K\Big(\frac{t-v_2}{b_n}\Big)K\Big(\frac{z-v_1}{b_n}\Big)\, dt\, dz \int_{\frac{t-1}{b_n}}^{\frac{t}{b_n}} K(\xi)K\Big(\frac{t-z}{b_n} - \xi\Big)\, d\xi \\
&\quad = b_n \int_{\Omega_n(\tau)}\int_{\Omega_n(\tau)} K\Big(\frac{t-v_2}{b_n}\Big)K\Big(\frac{z-v_1}{b_n}\Big)K_0\Big(\frac{z-t}{b_n}\Big)\, dt\, dz \\
&\quad \le c_{16} b_n^3, \quad K_0 = K * K.
\end{aligned}$$

Therefore

$$|A_n^{(4)}| \le c_{17}\,\frac{1}{b_n B_n^4}\int_0^1\int_0^1\int_0^1 \psi_n(u_2, v_1)\psi_n(u_2, v_2)\, du_2\, dv_1\, dv_2 + O\Big(\frac{N_n^4}{n b_n^2}\Big). \tag{18}$$

Applying the same operations to (18), we finally obtain

$$\begin{aligned}
|A_n^{(4)}| &\le c_{18}\,\frac{b_n^4}{b_n B_n^4} + O\Big(\frac{N_n^4}{n b_n^2}\Big) \\
&= O\Big(\frac{b_n^3 N_n^4}{b_n^2(\frac{N_n^2 B_n^2}{b_n})^2}\Big) + O\Big(\frac{N_n^4}{n b_n^2}\Big) = O(b_n N_n^4) + O\Big(\frac{N_n^4}{n b_n^2}\Big).
\end{aligned} \tag{19}$$

After combining relations (9), (10), (14) and (19) we establish that

$$2 \sum_{1\le k_1<k_2\le n} \mathbf{E}\,\xi_{k_1}^2\xi_{k_2}^2 \longrightarrow 1.$$

From this and (8) it follows that

$$\mathbf{E}\Big(\sum_{k=1}^n \xi_k^2 - 1\Big)^2 \longrightarrow 0 \ \text{ as } \ n \to \infty.$$

Thus

$$\frac{U_n - \mathbf{E}\, U_n}{B_n} \xrightarrow{d} N(0, 1). \tag{20}$$

The theorem is proved.

Let, further, $m_i = N_n$, $i = 1, \dots, n$. Then from Lemma 3 we have

$$\lim_{n\to\infty} \frac{N_n^2}{b_n} B_n^2 = \sigma^2(p), \tag{21}$$

Now let us find $\lim\limits_{n\to\infty} \mathbf{E}\, U_n$. We have

$$\operatorname{var} p_{1n}(x) = \frac{1}{N_n (n b_n)^2} \sum_{i=1}^{n} K^2\Big(\frac{x - x_i}{b_n}\Big) p(x_i)(1 - p_i(x)).$$

Hence, by virtue of Lemma 2, we find

$$\operatorname{var} p_{1n}(x) = \frac{1}{N_n} \left(\frac{1}{n b_n^2} \int_0^1 K^2\Big(\frac{x-u}{b_n}\Big) p(u)(1-p(u))\, du + O\Big(\frac{1}{(n b_n)^2}\Big)\right). \tag{22}$$

Further, taking into account that $\big[\frac{x-1}{b_n}, \frac{x}{b_n}\big] \supseteq [-\tau, \tau]$ for all $x \in \Omega_n(\tau)$, from (22) we find

$$\operatorname{var} p_{1n}(x) = \frac{1}{N_n} \left(\frac{1}{n b_n} p(x)(1 - p(x)) \int_{|x|\le\tau} K^2(u)\, du + O\Big(\frac{1}{n}\Big) + O\Big(\frac{1}{(n b_n)^2}\Big)\right).$$

Therefore

$$N_n \mathbf{E}\, U_n = \Delta + O(b_n) + O\Big(\frac{1}{n b_n}\Big), \tag{23}$$

$$\Delta = \int_0^1 p(x)(1 - p(x))\, du \int_{|x|\le\tau} K^2(u)\, du.$$

With (21) and (23) taken into account, from (20), as a corollary, we obtain

$$b_n^{-\frac{1}{2}} \frac{N_n U_n - \Delta}{\sigma(p)} \xrightarrow{d} N(0, 1). \tag{24}$$

Theorem 2 *Let $K(x) \in H(\tau)$, $p(x) \in C^{(2)}$. Let, further, $m_k = N_n$, $k = 1, \dots, n$, and $N_n \to \infty$. If $n N_n b_n^4 \to 0$, $\frac{N_n^4}{n b_n^2} \to 0$ and $N_n^4 b_n \to 0$ as $n \to \infty$, then*

$$b_n^{-\frac{1}{2}} \frac{T_n - \Delta}{\sigma(p)} \xrightarrow{d} N(0, 1),$$

Proof We have

$$T_n = N_n U_n + R_n^{(1)} + R_n^{(2)},$$

where

$$R_n^{(1)} = 2nb_n N_n \int\limits_{\Omega_n(\tau)} \big[p_{1n}(x) - \mathbf{E}\, p_{1n}(x)\big]\big[\mathbf{E}\, p_{1n}(x) - p_{2n}(x)p(x)\big]\, dx,$$

$$R_n^{(2)} = nb_n N_n \int\limits_{\Omega_n(\tau)} \big[\mathbf{E}\, p_{1n}(x) - p_{2n}(x)p(x)\big]^2\, dx.$$

It is easy to see that

$$\mathbf{E}\, p_{1n}(x) = \frac{1}{nb_n} \sum_{i=1}^{n} K\Big(\frac{x - x_i}{b_n}\Big) p(x_i).$$

Hence, by virtue of Lemma 2, it follows that

$$\mathbf{E}\, p_{1n}(x) = \frac{1}{b_n} \int\limits_0^1 K\Big(\frac{x - u}{b_n}\Big) p(u)\, du + O\Big(\frac{1}{nb_n}\Big)$$

$$= \int\limits_{\frac{x-1}{b_n}}^{\frac{x}{b_n}} K(t) p(x - b_n t)\, dt + O\Big(\frac{1}{nb_n}\Big). \qquad (25)$$

Since $p(x) \in C^{(2)}$ and $\big[\frac{x-1}{b_n}, \frac{x}{b_n}\big] \supseteq [-\tau, \tau]$ for all $x \in \Omega_n(\tau)$, from (25) we find

$$\mathbf{E}\, p_{1n}(x) = p(x) + O(b_n^2) + O\Big(\frac{1}{nb_n}\Big). \qquad (26)$$

Further, according to Lemma 2

$$p_{2n}(x) = \frac{1}{b_n} \int\limits_0^1 K\Big(\frac{x - u}{b_n}\Big)\, du + O\Big(\frac{1}{nb_n}\Big)$$

$$= \int\limits_{-\tau}^{\tau} K(u)\, du + O\Big(\frac{1}{nb_n}\Big) = 1 + O\Big(\frac{1}{nb_n}\Big). \qquad (27)$$

From (26) and (27) it follows that

$$b_n^{-\frac{1}{2}} R_n^{(2)} \leq c_{18}\Big(nN_n b_n^{\frac{9}{2}} + N_n b_n^{\frac{3}{2}} + \frac{\sqrt{N_n}}{nb_n^{\frac{3}{2}}}\Big) \longrightarrow 0. \tag{28}$$

Let us now estimate $b_n^{-\frac{1}{2}} \mathbf{E}\,|R_n^{(1)}|$. From (26) and (27) follows the inequality

$$b_n^{-\frac{1}{2}} \mathbf{E}\,|R_n^{(1)}| \leq c_{19} n N_n b_n^{\frac{1}{2}}\Big[b_n^2 + \frac{1}{nb_n}\Big] \int\limits_{\Omega_n(\tau)} \Big(\mathbf{E}\big[p_{1n}(x) - \mathbf{E}\,p_{1n}(x)\big]^2\Big)^{\frac{1}{2}} dx. \tag{29}$$

But, by Lemma 2, we have

$$\mathbf{E}\big[p_{1n}(x) - \mathbf{E}\,p_{1n}(x)\big]^2 = \frac{1}{N_n n b_n} p(x)(1-p(x)) \int\limits_{|x|\leq\tau} K^2(u)\,du + O\Big(\frac{1}{n}\Big) + O\Big(\frac{1}{(nb_n)^2}\Big).$$

From this and (29) we find that

$$b_n^{-\frac{1}{2}} \mathbf{E}\,|R_n^{(1)}| \leq c_{20}\Big(\sqrt{nN_n}\, b_n^2 + \frac{\sqrt{N_n}}{\sqrt{n}\, b_n}\Big) \longrightarrow 0. \tag{30}$$

The assertion of the theorem directly follows from (24) and relations (28) and (30).

Corollary 1 *Let $K(x)$ and $p(x)$ satisfy the conditions of the Theorem 2. Let, further, $m_k = N_0$, $k = 1, \dots, n$, $1 \leq N_0 < \infty$. If $nb_n^4 \to 0$, $nb_n^2 \to \infty$, then*

$$b_n^{-\frac{1}{2}} (\overline{T}_n - \Delta(p))\sigma^{-1}(p) \xrightarrow{d} N(0,1),$$

where

$$\overline{T}_n = nb_n N_0 \int\limits_{\Omega_n} \big(\widehat{p}_n(x) - p(x)\big)^2 p_{2n}^2(x)\,dx.$$

4 Application of the Statistic T_n for Hypothesis Testing

The assertion of Theorem 2 makes it possible to construct a criterion of the asymptotic level α, $0 < \alpha < 1$, for testing the hypothesis H_0, according to which $p(x) = p_0(x)$, $x \in \Omega_n(\tau)$. The critical region is established by the inequality

$$T_n \geq q_n(\alpha), \tag{31}$$

where

$$q_n(\alpha) = \Delta(p_0) + \lambda_\alpha \sqrt{b_n}\,\sigma(p_0),$$

$$\Delta(p_0) = \int_0^1 p_0(x)(1 - p_0(x))\,dx \int_{|x|\le\tau} K^2(x)\,dx,$$

$$\sigma^2(p_0) = 2\int_0^1 p_0^2(x)(1 - p_0^2(x))\,dx \int_{|u|\le 2\tau} K_0^2(u)\,du,$$

and λ_α is defined by the equality $\Phi(\lambda_\alpha) = 1 - \alpha$.

Now, let us analyze the asymptotic property of test criterion (31) (i.e. the behavior of the test power as $n \to \infty$). In the first place, we consider the question whether the criterion is consistent. The following assertion is valid.

Theorem 3 *Let all conditions of Theorem 2 be fulfilled. Then for $n \to \infty$,*

$$\Pi_n(p) = \mathbf{P}_{H_1}\left\{T_n \ge q_n(\alpha)\right\} \longrightarrow 1 \;\; as \;\; n \to \infty,$$

which means that the criterion defined in (31) is consistent against any alternative $H_1 : p(x) \ne p_0(x)$, $0 \le x \le 1$.

Proof Denote

$$z_n(p) = b_n^{-\frac{1}{2}}\left(nb_n N_n \int_{\Omega_n(\tau)} \left[\widehat{p}_n(x) - p(x)\right]^2 p_{2n}^2(x)\,dx - \Delta(p)\right)\sigma^{-1}(p).$$

It is not difficult to show that

$$\Pi_n(p) = \mathbf{P}_{H_1}\left\{z_n(p) \ge -nb_n^{\frac{1}{2}} N_n\left(\int_0^1 (p(x) - p_0(x))^2\,dx + o_{\mathbf{P}}(1)\right)\right\}.$$

Since $z_n(p)$ has an asymptotically normal distribution with the parameters (0, 1) for the hypothesis H_1 and $nb_n^{\frac{1}{2}} N_n \to \infty$, we have $\Pi_n(p) \to 1$ as $n \to \infty$.

Thus, for any fixed alternative the power of the tes criterion based on T_n tends to 1. However, if with the change of n the alternative changes and approaches the basic hepothesis H_0, then the test criterion power will not necessarily converge to 1. As an example, let us consider the sequence of Pitman type alternatives close to the hypothesis H_0:

$$H_1 : p_1^{(n)}(x) = p_0(x) + \gamma_n \varphi(x) + o(\gamma_n), \quad \gamma_n \to 0.$$

Theorem 4 *Let $p_0(x), \varphi(x) \in C^{(2)}$, and $K(x) \in H(\tau)$. If $b_n = n^{-\delta}$, $N_n = n^{\beta}$ and $\gamma_n = n^{-\frac{1+\beta}{2}+\frac{\delta}{4}}$ $(\beta + 1 < 4\delta,\ 4\beta + 2\delta < 1,\ 4\beta < \delta)$, then for the alternative H_1 the statistic $b_n^{-\frac{1}{2}}(T_n - \Delta(p_0))\sigma^{-1}(p_0)$ has a normal limit distribution with the parameters*

$$\left(\frac{1}{\sigma(p_0)}\int_0^1 \varphi^2(x)\,dx,\ 1\right),$$

i.e. the limit power of the test criterion is equal to

$$1 - \Phi\left(\lambda_\alpha - \frac{1}{\sigma(p_0)}\int_0^1 \varphi^2(u)\,du\right).$$

Proof Let write T_n in the form

$$\begin{aligned} T_n &= nb_nN_n\int_{\Omega(\tau)}\left(\widehat{p}(x)-p_1^{(n)}(x)\right)^2 p_{2n}^2(x)\,dx + nb_nN_n\int_{\Omega(\tau)}\left(p_1^{(n)}(x)-p_0(x)\right)^2 p_{2n}^2(x)\,dx \\ &\quad +2nb_nN_n\int_{\Omega_n(\tau)}\left(\widehat{p}_n(x)-p_1^{(n)}(x)\right)\left(p_1^{(n)}(x)-p_0(x)\right)p_{2n}^2(x)\,dx \\ &= T_n^* + A_1(n) + A_2(n). \end{aligned}$$

Since $p_{2n}(x) = 1 + O\left(\frac{1}{nb_n}\right)$ uniformly with respect to $x \in \Omega_n(\tau)$, we write

$$b_n^{-\frac{1}{2}}A_1(n) = \int_0^1 \varphi^2(u)\,du + o(1). \tag{32}$$

Denote

$$D_n = \int_{\Omega_n(\tau)}\left(p_{1n}(x) - \mathbf{E}_1 p_{1n}(x)\right)\varphi(x)p_{2n}(x)\,dx,$$

where $\mathbf{E}_1$ is the mathemetical expectation for the hypothesis H_1. Then

$$\begin{aligned} b_n^{-\frac{1}{2}}A_2(n) &= nb_n^{\frac{1}{2}}\gamma_nN_nD_n + nb_n^{\frac{1}{2}}\gamma_nN_n\left(b_n^2 + O\left(\frac{1}{nb_n}\right)\right) \\ &= nb_n^{\frac{1}{2}}\gamma_nN_nD_n + O(nb_n^{\frac{5}{2}}\gamma_nN_n) + O\left(\frac{\gamma_nN_n}{\sqrt{b_n}}\right). \end{aligned} \tag{33}$$

Now we show that

$$nb_n^{\frac{1}{2}}\gamma_nN_nD_n \xrightarrow{\mathbf{P}} 0.$$

Indeed,

$$\mathbf{E}\,|D_n| \le (\mathbf{E}\,D_n^2)^{\frac{1}{2}} = \left\{ \frac{1}{nb_n} \int\limits_{\Omega_n(\tau)} \int\limits_{\Omega_n(\tau)} p_{2n}(x) p_{2n}(y)\, dx\, dy \right.$$
$$\left. \times \frac{1}{nb_n} N_n^{-1} \sum_{i=1}^{n} K\Big(\frac{x - x_i}{b_n}\Big) K\Big(\frac{y - x_i}{b_n}\Big) p_1^{(n)}(x_i)(1 - p_1^{(n)}(x_i)) \right\}^{\frac{1}{2}}.$$

Further, using Lemma 2, we easily ascertain that

$$\mathbf{E}\,|D_n| \le \left\{ \frac{1}{nb_n} \int\limits_{\Omega_n(\tau)} \int\limits_{\Omega_n(\tau)} p_{2n}(x) p_{2n}(y)\, dx\, dy \right.$$
$$\left. \times \left[\frac{1}{b_n N_n} \int_0^1 K\Big(\frac{x-u}{b_n}\Big) K\Big(\frac{y-u}{b_n}\Big) p_1^{(n)}(u)(1-p_1^{(n)}(u))\, du + O\Big(\frac{1}{nb_n}\Big) \right] \right\}^{\frac{1}{2}}$$
$$\le c_{21} \frac{1}{\sqrt{nN_n}} \Big(1 + \frac{1}{nb_n^2}\Big).$$

Therefore

$$nb_n^{\frac{1}{2}} \gamma_n N_n \mathbf{E}\,|D_n| \le c_{22} n^{-\frac{\delta}{4}}.$$

Thus

$$nb_n^{\frac{1}{2}} \gamma_n N_n D_n \xrightarrow{\mathbf{P}} 0. \tag{34}$$

Further, the random variable $b_n^{-\frac{1}{2}}(T_n^* - \Delta(p_1^{(n)}))\sigma^{-1}(p_1^{(n)})$ is asymptotically normal with the mean 0 and the dispersion 1. From this and (32), (33), (34) the proof of Theorem 4 follows.

The proof of the next theorem is absolutely analogous to that of Theorem 4.

Theorem 5 *Let $p_0(x)$, $\varphi(x)$ and $K(x)$ satisfy the conditions of Theorem 4. If $b_n = n^{-\delta}$, $\gamma_n = n^{-\frac{1}{2}+\frac{\delta}{4}}$, $\frac{1}{4} < \delta < \frac{1}{2}$, then for the alternative $H_1 :\ p_1^{(n)}(x) = p_0(x) + \gamma_n\varphi(x)$ the statistic $b_n^{-\frac{1}{2}}(\overline{T}_n - \Delta(p_0))\sigma^{-1}(p_0)$ has a normal limit distribution with the parameters*

$$\left(\frac{N_0}{\sigma(p_0)} \int_0^1 \varphi^2(x)\, dx,\ 1 \right),$$

i.e.

$$\mathbf{P}_{H_1}(\overline{T}_n \ge q_n(\alpha)) \longrightarrow 1 - \Phi\left(\lambda_\alpha - \frac{N_0}{\sigma(p_0)} \int_0^1 \varphi^2(x)\, dx \right).$$

Remark 1 It should be emphasized that the behavior of the estimate $\widehat{p}_n(x)$ near the boundary of the interval [0, 1] is worse than within the interval $[\tau b_n, 1 - \tau b_n]$ (see [4]). That is why to avoid difficulties associated with this boundary effect, we consider the integral mean-square deviation on $\Omega_n(\tau)$. However it can be shown that under the conditions of Theorems 1 and 2 the results obtained above are valid for the modified estimate (see [1, 4]) of the function $p(x)$.

Remark 2 In this paper, some results, which were obtained by one of the authors in [7] and presented there without the proof, are generalized and refined.

Remark 3 Let x_i be the partition points of the interval [0, 1] chosen so that $H(x_j) = \frac{2j-1}{2n}$, $j = 1, \ldots, n$, where $H(x) = \int\limits_0^x h(u)\,du$, $h(u)$ is some known continuous distribution density on [0, 1]. In that case, the generalization of the results of this paper can be obtained by arguments analogous to those used above.

References

1. Absava, R.M., Nadaraya, E.A.: Some problems of theory of non-parametric estimation of functional characteristics of the distribution law of observations. Izd. Tbiliss. Univ, Tbilisi (2005) (in Russian)
2. Copas, J.B.: Plotting p against x. Appl. Statist. **32**(2), 25–31 (1983)
3. Efromovich, S.: Nonparametric Curve Estimation: Methods, Theory, and Applications. Springer Series in Statistics. Springer, New York (1999)
4. Hart, J.D., Wehrly, ThE: Kernel regression when the boundary region is large, with an application to testing the adequacy of polynomial models. J. Amer. Statist. Assoc. **87**(420), 1018–1024 (1992)
5. Lipcer, R.Sh., Shirjaev, A.N.: A functional central limit theorem for semimartingales. Teor. Veroyatnost. i Primenen. **25**(4), 683–703 (1980) (in Russian)
6. Nadaraya, E.A.: On a regression estimate. Teor. Verojatnost. i Primenen. **9**, 157–159 (1964). (in Russian)
7. Nadaraya, E.: Limit distribution of a quadratic deviation of a nonparametric estimate of the Bernoulli regression. Bull. Georgian National Acad. Sci. **173**(2), 221–224 (2006)
8. Nadaraya, E., Babilua, P., Sokhadze, G.: Estimation of a distribution function by an indirect sample. Ukr. Mat. Zh. **62**(12), 1642–1658 (2010); Ukr. Math. J. **62**(12), 1906–1924 (2010)
9. Okumura, H., Naito, K.: Weighted kernel estimators in nonparametric binomial regression. The International Conference on Recent Trends and Directions in Nonparametric Statistics. J. Nonparametr. Stat. **16**(1–2), 39–62 (2004)
10. Staniswalis, J.G., Cooper, V.: Kernel estimates of dose response. Biometrics **44**(4), 1103–1119 (1988)
11. Watson, G.S.: Smooth regression analysis. Sankhya Ser. A **26**, 359–372 (1964)

Scaling Property for Fragmentation Processes Related to Avalanches

Lucian Beznea, Madalina Deaconu, and Oana Lupaşcu-Stamate

Abstract We emphasize a scaling property for the continuous time fragmentation processes related to a stochastic model for the fragmentation phase of an avalanche. We present numerical results that confirm the validity of the scaling property for our model, based on the appropriate stochastic differential equation of fragmentation and on a fractal property of the solution.

Keywords Scaling · Fragmentation stochastic differential equation · Jump process · Monte Carlo method

Mathematics Subject Classification 39A13 · 65C30 · 60J45 · 65C05 · 60J35

1 Introduction

In the papers [1–3] we studied binary fragmentation processes (and associated non-local branching processes, cf. [4]) of an infinite particles system, including a numerical approach for the time evolution of the fragmentation phase of an avalanche. A fractal property was emphasized for the process related to the avalanches.

L. Beznea (✉)
Simion Stoilow Institute of Mathematics of the Romanian Academy, Centre Francophone en Mathématique de Bucarest, P.O. Box 1-764, 014700 Bucharest, Romania
e-mail: lucian.beznea@imar.ro

Faculty of Mathematics and Computer Science, University of Bucharest, Bucharest, Romania

M. Deaconu
Université de Lorraine, CNRS, Inria, IECL, 54000 Nancy, France
e-mail: Madalina.Deaconu@inria.fr

O. Lupaşcu-Stamate
Gheorghe Mihoc–Caius Iacob Institute of Mathematical Statistics and Applied Mathematics of the Romanian Academy, Calea 13 Septembrie 13, 050711 Bucharest, Romania
e-mail: oana.lupascu_stamate@yahoo.com

G. Jaiani and D. Natroshvili (eds.), *Applications of Mathematics and Informatics in Natural Sciences and Engineering*, Springer Proceedings in Mathematics & Statistics 334, https://doi.org/10.1007/978-3-030-56356-1_3

In this paper we prove that a scaling property holds for the above mentioned process and we present numerical results that confirm the validity of this property.

The study of the scaling property is closely related to the study of the self-similarity property. In this direction some results were recently obtained by using deterministic approaches. In particular in [5] this study is performed by studying the asymptotic behaviour of the first eigenvalue, as it represents the asymptotic growth of the solution. In the presence of a transport term it is shown that this behaviour depends on wether transport dominates fragmentation or not. This equation has some applications in biology and medicine. Previously in [7] the self-similarity property was used for the coagulation-fragmentation equation to obtain the existence of a stationary solution for any given mass. In [10] the fragmentation is used to model cell-division and the authors prove the existence of a stable steady distribution.

The paper is organised as follows.

In Sect. 2 we present a general result, characterizing (in Theorem 1) the scaling property of Markov process in terms of the transition function, the associated resolvent of kernels, and of the generator. It is pointed out also the case of a pure jump process.

In Sect. 3 it is given the main application, by proving (in Corollary 1) that the weak solution of the stochastic differential equation of fragmentation for avalanches has a scaling property. As it was already mentioned, it is a second specific property emphasized for this SDE of fragmentation, the first one being the fractal property proved in [2], and for the reader convenience we presented it at the end of the section.

Finally, in Sect. 4 we discuss the numerical results, obtained by Monte Carlo simulation, that confirm the validity of the scaling property we proved.

2 Scaling Property for Jump Processes

Let E be Lusin topological space (i.e., E is homeomorphic to a Borel subset of a compact metric space) with Borel σ-algebra $\mathscr{B}(E)$. We denote by $p\mathscr{B}(E)$ (resp. $b\mathscr{B}(E)$) the set of all positive Borel measurable functions on E (resp. the set of all bounded real-valued Borel measurable functions on E).

Let $X = (\Omega, \mathscr{F}, \mathscr{F}_t, X_t, \theta_t, \mathbb{P}^x, \zeta)$ be the right Markov process on E having $(P_t)_{t\geqslant 0}$ as transition function, $P_t f(x) = \mathbb{E}^x(f(X_t), t < \zeta)$, $t \geqslant 0$, $f \in p\mathscr{B}(E)$. Let further $(U_\alpha)_{\alpha>0}$ be the associated sub-Markovian resolvent of kernels, $U_\alpha f := \int_0^\infty e^{-\alpha t} P_t f \mathrm{d}t$.

We consider the generator $(L, \mathscr{D}(L))$ of X as follows; cf. [6] pag. 55, and [9]. Let

$$\mathscr{B}_o := \{f \in b\mathscr{B}(E) : \lim_{t\searrow 0} P_t f = f \text{ pointwise on } E\}$$

and $\mathscr{D}(L)$ be the set of all $f \in \mathscr{B}_o$ such that $\left(\frac{P_t f(x)-f(x)}{t}\right)_{t,x}$ is bounded for $x \in E$ and t in a neighbourhood of zero, there exists $\lim_{t\searrow 0} \frac{P_t f - f}{t}$ pointwise and the above limit is an element of $\mathscr{B}_o$. Define the linear operator $L : \mathscr{D}(L) \longrightarrow b\mathscr{B}(E)$ as

$$Lf(x) := \lim_{t \searrow 0} \frac{P_t f(x) - f(x)}{t}, \quad f \in \mathscr{D}(L), \ x \in E.$$

The operator $(L, \mathscr{D}(L))$ is called the *weak generator* of X. Recall that $\mathscr{D}(L) = U_\alpha(\mathscr{B}_o)$ for all $\alpha > 0$, and if $f = U_\alpha g$, with $g \in \mathscr{B}_o$, then $(\alpha - L)f = g$.

We present now the classical construction of a jump process (see, e.g., [8], page 163), as we need it for the fragmentations processes related to avalanches.

Let N be a bounded kernel on E and denote by $\lambda(x)$ the total mass of the measure N_x, $x \in E$, $\lambda(x) := N1(x) \in E$. We set

$$\lambda_o := ||N1||_\infty \text{ and } N' := \frac{1}{\lambda_o} N + (1 - \frac{\lambda}{\lambda_o})I,$$

and define the bounded linear operator $\widetilde{N}$ on $b\mathscr{B}(E)$ as

$$\widetilde{N} f(x) = \lambda_o \int_E [f(y) - f(x)] N'_x(\mathrm{d}y) \text{ for all } f \in b\mathscr{B}(E) \text{ and } x \in E.$$

Then $\widetilde{N} = N - \lambda I = \lambda_o(N' - I)$ and it is the generator of a C_0–semigroup $(P_t)_{t \geqslant 0}$ on $b\mathscr{B}(E)$,

$$P_t := \mathrm{e}^{t\widetilde{N}}, \ t \geqslant 0.$$

Each P_t is a Markovian kernel on E, more precisely, $P_t f = \mathrm{e}^{-t\lambda_o} \sum_{k \geqslant 0} \frac{(\lambda_o t)^k}{k!} N'^k f$, where $N'^k := \underbrace{N' \circ \ldots \circ N'}_{k \text{ times}}$. The operator $\widetilde{N}$ is the generator of a (continuous time) pure jump Markov process $X = (X_t)_{t \geqslant 0}$ with state space E. Clearly, $\widetilde{N}$ is the weak generator of $(P_t)_{t \geqslant 0}$, with $\mathscr{D}(\widetilde{N}) = b\mathscr{B}(E)$.

The scaling property. Assume that E is a star-convex subset of $\mathbb{R}^d$, $d \geqslant 1$, i.e. there exists an x_o in E such that for all x in E the line segment from x_o to x is in E. For simplicity we suppose that $x_o = 0$. For a real-valued function f on E and $s \in (0, 1)$ we denote be f_s the function on E defined as $f_s(x) := f(sx)$, $x \in E$.

Let $n \in \mathbb{Z}$. A linear operator $(L, \mathscr{D}(L))$ on $b\mathscr{B}(E)$ is called *homogeneous of degree n* provided that for every $s \in (0, 1)$ and $f \in \mathscr{D}(L)$ one has $f_s \in \mathscr{D}(L)$ and $(Lf)_s = s^n L(f_s)$.

Clearly, the Laplace operator (in a star-convex subset of $\mathbb{R}^d$) is homogeneous of degree -2. In the next section we shall give examples of operators related to fragmentation processes, satisfying such a scaling property.

We can state now the main result of this section.

Theorem 1 *Let $n \in \mathbb{Z}$, $(P_t)_{t \geqslant 0}$ be the transition function of a right (Markov) process $(X, \mathbb{P}^x)$ with state space E, let $(L, \mathscr{D}(L))$ be the weak generator of $(P_t)_{t \geqslant 0}$, and $(U_\alpha)_{\alpha > 0}$ the associated resolvent.*

(i) The following assertions are equivalent.

$(i.a)$ *The transition function* $(P_t)_{t\geqslant 0}$ *satisfies*

$$(P_t f)_s = P_{ts^n}(f_s) \text{ for all } f \in bp\mathscr{B}(E), s \in (0,1), \text{ and } t \geqslant 0. \tag{1}$$

$(i.b)$ *The resolvent family* $(U_\alpha)_{\alpha>0}$ *satisfies*

$$s^n (U_\alpha f)_s = U_{\frac{\alpha}{s^n}}(f_s) \text{ for all } f \in bp\mathscr{B}(E), s \in (0,1), \text{ and } \alpha > 0 \tag{2}$$

$(i.c)$ *The weak generator* $(L, \mathscr{D}(L))$ *is homogeneous of degree* n.
$(i.d)$ *The process* X *has the following scaling property:*

$$\mathbb{E}^{sx}(X_t \in A) = \mathbb{E}^x(X_{ts^n} \in \frac{1}{s}A) \text{ for all } A \in \mathscr{B}(E), x \in E, s \in (0,1), \text{ and } t \geqslant 0. \tag{3}$$

(ii) *Assume that* N *is a kernel on* E *which is homogeneous of degree* n. *Then the pure jump Markov process having the generator* $\widetilde{N}$ *has the scaling property (3).*

Proof We clearly have $(i.a) \iff (i.d)$ because $(P_t)_{t\geqslant 0}$ is the transition function of X.

$(i.a) \Longrightarrow (i.b)$. We have $(U_\alpha f)_s = \int_0^\infty e^{-\alpha t}(P_t f)_s \mathrm{d}t = \int_0^\infty e^{-\alpha t} P_{ts^n}(f_s)\mathrm{d}t = \frac{1}{s^n} U_{\frac{\alpha}{s^n}}(f_s)$, where we used the hypothesis $(i.a)$ to get the second equality.

$(i.b) \Longrightarrow (i.a)$. With the same computation as before, we get from $(i.b)$ that for all $\alpha > 0$ we have $\int_0^\infty e^{-\alpha t}(P_t f)_s \mathrm{d}t = \int_0^\infty e^{-\alpha t} P_{ts^n}(f_s)\mathrm{d}t$. Since any bounded β-level excessive function belongs to $\mathscr{B}_o$, using a monotone class argument, in order to prove (1) we may assume that $f \in \mathscr{B}_o$ and therefore the real-valued functions $t \longmapsto (P_t f)_s(x)$ and $t \longmapsto P_{ts^n}(f_s)(x)$ are both right continuous on $[0,\infty)$ for every $x \in E$. By the uniqueness property of the Laplace transform we conclude now that (1) holds.

$(i.a) \Longrightarrow (i.c)$. Let $f \in \mathscr{D}(L)$. Observe first that if $g \in \mathscr{B}_o$ then for all $s \in (0,1)$ we have $g_s \in \mathscr{B}_o$ because $\lim_{t\searrow 0} P_t(g_s) = \lim_{t\searrow 0} P_{ts^n} g = g$. Consequently, if $f \in \mathscr{D}(L)$ then $(Lf)_s \in \mathscr{B}_o$ and we have pointwise $(Lf)_s = \lim_{t\searrow 0} \frac{(P_t f)_s - f_s}{t} = s^n \lim_{t\searrow 0} \frac{P_{ts^n}(f_s) - f_s}{ts^n} = s^n L(f_s)$. As a consequence f_s also belongs to $\mathscr{D}(L)$, hence $(L, \mathscr{D}(L))$ is homogeneous of degree n.

$(i.c) \Longrightarrow (i.b)$. As before, in order to prove (2), we may suppose that $f \in \mathscr{B}_o$ and let $g := U_\alpha f \in \mathscr{D}(L)$. Because g_s also belongs to $\mathscr{D}(L)$, there exists $h \in \mathscr{B}_o$ such that $g_s = U_{\frac{\alpha}{s^n}} h$. We have $(Lg)_s = \alpha(U_\alpha f)_s - f_s = \alpha g_s - f_s = \alpha U_{\frac{\alpha}{s^n}} h - f_s$. We have also $L(g_s) = L(U_{\frac{\alpha}{s^n}} h) = \frac{\alpha}{s^n} U_{\frac{\alpha}{s^n}} h - h$. The equality $(Lg)_s = s^n L(g_s)$ is therefore equivalent with $\alpha U_{\frac{\alpha}{s^n}} h - f_s = \alpha U_{\frac{\alpha}{s^n}} h - s^n h$. Hence $f_s = s^n h$, $U_{\frac{\alpha}{s^n}}(f_s) = s^n U_{\frac{\alpha}{s^n}} h = s^n g_s = s^n (U_\alpha f)_s$.

(ii) Observe first that the hypothesis on N implies that $\lambda(sx) = (N1)_s(x) = s^n N1(x) = s^n \lambda(x)$ for all $x \in E$ and $s \in (0,1]$. It follows that the kernel λI is homogeneous of degree n and we deduce that $\widetilde{N} = N - \lambda I$ is also homogeneous of degree n. The scaling property of X is now a consequence of the equivalence $(i.c) \iff (i.d)$, since $\widetilde{N}$ is its generator. □

Remark 1 By Theorem 1 if follows that:
If N is a kernel on E which is homogeneous of degree n, then the induced semigroup $P_t = e^{t\widetilde{N}}, t \geqslant 0$, *satisfies the scaling property* (1).
One can give a direct, alternative proof for this assertion. Indeed, observe first that the hypothesis on N implies that $\lambda(sx) = (N1)_s(x) = s^n N1(x) = s^n\lambda(x)$ for all $x \in E$ and $s \in (0, 1]$. It results that the kernel λI is homogeneous of degree n and we deduce that $\widetilde{N} = N - \lambda I$ is also homogeneous of degree n. Therefore $(\widetilde{N}^k f)_s = s^{kn}\widetilde{N}^k(f_s)$ for any $k \in \mathbb{N}^*$ and we conclude that $(P_t f)_s = \sum_{k\geqslant 0} \frac{t^k}{k!}(\widetilde{N}^k f)_s = \sum_{k\geqslant 0} \frac{(ts^n)^k}{k!}\widetilde{N}^k(f_s) = P_{ts^n}(f_s)$.

3 Scaling Property for the SDE of Fragmentation

In this section we consider the framework from [2].

Discontinuous fragmentation kernels for avalanches. We describe first a binary fragmentation model. Consider an infinite system of particles, each particle being characterized by its mass. As time evolves the particles perform fragmentation, that is one particle can split into two smaller particles by conserving the total mass. Let F be a *fragmentation kernel*, that is, a symmetric function $F : (0, 1]^2 \longrightarrow \mathbb{R}_+$. Here $F(x, y)$ represents the rate of fragmentation of a particle of size $x + y$ into two particles of sizes x and y.

The following assumption is suggested by the so called *rupture properties*, emphasized in the deterministic modelling of the snow avalanches:

(H) There exists a function $\Phi : (0, \infty) \longrightarrow (0, \infty)$ such that $F(x, y) = \Phi\left(\frac{x}{y}\right)$ for all $x, y > 0$.

Since the fragmentation kernel F is assumed to be a symmetric function, we have $\Phi(z) = \Phi\left(\frac{1}{z}\right)$ for all $z > 0$. An example is as follows. Fix a "ratio" r, $0 < r < 1$, and consider the fragmentation kernel $F^r : [0, 1]^2 \longrightarrow \mathbb{R}_+$, defined as $F^r(x, y) := \frac{1}{2}(\delta_r(\frac{x}{y}) + \delta_{1/r}(\frac{x}{y}))$, if $x, y > 0$, and $F^r(x, y) := 0$ if $xy = 0$.

One can see that the fragmentation kernel F^r satisfies condition (H), more precisely we have $F^r(x, y) = \Phi^r(\frac{x}{y})$ for all $x, y > 0$, where $\Phi^r : (0, \infty) \longrightarrow (0, \infty)$ is defined as $\Phi^r(z) := \frac{1}{2}(\delta_r(z) + \delta_{1/r}(z))$, $z > 0$. Clearly, the function Φ^r is not continuous. By approximating the function Φ^r with a convenient sequence of continuous functions, one can see that the kernel N^{F^r} associated with F^r is given by the following linear combination of Dirac measures:

$$N_x^{F^r} := \lambda_o(\beta x\delta_{\beta x} + (1 - \beta)x\delta_{(1-\beta)x}), \tag{4}$$

where $\lambda_o := \frac{\beta^2+(1-\beta)^2}{4}$ with $\beta := \frac{r}{1+r}$. In this case the kernel N^{F^r} is no more Markovian and has no density with respect to the Lebesgue measure.

The corresponding stochastic differential equation of fragmentation. To emphasize the stochastic differential equation of fragmentation which is related to our stochastic model for the avalanches, we consider the kernel N^{F^r} on $E := [0, 1]$ and the associated pure jump process $X = (X_t)_{t \geqslant 0}$ with state space E.

We state now the *stochastic differential equation of fragmentation for avalanches*:

$$X_t = X_0 - \int_0^t \int_0^1 \Big((1-\beta) X_{\alpha -} \mathbb{1}_{[\frac{s}{\beta \lambda_o} < X_{\alpha -} \leqslant 1]} + \beta X_{\alpha -} \mathbb{1}_{[\frac{s}{\lambda_o} < X_{\alpha -} \leqslant \frac{s}{\beta \lambda_o}]} \Big) \, p(\mathrm{d}\alpha, \mathrm{d}s), \tag{5}$$

where $p(\mathrm{d}\alpha, \mathrm{d}s)$ is a Poisson measure with intensity $q := \mathrm{d}\alpha \mathrm{d}s$.

Recall that the solution X of (5) describes the time evolution of the size of a typical particle as follows. At some exponential random instants of parameter λ_0, either, with probability $1 - X$, no fragmentation occurs for the typical particle, or else, it breaks into two smaller particles: we subtract $(1 - \beta)X$ from X with probability βX, or βX with probability $(1 - \beta)X$. The conditions on the particle size are induced by the specific property of an avalanche, depending on β.

The existence of the weak solution to the Eq. (5) was proved in [2]. The next corollary shows that this solution satisfies the claimed scaling property.

Corollary 1 *The weak solution of the stochastic differential equation of fragmentation for avalanches (5), with the initial distribution δ_x, $x \in E$, is equal in distribution with $(X, \mathbb{P}^x)$ and the following scaling property holds:*

$$\mathbb{E}^{sx}(X_t \in A) = \mathbb{E}^x(X_{ts} \in \frac{1}{s} A) \ \textit{for every } x \in E, t \geqslant 0, A \in \mathscr{B}(E), \ \textit{and } s \in (0, 1].$$

Proof We show first that the kernel N^{F^r} is homogeneous of degree one ($n = 1$). It is sufficient to show that a kernel K of the form $Kf(x) := xf(\beta x)$ has this property. We have indeed $Kf(sx) = sxf(\beta sx) = sK(f_s)(x)$.

The scaling property follows now by assertion (ii) of Theorem 1 (see also Remark 1) because we know that the generator of X is $\widetilde{N^{F^r}}$. □

The fractal property. We consider a sequence $(d_n)_{n \geqslant 1}$ such that $d_1 < \beta \leqslant 1/2$ and $d_{n+1}/d_n < \beta$ for all $n \geqslant 1$. Let $n \geqslant 1$ be fixed and define

$$E_n := [d_n, 1], \, E'_n := [d_{n+1}, d_n), \text{ and } E'_0 = E_1.$$

Then clearly $E_n = \bigcup_{k=1}^n E'_{k-1}$.

The kernel N^{F^r} given by (4) is used to define the kernel N^r_n on E_n as

$$N^r_n f := \sum_{k=1}^n \mathbb{1}_{E'_{k-1}} N^{F^r}(f \mathbb{1}_{E'_{k-1}}) \text{ for all } f \in pb\mathscr{B}(E_n).$$

Further, we consider the first order integral operator $\mathscr{F}^r_n$,

$$\mathscr{F}^r_n f(x) := \widetilde{N^r_n} f(x) = \int_{E_n} [f(y) - f(x)](N^r_n)_x(\mathrm{d}y) \text{ for all } f \in pb\mathscr{B}(E_n) \text{ and } x \in E_n.$$

The operator $\mathscr{F}_n^r$ is the generator of a (continuous time) jump Markov process $X^{r,n} = (X_t^{r,n})_{t \geqslant 0}$. Its transition function is $P_t^{r,n} := e^{\mathscr{F}_n^r t}, t \geqslant 0$.

For every $x \in E$ let

$$E_{\beta,x} := \{\beta^i(1-\beta)^j x : i, j \in \mathbb{N}\} \cup \{0\} \text{ and } E_{\beta,x,n} := E_{\beta,x} \cap E_n \text{ for } n \geqslant 1.$$

We can state now the fractal property of the process $X^{r,n}$, proved in [2].

Theorem 2 *If $n \geqslant 1$ then the following assertions hold for the Markov process $X^{r,n}$ with state space E_n and transition function $(P_t^{r,n})_{t \geqslant 0}$.*

(i) If $t \geqslant 0$ and $x \in E_n$ then $P_t^{r,n}(\mathbb{1}_{(x,1]})(x) = 0$.
(ii) For every $\phi \in pb\mathscr{B}(E_n)$ and each probability ν on E_n, the process $\phi(X_t^{r,n}) - \int_0^t \mathscr{F}_n^r \phi(X_s^{r,n})\mathrm{d}s$, $t \geqslant 0$, is a martingale under $\mathbb{P}^\nu$, with respect to the natural filtration of $X^{r,n}$.
(iii) If $x \in E_n$ then $\mathbb{P}^x$-a.s. $X_t^{r,n} \in E_{\beta,x,n}$ for all $t \geqslant 0$.

4 Numerical Results

Let $A \subset [0, 1]$ be a fixed set and $x \in [0, 1]$. By Corollary 1 for all $n \in \mathbb{N}^*$ and all time $t > 0$ we have the following scaling property:

$$\mathbb{E}^x(X_t \in A) = \mathbb{E}^{\frac{1}{n}x}\left(X_{nt} \in \frac{1}{n}A\right) \text{ for all } t \geqslant 0. \tag{6}$$

The relation (6) indicates that the probability that the process starting from x is in the set A at time t is exactly the probability that the process starting from x/n, belongs to the smaller set A/n at time nt. The key point of the equality (6) is that it depends on n only on the right hand side.

To test numerically the relation (6) we use a Monte Carlo simulation for the stochastic differential equation of fragmentation given by (5).

We fix a set $A \subset [0, 1]$, a point $x \in [0, 1]$, a final time $T \in \mathbb{R}_+^*$ and $n \in \mathbb{N}^*$. In the first step, we sample values of X_T starting from x as a solution of the corresponding stochastic differential equation of fragmentation with the discontinuous kernel F^r, by using the algorithm developed and implemented in [3]. For the reader's convenience we recall it below, we can remark the fractal property of the resulting fragments after the splitting, property which holds according to assertion (iii) of Theorem 2. In the second step we compute the probability that the samples of X_T belong to the set A. Then, we compare it with the probability that the process X_{nT}, starting from x/n, belongs to $\frac{1}{n}A$.

We fix the parameter $\beta < \frac{1}{2}$ and a final time T.

Algorithm
Step 0: `Sampling the initial particle` $X_0 \sim Q_0$
Step p: `Sampling a random variable` $S_p \sim \text{Exp}(\lambda_0)$
`Set` $T_p = T_{p-1} + S_p$
`Set` $X_t = X_{p-1}$ `for each` $t \in [T_{p-1}, T_p)$
`Set` $X_p = \beta X_{p-1}$ `with probability` βX_{p-1},

Table 1 Monte Carlo estimators for 10^4 simulations for $x = 1,\ \beta = \frac{1}{6},\ T = 20$ fixed, and different values of n

n	$\mathbb{E}^x(X_t \in A)$	$\mathbb{E}^{\frac{1}{n}x}\left(X_t \in \frac{1}{n}A\right)$
3	1	0.9998
10	1	1
20	1	1
30	1	1
40	1	1
50	1	1

Table 2 Monte Carlo estimators for 10^4 simulations for $x = 1, n = 3,\ T = 20$ fixed, and different values of β

n	$\mathbb{E}^x(X_t \in A)$	$\mathbb{E}^{\frac{1}{n}x}\left(X_t \in \frac{1}{n}A\right)$
$\frac{1}{6}$	1	0.9998
$\frac{1}{3}$	1	1
$\frac{1}{9}$	1	1

Table 3 Monte Carlo estimators for 10^4 simulations for $x = 1, n = 3,\ \beta = \frac{1}{6}$ fixed, and different values of T

n	$\mathbb{E}^x(X_t \in A)$	$\mathbb{E}^{\frac{1}{n}x}\left(X_t \in \frac{1}{n}A\right)$
20	1	1
30	1	1

$X_p = (1-\beta)X_{p-1}$ `with probability` $(1-\beta)X_{p-1}$,
`or` $X_p = X_{p-1}$ `with probability` $1 - X_{p-1}$

Stop: `When` $T_p > T$.
Outcome: The approximated particle mass at time T, X_{p-1}.

To implement the above Monte Carlo simulation associated to the relation (6), we fix the set A a union of disjoint intervals, $A = [0, \frac{1}{4}] \cup [\frac{1}{2}, \frac{3}{4}]$, the starting point $x = 1$, that does not belong to A. We consider the Monte Carlo parameter 10^4. Notice that the fractal character of the particles is encoding in the ratio β.

In Table 1 we give the Monte Carlo estimator for each one of the terms of relation (6), for $\beta = \frac{1}{6},\ T = 20$ fixed, and different values of n. In Table 2 we illustrate the Monte Carlo estimator of the each term of relation (6) for $n = 3,\ T = 20$ fixed, and different values of β. In Table 3 is written down the Monte Carlo estimator of the each term of relation (6), for different values of n, with parameter 10^4.

We represent in Fig. 1 the evolution in time of $t \longmapsto \mathbb{E}^x(X_t \in A)$ the red color and $t \longmapsto \mathbb{E}^{\frac{1}{n}x}\left(X_{nt} \in \frac{1}{n}A\right)$ the blue color in the time interval $t \in [50, 100]$ for β, A, x chosen above, and $n = 3$. Remark that in large time the red trajectory is very close to the blue one, that suggest the validity of the relation (6).

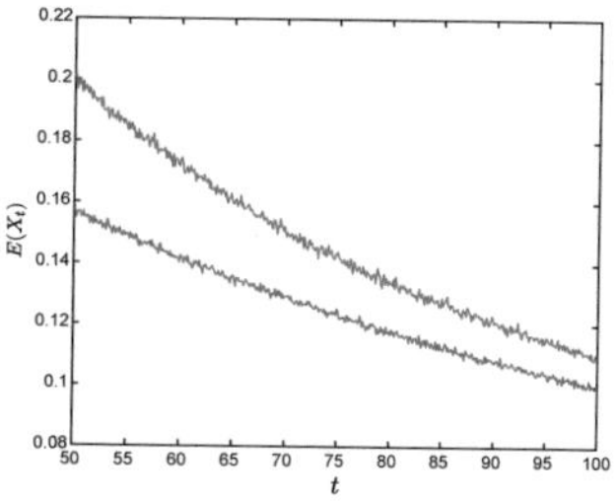

Fig. 1 The path of Monte Carlo approximation for $t \longmapsto \mathbb{E}^x(X_t \in A)$, the red color, and $t \longmapsto \mathbb{E}^{\frac{1}{n}x}\left(X_{nt} \in \frac{1}{n}A\right)$, the blue color, for $\beta = \frac{1}{6}$, $A = [0, \frac{1}{4}] \cup [\frac{1}{2}, \frac{3}{4}]$, $n = 3$, the Monte Carlo parameter 10^4, and the Euler step 10^{-3}

Acknowledgements For the third named author this work was supported by a grant of the Romanian National Authority for Scientific Research and Innovation, CNCS-UEFISCDI, project number PN-III-P1-1.1-PD- 2016-0293, within PNCDI III. Support from GDRI ECO-Math is kindly acknowledged.

References

1. Beznea, L., Deaconu, M., Lupaşcu, O.: Branching processes for the fragmentation equation. Stoch. Process. Their Appl. **25**, 1861–1885 (2015)
2. Beznea, L., Deaconu, M., Lupaşcu, O.: Stochastic equation of fragmentation and branching processes related to avalanches. J. Stat. Phys. **162**, 824–841 (2016)
3. Beznea, L., Deaconu, M., Lupaşcu-Stamate, O.: Numerical approach for stochastic differential equations of fragmentation; application to avalanches. Math. Comput. Simul. **160**, 111–125 (2019)
4. Beznea, L., Lupaşcu, O.: Measure-valued discrete branching Markov processes. Trans. Am. Math. Soc. **368**, 5153–5176 (2016)
5. Calvez, V., Doumic, M., Gabriel, P.: Self-similarity in a general aggregation–fragmentation problem. Application to fitness analysis. J. Math. Pures Appl. **98**, 1–27 (2012)
6. Dynkin, E.B.: Markov Processes, vol. I. Springer, Berlin (1965)
7. Escobedo, M., Mischler, S., Rodriguez Ricard, M.: On self-similarity and stationary problem for fragmentation and coagulation models. Ann. I. H. Poincaré. **22**, 9–125 (2005)
8. Ethier, N.S., Kurtz, T.G.: Markov Processes–Characterization and Convergence. Willey-Interscience, Hoboken (1986)
9. Fitzsimmons, P.J.: Construction and regularity of measure-valued Markov branching processes. Isr. J. Math. **64**, 337–361 (1988)
10. Perthame, B., Ryzhik, L.: Exponential decay for the fragmentation or cell-division equation. J. Differ. Equ. **210**, 155–177 (2005)

Conflict Resolution Models and Resource Minimization Problems

Temur Chilachava and George Pochkhua

Abstract Nonlinear mathematical models of economic cooperation between two politically (non-military confrontation) mutually opposing sides (two countries or a country and its legal region) are proposed, which consider economic cooperation between parts of the population of the sides, aimed at rapprochement of the sides and peaceful settlement of conflicts. Qualitatively different four mathematical models are considered: in the first case, the process of economic cooperation is free from political pressure; in the second case, the governments of both sides interfere with the process of economic cooperation; in the third case, the governments of both sides encourage the process of economic cooperation; in the fourth case, the government of first side interferes, and the government of the second side promotes cooperation. With some dependencies between constant coefficients of the first model, the first integrals and exact analytical solutions are found. A theorem has been proven to optimize (minimize) the financial resources by which economic cooperation can peacefully resolve political conflict. In the case of variable coefficients, computer simulations were performed for all four mathematical models using the MATLAB software environment to solve numerically the Cauchy problem for nonlinear dynamic systems. Minimum values of management parameters (optimization of financial resources) are found, at which conflicts can be resolved.

1 Introduction

Synergetics gave a powerful push using of mathematical models in social sciences: sociology, history, demography, political science, conflicting science, etc. Creation of mathematical models is more original in social sphere, because, they are more difficult to substantiate [1–6].

T. Chilachava (✉) · G. Pochkhua
Sokhumi State University, 61 Politkovskaya Str., 0186 Tbilisi, Georgia
e-mail: temo_chilachava@yahoo.com

G. Pochkhua
e-mail: gia.pochkhua@gmail.com

G. Jaiani and D. Natroshvili (eds.), *Applications of Mathematics and Informatics in Natural Sciences and Engineering*, Springer Proceedings in Mathematics & Statistics 334, https://doi.org/10.1007/978-3-030-56356-1_4

In 2005, mathematicians Robert Aumann and Thomas Schelling won the Nobel Prize in Economics for the scientific work cycle Understanding of the problems of the conflict and cooperation through the game theory.

Regarding to the conflict, the "repeated game principle presents another important methodological aspect of mathematical modeling (game theory). According to this principle: the long-term relationship of subjects in competition can generate cooperation between them, for which, there cannot be found a sufficient basis in case of one time relationship (contact). In other words, long-term relationship generates common interests and preconditions for cooperation [7–9].

Lee Kuan Yew, author of the Singaporean "Economic Miracle", noted: "If you want economic growth, do not break out the war with neighbors, establish trade relations with them, instead".

Taking into consideration the existing political conflict regions in the world, we consider efficient mathematical modeling and the corresponding computer simulations very perspective in determining conditions leading to the solving of conflicts (cf. [10–12]).

2 Description of Mathematical Models

The nonlinear mathematical model (the dynamic system) of economic cooperation between two political warring sides offered by us, read as:

$$\begin{cases} \frac{dN_1(t)}{dt} = -\alpha_1(t)\,[a(t) - N_1(t)]\,[b(t) - N_2(t)] + \beta_1(t)N_1(t)N_2(t) + F_1(t, N_1(t)) \\ \frac{dN_2(t)}{dt} = -\alpha_2(t)\,[a(t) - N_1(t)]\,[b(t) - N_2(t)] + \beta_2(t)N_1(t)N_2(t) + F_2(t, N_2(t)) \end{cases}, \quad (1)$$

$$N_1(0) = N_{10}, \; N_2(0) = N_{20}. \quad (2)$$

Depending on the functions $F_1(t, N_1(t))$, $F_2(t, N_2(t))$, we will get four qualitatively different social processes, leading respectively to four different mathematical models:

The first model (the process of economic cooperation is free from political pressure):

$$F_1(t, N_1(t)) \equiv 0, \; F_2(t, N_2(t)) \equiv 0. \quad (3)$$

The second model (the governments of both sides interfere with the process of economic cooperation, influencing various levels of pressure upon the citizens inclined to mutual economic cooperation):

$$F_1(t, N_1(t)) = -\delta_1(t)N_1(t), \; F_2(t, N_2(t)) = -\delta_2(t)N_2(t). \quad (4)$$

The third model (the governments of both sides encourage the process of economic cooperation):

$$F_1(t, N_1(t)) = \gamma_1(t)\,[a(t) - N_1(t)]\,,\ F_2(t, N_2(t)) = \gamma_2(t)\,[b(t) - N_2(t)]\,. \quad (5)$$

The four model (the government of the first side interferes, and the government of the second side promote cooperation):

$$F_1(t, N_1(t)) = -\delta_1(t)N_1(t)),\ F_2(t, N_2(t)) = \gamma_2(t)\,[b(t) - N_2(t)]\,. \quad (6)$$

In mathematical models (1)–(6), the following notation are employed:

$N_1(t)$—number of the citizens of the first side in time-point t, wishing or already being in economic cooperation and inclined to the subsequent peaceful resolution of the conflict;

$N_2(t)$—number of the citizens of the second side in time-point t, wishing or already being in economic cooperation and inclined to the subsequent peaceful resolution of the conflict;

$\alpha_1(t), \alpha_2(t)$—coefficients of aggression (alienation) of the sides;

$\beta_1(t), \beta_2(t)$—coefficients of cooperation of the sides;

$\delta_1(t), \delta_2(t)$—coefficients of coercion to aggression (alienation) of the sides (model 2, model 4);

$\gamma_1(t), \gamma_2(t)$—coefficients of coercion to cooperation of the sides (model 3, model 4);

$a(t), b(t)$—the population according to the first and second sides in time-point t;

$N_1, N_2 \in C^1[0, T]$;

T—time interval for model (conflict) consideration.

We assume that weak condition of conflict resolution are (in the mathematical model we assume that the conflict is resolved if at the same time more than half of the population of both sides support the process of economic cooperation, which promotes political reconciliation; simple majority of the population):

$$\begin{cases} \frac{a(t)}{2} < N_1(t) \le a(t) \\ \frac{b(t)}{2} < N_2(t) \le b(t) \end{cases}, \quad t \ge t_1, \quad (7)$$

and strong condition of conflict resolution are (in the mathematical model we assume that the conflict is resolved if at the same time more than two thirds of the population of both sides support the process of economic cooperation, which promotes political reconciliation; the qualified most of the population):

$$\begin{cases} \frac{2a(t)}{3} < N_1(t) \le a(t) \\ \frac{2b(t)}{3} < N_2(t) \le b(t) \end{cases}, \quad t \ge t_1. \quad (8)$$

3 First Mathematical Model in the Case of Constant Coefficients

Let's consider a special case of constant coefficients of the first mathematical model (1)–(3):

$$\begin{gathered}\alpha_i(t) = \alpha_i = const > 0,\ \beta_i(t) = \beta_i = const > 0, i = 1, 2.,\\ a(t) = a = const,\ b(t) = b = const.\end{gathered} \tag{9}$$

Here β_1 and β_2 are coefficients (factors) of cooperation of the sides, depending on financial support (investments) of international peacekeeping organizations, promoting the process of economic cooperation of the sides, i.e. are parameters of management.

It is natural and interesting to find the minimum value of these factors at which the conflict can be resolved, i.e. to find the minimum values of external investments that facilitate the process of cooperation between the sides, in order to resolve the conflict.

The following theorem specifies conditions for model coefficient values and initial data under which a political conflict is resolved.

Theorem 1 *If the following conditions hold*

$$\frac{\beta_1}{\alpha_1} = \frac{\beta_2}{\alpha_2} = \frac{1}{p} > 1, \tag{10}$$

$$\beta_1 > \alpha_1 \left(1 + \frac{a}{N_{10}} \cdot \frac{b}{N_{20}} - \frac{a}{N_{10}} - \frac{b}{N_{20}}\right), \tag{11}$$

then the exact analytical solution to Cauchy's problem

$$\begin{cases} \frac{dN_1(t)}{dt} = -\alpha_1 [a - N_1(t)] [b - N_2(t)] + \beta_1 N_1(t) N_2(t) \\ \frac{dN_2(t)}{dt} = -\alpha_2 [a - N_1(t)] [b - N_2(t)] + \beta_2 N_1(t) N_2(t) \end{cases}, \tag{12}$$

$$N_1(0) = N_{10},\ N_2(0) = N_{20},$$

meets conditions

$$\begin{cases} \frac{a}{2} < N_1(t_1) \le a \\ \frac{b}{2} < N_2(t_1) \le b \end{cases}, \tag{13}$$

when

$$t_1 = \max\left\{ \frac{1}{\sqrt{\varepsilon^2 + 4\delta^2(1-p)\beta_2}} \ln\left[\frac{\frac{a}{2} - N_{13}}{\frac{a}{2} - N_{14}} \frac{N_{10} - N_{14}}{N_{10} - N_{13}}\right];\right.$$

$$\frac{1}{\sqrt{\varepsilon^2+4\delta^2(1-p)\beta_2}}\ln\left[\frac{s-N_{13}}{s-N_{14}}\frac{N_{10}-N_{14}}{N_{10}-N_{13}}\right]\Bigg\},\ s=\frac{\beta_1}{\beta_2}\left(\frac{b}{2}-N_{20}\right)+N_{10},$$

$$\delta^2\equiv pa[\beta_2N_{10}+(b-N_{20})\beta_1],\qquad \varepsilon\equiv\beta_1pb+\beta_2pa+\beta_1(1-p)N_{20}-(1-p)\beta_2N_{10}, \tag{14}$$

$$N_{13}=-\frac{\varepsilon}{2(1-p)\beta_2}+\frac{\sqrt{\varepsilon^2+4\delta^2(1-p)\beta_2}}{2\,(1-p)\,\beta_2}>0,$$

$$N_{14}=-\frac{\varepsilon}{2(1-p)\beta_2}-\frac{\sqrt{\varepsilon^2+4\delta^2(1-p)\beta_2}}{2\,(1-p)\,\beta_2}<0.$$

Proof From (10), (12), it is easy to find the first integral of the system of nonlinear differential equations (12)

$$N_2(t)=N_{20}+\frac{\beta_2}{\beta_1}(N_1(t)-N_{10}). \tag{15}$$

Substituting (15) in the first equation of system (12), we get

$$\begin{gathered}\tfrac{dN_1(t)}{dt}=-pa\,[\beta_1\,(b-N_{20})+\beta_2N_{10}]+\\ +[\beta_1pb+\beta_2pa+\beta_1\,(1-p)\,N_{20}-(1-p)\,\beta_2N_{10}]\,N_1+\\ +(1-p)\,\beta_2N_1^2,\end{gathered} \tag{16}$$

$$N_1(0)=N_{10}.$$

Introduce the notation:

$$\begin{gathered}\delta^2\equiv pa[\beta_2N_{10}+(b-N_{20})\beta_1],\\ \varepsilon\equiv\beta_1pb+\beta_2pa-\beta_1(p-1)N_{20}+(p-1)\beta_2N_{10}>0.\end{gathered} \tag{17}$$

Then Cauchy's problem (16) can be rewritten in the following form

$$\frac{dN_1(t)}{dt}=(1-p)\,\beta_2\left\{\left[N_1+\frac{\varepsilon}{2\,(1-p)\,\beta_2}\right]^2-\frac{\varepsilon^2+4\delta^2\,(1-p)\,\beta_2}{4\,(p-1)^2\,\beta_2^2}\right\}. \tag{18}$$

Further, let

$$N_{13}\equiv-\frac{\varepsilon}{2(1-p)\beta_2}+\frac{\sqrt{\varepsilon^2+4\delta^2(1-p)\beta_2}}{2(1-p)\beta_2}>0, \tag{19}$$

$$N_{14}\equiv-\frac{\varepsilon}{2(1-p)\beta_2}-\frac{\sqrt{\varepsilon^2+4\delta^2(1-p)\beta_2}}{2(1-p)\beta_2}<0. \tag{20}$$

Then from (18) we get

$$\frac{dN_1(t)}{dt} = (1-p)\,\beta_2\,[N_1(t) - N_{13}]\,[N_1(t) - N_{14}]\,. \tag{21}$$

Taking into account initial condition (16), one can find the exact solution of equation (21),

$$N_1(t) = \frac{N_{13}(N_{10} - N_{14}) - N_{14}(N_{10} - N_{13})\exp\{t\sqrt{\varepsilon^2 + 4\delta^2(1-p)\beta_2}\}}{N_{10} - N_{14} - (N_{10} - N_{13})\exp\{t\sqrt{\varepsilon^2 + 4\delta^2(1-p)\beta_2}\}}. \tag{22}$$

Analysis of the exact solution (15), (22) shows that the following relations hold:

$$N_1(t_*) = \frac{a}{2},\ t_* = \frac{1}{\sqrt{\varepsilon^2 + 4\delta^2(1-p)\beta_2}} \ln\left[\frac{\frac{a}{2} - N_{13}}{\frac{a}{2} - N_{14}}\frac{N_{10} - N_{14}}{N_{10} - N_{13}}\right], \tag{23}$$

$$N_1(t) > \frac{a}{2},\ t > t_*,$$

$$N_2(t_{**}) = \frac{b}{2},\ t_{**} = \frac{1}{\sqrt{\varepsilon^2 + 4\delta^2(1-p)\beta_2}} \ln\left[\frac{s - N_{13}}{s - N_{14}}\frac{N_{10} - N_{14}}{N_{10} - N_{13}}\right], \tag{24}$$

$$s = \frac{\beta_1}{\beta_2}\left(\frac{b}{2} - N_{20}\right) + N_{10},\ N_1(t) > \frac{a}{2},\ t > t_{**}.$$

At the same time, in accordance with (21), it is natural to assume that the initial condition satisfies the inequality

$$N_{10} > N_{13}, \tag{25}$$

which automatically guarantee remove positivity of expressions within logarithms (23), (24).

Taking into account (17) and (19), inequality (25) leads to the following inequality

$$p < \frac{N_{10}N_{20}}{ab - aN_{20} - bN_{10} + N_{10}N_{20}},$$

which, in turn, is equivalent to the inequality (11)

$$\beta_1 > \alpha_1\left(1 + \frac{a}{N_{10}}\cdot\frac{b}{N_{20}} - \frac{a}{N_{10}} - \frac{b}{N_{20}}\right).$$

Thus, when

$$t = t_1 \equiv \max\{t_*; t_{**}\},$$

$$t_1 = \max\left\{\frac{1}{\sqrt{\varepsilon^2 + 4\delta^2(1-p)\beta_2}} \ln\left[\frac{\frac{a}{2} - N_{13}}{\frac{a}{2} - N_{14}}\frac{N_{10} - N_{14}}{N_{10} - N_{13}}\right];\right.$$

$$\left.\frac{1}{\sqrt{\varepsilon^2+4\delta^2(1-p)\beta_2}}\ln\left[\frac{s-N_{13}}{s-N_{14}}\frac{N_{10}-N_{14}}{N_{10}-N_{13}}\right]\right\},$$

the system of inequalities (13) holds, which completes the proof.

The above theorem, for the first mathematical model in the case of constant model coefficients, allows for fixed initial conditions (population of the sides (zero demographic factors); the initial population of the sides prone to economic cooperation; factors of aggressiveness of parts of the population of the sides preventing economic cooperation) to obtain minimum values for parameters of management, in which political conflict will be resolved

$$\begin{aligned}\beta_1 > \beta_{1\min} = \alpha_1\left(1+\tfrac{a}{N_{10}}\cdot\tfrac{b}{N_{20}}-\tfrac{a}{N_{10}}-\tfrac{b}{N_{20}}\right),\\ \beta_2 > \beta_{2\min} = \alpha_2\left(1+\tfrac{a}{N_{10}}\cdot\tfrac{b}{N_{20}}-\tfrac{a}{N_{10}}-\tfrac{b}{N_{20}}\right).\end{aligned} \tag{26}$$

4 Computer Modeling in the Case of Variable Model Coefficients. Optimization of Management Parameters

Let us consider mathematical models (1)–(6) in the case of variable coefficients taking into account non-zero demographic factors of population of the sides.

In mathematical models (1)–(6) with variable coefficients, we assume that the coefficients are exponential functions and we perform the appropriate computer modeling.

The calculations are performed during the mathematical models review period $t \in [0, T]$.

Computer modeling (simulations), depending on the variable coefficients of the mathematical models, produces two different results (for example, with weak conflict resolution):

There exists time $t_1 : 0 < t_1 \leq T$ which system (7) is completed (the conflict is resolved);

The system (7) is not completed for $t \in [0, T]$ (the conflict is not resolved).

Below we use the following exponential functions:

$$\begin{gathered} a(t)=a_0e^{n_1\frac{t}{T}},\ b(t)=b_0e^{n_2\frac{t}{T}},\ \alpha_1(t)=\alpha_{10}e^{n_3\frac{t}{T}},\ \alpha_2(t)=\alpha_{20}e^{n_4\frac{t}{T}},\\ \beta_1(t)=\beta_{10}e^{n_5\frac{t}{T}},\ \beta_2(t)=\beta_{20}e^{n_6\frac{t}{T}},\ \delta_1(t)=\delta_{10}e^{n_7\frac{t}{T}},\ \delta_2(t)=\delta_{20}e^{n_8\frac{t}{T}},\\ \gamma_1(t)=\gamma_{10}e^{n_9\frac{t}{T}},\ \gamma_2(t)=\gamma_{20}e^{n_{10}\frac{t}{T}},\end{gathered} \tag{27}$$

where

$a_0 = 2\cdot 10^5$—the population of the first side (at start point in time, start of process),

$b_0 = 4 \cdot 10^6$—the population of the second side (at start point in time, start of process),

$N_{10} = 2 \cdot 10^4$—10% of the population of the first side (at start point in time, start of process),

$N_{20} = 8 \cdot 10^5$—20% of the population of the second side (at start point in time, start of process),

n_1, n_2—the coefficients of demographic factors,

n_3, n_4—the coefficients of aggression,

n_5, n_6—the coefficients of cooperation,

n_7, n_8—coefficients of coercion to aggression (alienation) of the sides,

n_9, n_{10}—coefficients of coercion to cooperation of the sides.

For clarity, we assume either $T = 120$ or $T = 240$ months.

Below we present some numerical results of computer simulation and management parameter optimization for the above described four mathematical models (considered weak conditions (7) for conflict resolution).

Model 1

Case 1.1:

For fixed values of the parameters (see (27))

$$\alpha_{10} = 4 \cdot 10^{-11}, \alpha_{20} = 1 \cdot 10^{-11}, \ \beta_{20} = 9 \cdot 10^{-8},$$

$$n_1 = 0.1, \ n_2 = 0.2, \ n_3 = 1, \ n_4 = 1, \ n_5 = 1.2, \ n_6 = 1.3,$$

we have found minimum value $\beta_{10} = 0.726 \cdot 10^{-8}$ for which the conflict is resolved.

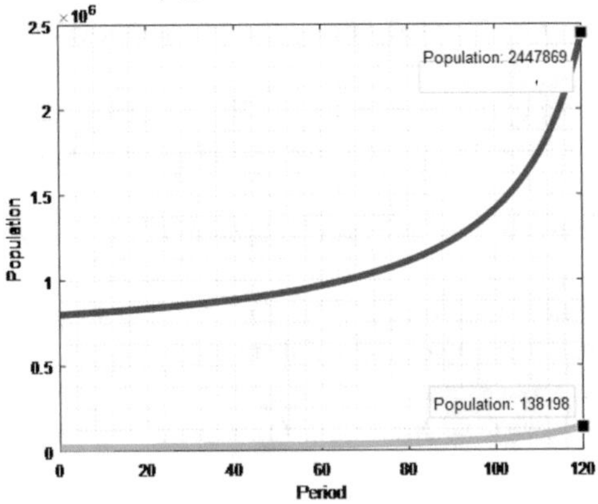

Case 1.2:

For fixed values of the parameters (see (27))

$$\alpha_{10} = 4 \cdot 10^{-11}, \alpha_{20} = 1 \cdot 10^{-11}, \ \beta_{10} = 1 \cdot 10^{-8},$$

$$n_1 = 0.1, \ n_2 = 0.2, \ n_3 = 1, \ n_4 = 1, \ n_5 = 1.2, \ n_6 = 1.3,$$

we have found minimum value $\beta_{20} = 7.358 \cdot 10^{-8}$ for which the conflict is resolved $(T = 114)$.

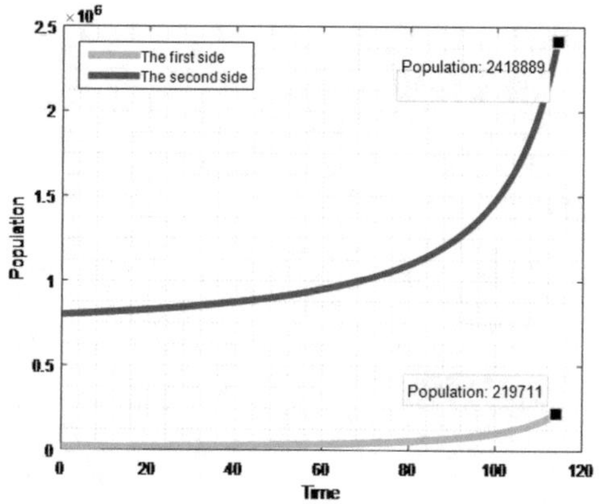

Model 2

Case 2.1:

For fixed values of the parameters (see (27))

$$\alpha_{10} = 4 \cdot 10^{-11}, \alpha_{20} = 1 \cdot 10^{-11},\ \beta_{20} = 8 \cdot 10^{-8},\ \delta_{10} = 2 \cdot 10^{-4},\ \delta_{20} = 3 \cdot 10^{-4},$$

$$n_1 = 0.1,\ n_2 = 0.2,\ n_3 = 1,\ n_4 = 1,\ n_5 = 1.2,\ n_6 = 1.3,\ n_7 = 0.3,\ n_8 = 0.5,$$

we have found minimum value $\beta_{10} = 0.858 \cdot 10^{-8}$ for which the conflict is resolved.

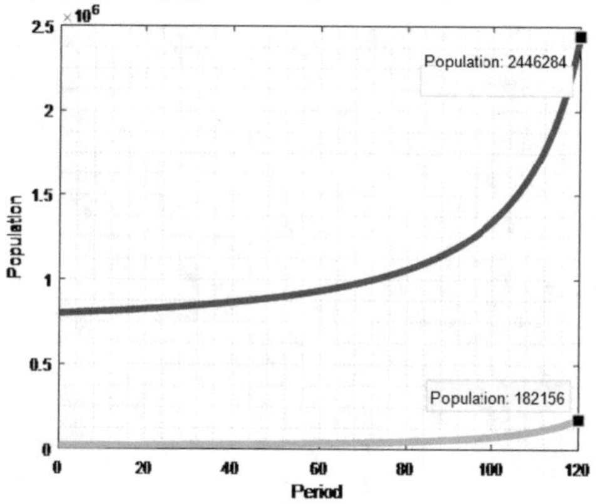

Case 2.2:

For fixed values of the parameters (see (27))

$$\alpha_{10} = 4 \cdot 10^{-11}, \alpha_{20} = 1 \cdot 10^{-11},\ \beta_{10} = 1 \cdot 10^{-8},\ \delta_{10} = 2 \cdot 10^{-4},\ \delta_{20} = 3 \cdot 10^{-4},$$

$$n_1 = 0.1,\ n_2 = 0.2,\ n_3 = 1,\ n_4 = 1,\ n_5 = 1.2,\ n_6 = 1.3,\ n_7 = 0.3,\ n_8 = 0.5,$$

we have found minimum value $\beta_{20} = 7.656 \cdot 10^{-8}$ for which the conflict is resolved $(T = 115)$.

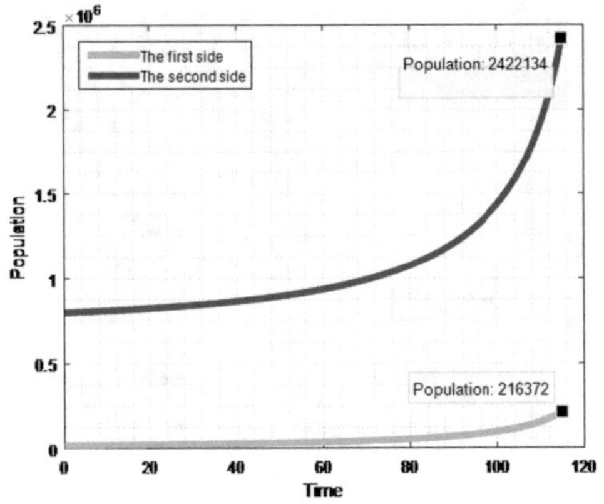

Model 3

Case 3.1:

For fixed values of the parameters (see (27))

$$\alpha_{10} = 5 \cdot 10^{-11}, \alpha_{20} = 1 \cdot 10^{-11}, \ \beta_{10} = 1 \cdot 10^{-8}, \ \beta_{20} = 8 \cdot 10^{-8}, \ \gamma_{20} = 4 \cdot 10^{-5},$$

$$n_1 = 0.1, \ n_2 = 0.2, \ n_3 = 1, \ n_4 = 1, \ n_5 = 1, \ n_6 = 1, \ n_9 = 1.2, \ n_{10} = 1.4,$$

we have found minimum value $\gamma_{10} = 6.550 \cdot 10^{-5}$ for which the conflict is resolved.

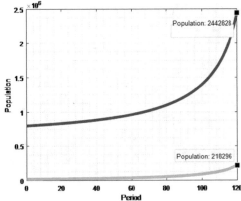

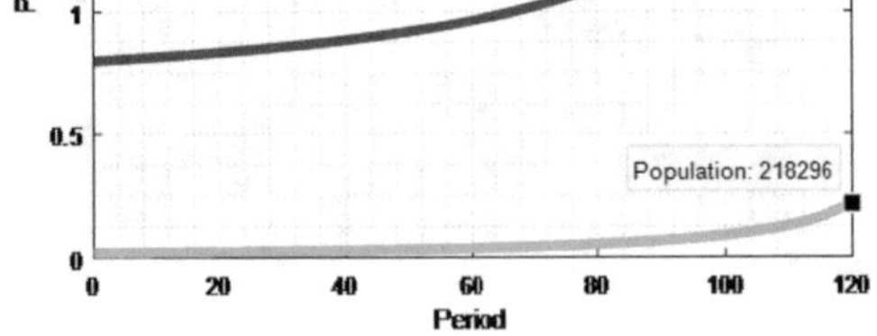

Case 3.2:

For fixed values of the parameters (see (27))

$$\alpha_{10} = 5 \cdot 10^{-11}, \alpha_{20} = 1 \cdot 10^{-11}, \ \beta_{10} = 1 \cdot 10^{-8}, \ \beta_{20} = 8 \cdot 10^{-8}, \ \gamma_{10} = 6 \cdot 10^{-5},$$

$$n_1 = 0.1, \ n_2 = 0.2, \ n_3 = 1, \ n_4 = 1, \ n_5 = 1, \ n_6 = 1, \ n_9 = 1.2, \ n_{10} = 1.4,$$

we have found minimum value $\gamma_{20} = 4.398 \cdot 10^{-5}$ for which the conflict is resolved.

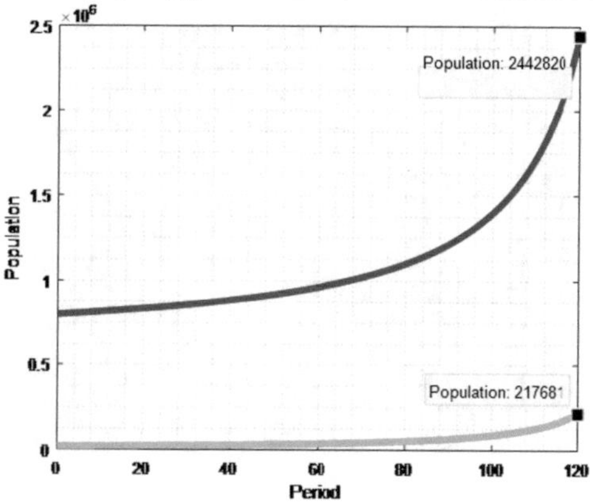

Model 4

Case 4.1:

For fixed values of the parameters (see (27))

$$\alpha_{10} = 4 \cdot 10^{-11}, \alpha_{20} = 2 \cdot 10^{-11}, \ \beta_{10} = 1 \cdot 10^{-8}, \ \beta_{20} = 8 \cdot 10^{-8}, \ \delta_{10} = 3 \cdot 10^{-5},$$

$$n_1 = 0.1, \ n_2 = 0.2, \ n_3 = 1, \ n_4 = 1, \ n_5 = 1, \ n_6 = 1, \ n_7 = 1, \ n_{10} = 1.5,$$

we have found minimum value $\gamma_{20} = 6.728 \cdot 10^{-5}$ for which the conflict is resolved.

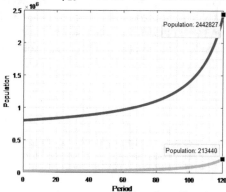

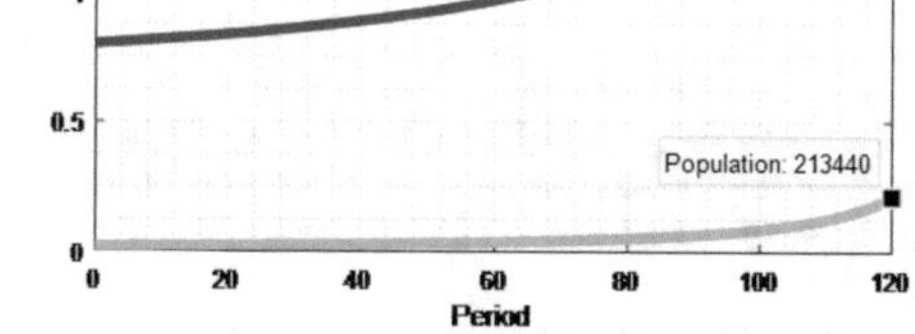

Case 4.2:

For fixed values of the parameters (see (27))

$$\alpha_{10} = 4 \cdot 10^{-11}, \alpha_{20} = 2 \cdot 10^{-11}, \ \beta_{10} = 1 \cdot 10^{-8}, \ \beta_{20} = 8 \cdot 10^{-8}, \ \delta_{10} = 3 \cdot 10^{-5},$$

$$n_1 = 0.1, \ n_2 = 0.2, \ n_3 = 1, \ n_4 = 1, \ n_5 = 1, \ n_6 = 1, \ n_7 = 1, \ n_{10} = 2,$$

we have found minimum value $\gamma_{20} = 5.288 \cdot 10^{-5}$ for which the conflict is resolved.

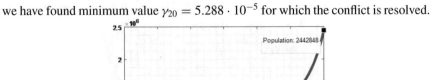

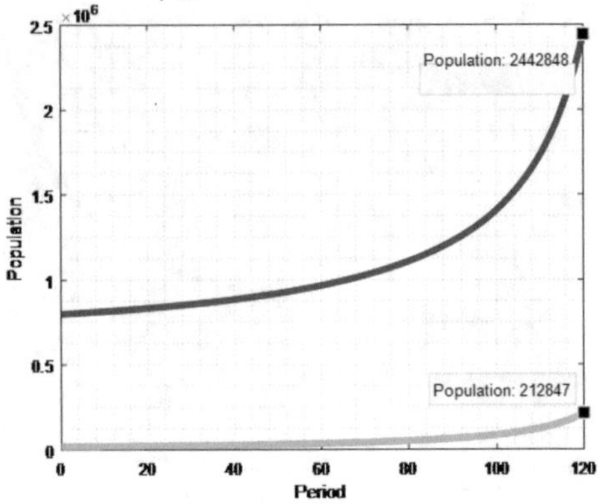

5 Conclusion

Some dependence between constant coefficients of the first mathematical model (the process of economic cooperation is free from political pressure), conditions on management parameters are analytically found to optimize (minimize) financial resources under which economic cooperation can peacefully resolve political conflict.

In the case of models with variable coefficients, for all four mathematical models described in the main text, with the help of computer simulations minimum values of management parameters (optimization of financial resources) have been found for which the conflicts can be resolved (more than half of the population of both sides support an economic cooperation process that promotes political reconciliation).

References

1. Chilachava, T., Kereselidze, N.: Optimizing problem of mathematical model of preventive information warfare, information and computer technologies, Theory and Practice: Proceedings of the International Scientific Conference ICTMC- 2010, USA, Imprint: Nova, pp. 525–529 (2012)
2. Chilachava, T., Kereselidze, N.: Mathematical modeling of information warfare. Inf. Warfare **1**(17), 28–35 (2011)
3. Chilachava, T.: Nonlinear mathematical model of dynamics of voters of two political subjects. Seminar of I.Vekua Institute of Applied Mathematics, Reports, **39**, 13–22 (2013)
4. Chilachava, T.: About some exact solutions of nonlinear system of the differential equations describing three-party elections. Appl. Math. Inf. Mech. **21**(1), 60–75 (2016)
5. Chilachava, T.: Research of the dynamic system describing globalization process. In: Springer Proceedings in Mathematics & Statistics, Mathematics, Informatics and their Applications in Natural Sciences and Engineering AMINSE 2017, vol. 276, pp. 67–78 (2019)

6. Chilachava, T., Pinelas, S., Pochkhua, G.: Research of three-level assimilation processes. Differential and Difference Equations with Applications. Springer Proceedings in Mathematics & Statistics (2020)
7. Aumann, R., Katznelson, Y., Radner, R., Rosenthal, R., Weiss, B.: Approximate purification of mixed strategies. Math. Oper. Res. INFORMS **8**(3), 327–341 (1983)
8. Aumann, R., Hart, S.: Handbook of Game Theory with Economic Applications. Handbook of Game Theory with Economic Applications, 1st edn., vol. 2, No. 2. Elsevier, Amsterdam (1994)
9. Aumann, R.: Repeated Games with Incomplete Information, 1st edn., vol. 1. The MIT Press, Cambridge (1995)
10. Chilachava, T., Pochkhua, G.: Research of the dynamic system describing mathematical model of settlement of the conflicts by means of economic cooperation. GESJ: Comput. Sci. Telecommun. **3**(55), 18–26 (2018)
11. Chilachava, T., Pochkhua, G.: About a possibility of resolution of conflict by means of economic cooperation, Problems of management of safety of difficult systems., vol. 69–74 (2018)
12. Chilachava, T., Pochkhua, G.: Research of the nonlinear dynamic system describing mathematical model of settlement of the conflicts by means of economic cooperation. In: 8th International Conference on Applied Analysis and Mathematical Modelling, ICAAMM 2019, Proceedings Book, pp. 183–187 (2019)

Modeling of Extreme Events and Regional Climate Variability on the Territory of the Caucasus (Georgia)

Teimurazi Davitashvili, Inga Samkharadze, and Meri Sharikadze

Abstract Currently, the problem of climate change is an urgent issue in the South Caucasus region, as well as in Georgia, where increased trends of average annual temperature with heavy precipitation, hail, floods and drought have become more frequent. To prevent the consequences of these events (accidents) in a timely manner, it is necessary to take more effective steps in providing scientific information on extreme events a regional and local scale, the official environmental authorities, society and the scientific community. In this work, on the one hand, a comparative study of three cumulus parameterization and five microphysical schemes of the Weather Research and Forecast (WRF) v.3.6 model, is carried out for four exceptional precipitation phenomena that have occurred in Georgia (Tbilisi) in the summer of 2015 and 2016. On the other hand, the Real-time Environmental Applications and Display System (READY) is used to study these phenomena. Predicted events are evaluated by a thorough examination of weather radar data. Some results of numerical calculations based on the WRF and READY systems are presented and analyzed.

Keywords WRF · READY · Showers · Georgia

T. Davitashvili (✉) · M. Sharikadze
I. Vekua Institute of Applied Mathematics Iv. Javakhishvili Tbilisi State University, Tbilisi, Georgia
e-mail: teimuraz.davitashvili@tsu.ge

M. Sharikadze
e-mail: meri.sharikadze@tsu.ge

I. Samkharadze
Faculty of Exact and Natural Sciences, Hydrometeorological Institute of Georgian Technical University, Iv. Javakhishvili Tbilisi State University, Tbilisi, Georgia
e-mail: inga.samkharadze562@ens.tsu.edu.ge

G. Jaiani and D. Natroshvili (eds.), *Applications of Mathematics and Informatics in Natural Sciences and Engineering*, Springer Proceedings in Mathematics & Statistics 334, https://doi.org/10.1007/978-3-030-56356-1_5

1 Introduction

For the last four decades the number of the extreme weather events and natural hazards (floods, landslides, storms, heavy showers, hails) has significantly increased on the territory of Georgia [24]. How it seems for prevention of accidents it is necessary to take more efficient steps in provision with scientific information (regional and local scale extreme weather prediction, climate change tendencies) against the freaks of nature. The question of studying formation of hazardous precipitations on the background of modern climate change based on the regional weather forecasting models such as Weather Research and Forecast (WRF) model and on the modern sounding technologies is an urgent issue for Georgia. In a numerical model like WRF, the Microphysics (MP) and Cumulus Parameterization schemes (CPSs) are mainly responsible for precipitation generation [13]. The subgrid-scale convective process and shallow clouds are managed by the cumulus parameterization [25]. Significant MP and CPSs options are especially highlighted in those studies that deal with the warm season convective events predictions over the mountains territories [7]. Problem of the MP and CPSs option for warm season precipitation prediction has been widely explored in the scientific literature ([8, 11, 13, 15, 17, 22, 31, 32, 37] among others). Generally the global models are well characterizing the large scale weather systems above the Caucasus Mountains region, but not enough the fine scale atmosphere processes which associate with local terrain and land cover. To capture these smaller scale features of atmosphere processes, a simulation with sufficiently high spatial resolution of the local complex terrain and the heterogeneous land surface of the complex territory is necessary [5, 6, 6]. A numerical calculations by WRF model the convective cumulus schemes mainly are used in the coarse size grid meshes [27] and that is why a series of numerical simulations have been conducted without use of cumulus schemes at high-resolution (less than 10 km) grid spacing (Nasrollahi et al. 2007). However, it has been demonstrated that using a cumulus scheme at a 9-km grid size had improved the results of simulations of the early rapid intensification of Hurricane Emily, while the effect of using such parameterization at 3-km grid size had only a small impact on the results [14]. Besides, some of the previous studies also showed that application of the cumulus schemes in the numerical simulations at a horizontal resolution of 9 km [9] and even 6 km [16] had improved the results of numerical simulations. Our previous studies also showed that the simulations of inner-massive cumulus processes over the complex territory of Caucasus were influenced by choosing physical processes in WRF model [7], by the interactions among the physical processes and by the model horizontal grid spacing [6]. Furthermore, sounding is a convenient tool for assessing the state of the atmosphere. Indeed, many countries developed their own radiosonde systems as a matter of pride [12]. To improve weather forecasts (with an emphasis on weather hazard events), an Atmospheric Sounding Program (ASP) was developed is a convenient tool for displaying vertical profiles of thermodynamic quantities (temperature, humidity, and pressure) recorded by a radiosonde and based on vertical atmospheric profiles to predict the type of precipitation, the evolution of the boundary layer (clouds and temperatures) and the type of convection [19]. Fully automated sounding equipment reduces operating costs and provides greater flexibility in site selection and observation schedule, provides full coverage of upper-air meteorological observations [34]. Proximity sounding analysis has long been a tool for determining the environmental conditions associated with different types of weather phenomena and for distinguishing between them [4]. For example, more than 65 derechos (long-lived, widespread damaging convective windstorms), were identified during the years 1983 to 1993 accompanied by 115 proximity soundings [10]. ASP has been widely used: for comparison of radiosonde and COSMIC data [29]; in order to study the circulation of the upper troposphere and lower stratosphere [3]; for comparisons of convective available potential energy (CAPE) with standard instability indices and for evaluating the convective potential of the atmosphere such as the lifted index [1, 2]. In this study, on the one hand, we intended to determine some acceptable, possible combinations of MP and CPSs schemes of the WRF v.3.6.1 model for predicting convective phenomena (precipitation, hail) of the warm season over the territory of Georgia. On the other hand, the data from the sounding system and the meteorological radar were used to further evaluate the results of the WRF simulation.

2 Problem Formulation

The study of the formation and prediction of strong convection (hazardous precipitation, hail) based on numerical models is an urgent problem for mountainous countries and for the territory of Georgia. But it should be noted that in the summer, short-term forecasts of heavy rainfall using numerical models (WRF, etc.) are challenging, especially in mountainous areas [7]. Since MP and CPSs schemes result in significant variability in precipitation prediction in the WRF model, not to a lesser extent, a choice among of physical schemes in the WRF v.3.6.1 model is very much necessary for showers, hails and as a consequence floods properly prediction on the territory of Georgia. Since this study also seeks to support operation works at the Georgian Environmental Agency (GEA) in short term local scale weather forecasts (where the WRF model is used too) the first step in this process is to evaluate through analysis of MP and CPSs schemes option for predicting extreme summer precipitation in the WRF v.3.6 model. Four sets of high-quality observations were used to evaluate the accuracy of forecasts of heavy summer precipitation. To this end, the dependence of forecasts on the resolution of grid points and a comparative study of the MP and CPS options for predicting precipitation in the complex territory of Tbilisi was studied. Namely, four cases of convective events (June 13, 2015, June 21, 23 and August 2, 2016) that took place in Tbilisi and various combinations between the three CPS schemes Kain–Fritsch, Grell–Devenyi ensemble and Betts–Miller–Janjic) and five MP (WSM6, Purdue Lin, Thompson, Morrison 2-Moment and Goddard) were studied.

2.1 Data and Methodology

For analyzing and assessment of forecasting results we were supported by the weather radar data and by the Global Data Assimilation System (GDAS). Computations were executed by the computer system with working nodes (16 cores+, 32Gb Ram on each) located in the Georgian Research and Educational Networks Association (GRENA) which in its normal course was connected to the European GRID infrastructure. Therefore it was a good opportunity for running model on larger number of CPUs and storing large amount of data on the grid storage elements.

2.1.1 Real-Time Environmental Applications and Display System

Air masses vertical movements are important in the development of ongoing atmospheric processes in the mountainous regions. Indeed, convection of air masses is intensified under the influence of relief that causes formation of intense convective clouds on the territories with complex orography. Forecasting dangerous, local convective processes with numeral methods within short timeframes mostly do not have good results. That is why it is essential to develop different methods together with the numeral methods of weather forecasting. Namely, in order to estimate possible quality of convective processes, we have to study the thermodynamic condition of atmosphere and establish the quality of atmospheric instability with the purpose of forecasting possible dangerous convective local processes. In order to estimate quality of atmospheric instability it is essential to study the vertical structure of atmosphere. The state of the atmosphere by stratification can be unstable, unsustainable, and indistinguishable. In the case of stable stratification, the atmosphere has the ability to maintain or stop vertical movements. The particle method is one of the most widely used methods to measure the degree of convection in the atmosphere, assuming that at the beginning the particle and the surrounding atmosphere have the same temperatures and do not heat up to the environment when the particle ascends, that is, the process proceeds adiabatically. Since the particle is not initially saturated, so it moves to the Lifting Condensation Level (LCL) dry, and at the LCL air parcel's temperature and dew point are equal, water begins to condense out, movement becomes slower and after the LCL the air parcel's temperature follows the moist adiabat as it rises. At the Level of Free Convection (LFC), the temperature of the air equals that of the environment and above the LEC, the air parcel temperature is higher than that of the environment such that if the air rises above the LFC, it will continue to rise to the level Z_{max} where its final maximum updraft strength w will be at Z_{max} [21, 35]:

$$w(Z_{max}) = \sqrt{2CAPE} \tag{1}$$

where $CAPE$ (Convective Available Potential Energy) is the work of per unit mass done on the parcel as it rises (J/kg), represents the vertically integrated positive buoyancy of a parcel experiencing adiabatic ascent and has the following form [21]:

$$CAPE = \int_{Z_{LPC}}^{Z_{max}} g \frac{T_{parcel} - T_{env}}{T_{env}} dZ \tag{2}$$

where T_{parcel} is the temperature of the parcel and T_{env} is the temperature of the environment, g is gravity acceleration, Z_{max} is the height of the equilibrium level (an atmospheric layer, above which the ambient temperature is greater than the particle temperature and generally during a strong convection it is close to the height of the tropopause).

Equation (2) is called the "Positive Energy Above LFC" on the Skew-T program, and is calculated using potential temperature. Equation (2) provides an estimate of maximum updraft strength in convective storms according to (1). CAPE is a fundamental indicator of potential intensity of deep, moist convection, since it is proportional to the energy available for a rising parcel and therefore (2) combines a significant part of the thermodynamic information contained in the sounding. The Convective Inhibition energy (CIN) is the amount of work required to lift a parcel through a layer that is warmer than the parcel and a very similar to CAPE has the following form:

$$CIN = \int_{Z_{max}}^{Z_{LPC}} g \frac{T_{parcel} - T_{env}}{T_{env}} dZ \tag{3}$$

where Z_{ML} is the mixed layer depth (the height of the surface) and all other variables are the same as in the CAPE calculation. Equation (3) is a measure of the "negative area" on a sounding between the surface and the LFC. CIN is the amount of work required to raise a parcel upward sufficient to overcome negative buoyancy. This negative area is often referred to as a "lid" or "cap" [21, 35]. Nowadays, operation of radiosonde relates to some financial expenses and it is not almost in use in some developing countries (among them even in Georgia). In order to effectively and timely forecast ongoing atmospheric processes, the National Oceanic and Atmospheric Administration's (NOAA) Air Resources Laboratory (ARL) has created Real-time Environmental Applications and Display System (READY, http://www.ready.noaa.gov). For the purpose of evaluation thermodynamic condition of the atmosphere by extreme days discussed in this paper, archive of READY System [23, 28] have been applied. Calculated results have been compared with real data obtained from Sighnaghi weather radar (Meteor 735CDP 10-Doppler Weather Radar) monitored by Military Scientific Technical Center "DELTA" located in the Village of Chotori, Sighnaghi Municipality, through that on entire Kakheti region clouds are being observed. So in this study, WRF v.3.6 model's different microphysics and convective scheme components options with nested grid resolutions in the range from 2.2 to 19.8 km with an emphasis on 6.6 and 2.2 km mesh sizes simulations against READY system and radar observation data are studied.

2.1.2 WRF Model and Simulation Design

The WRF is a real-time numerical weather forecasting system in which the atmosphere modeling system includes the development and research of data assimilation, research on parameterized physics, regional climate modeling, air quality modeling, atmosphere-ocean coupling and idealized simulations [26]. In WRF there are two dynamics solvers: the Advanced Research WRF (ARW) solver (Eulerian mass or "Em" solver) developed primarily at the National Center for Atmospheric Research's, and the NMM (Nonhydrostatic Mesoscale Model) solver developed at National Centers for Environmental Prediction. The ARW dynamics solver is based on compressible, non-hydrostatic Euler equations, which in the vertical mass coordinate following the terrain (Laprise 1992) in the flux form using variables that have conservation properties [20] have the following form [26]:

$$\frac{\partial U}{\partial t} + (\nabla \cdot Vu) - \frac{\partial}{\partial x}\left(P\frac{\partial \phi}{\partial \eta}\right) + \frac{\partial}{\partial \eta}\left(P\frac{\partial \phi}{\partial x}\right) = F_U, \tag{4}$$

$$\frac{\partial V}{\partial t} + (\nabla \cdot Vv) - \frac{\partial}{\partial y}\left(P\frac{\partial \phi}{\partial \eta}\right) + \frac{\partial}{\partial \eta}\left(P\frac{\partial \phi}{\partial y}\right) = F_V, \tag{5}$$

$$\frac{\partial W}{\partial t} + (\nabla \cdot V_w) - g\left(\frac{\partial P}{\partial \eta} - \eta\right) = F_w \tag{6}$$

$$\frac{\partial \theta}{\partial t} + (\nabla \cdot \theta) = F_\theta, \tag{7}$$

$$\frac{\partial \mu}{\partial t} + (\nabla \cdot V) = 0, \tag{8}$$

$$\frac{\partial \phi}{\partial t} + [V \cdot \nabla \phi - gW] = 0, \tag{9}$$

along with the diagnostic relation for the inverse density

$$\frac{\partial \phi}{\partial \eta} = -\alpha\mu, \tag{10}$$

and the equation of state

$$P = P_0(R_d\theta/P_0\alpha)^\gamma \tag{11}$$

In (4)–(9) the following notation are employed,

$$\nabla \cdot Va = \frac{\partial(Ua)}{\partial x} + \frac{\partial(Va)}{\partial y} + \frac{\partial(\Omega a)}{\partial \eta}$$

$$V \cdot \nabla a = U\frac{\partial a}{\partial x} + V\frac{\partial a}{\partial y} + \Omega\frac{\partial a}{\partial \eta}$$

where η is terrain-following hydrostatic-pressure vertical coordinate, defined as $\eta = \left(P_h - P_{h_t}/\mu\right)$, $\mu = P_{h_S} - P_{h_t}$, and $\mu(x, y)$ represents the mass per unit area within the column in the model domain at (x, y), P_h is the hydrostatic component of the pressure, and P_{h_S} and P_{h_t} refer to values along the surface and top boundaries, respectively. $V = \mu v = (U, V, W)$, $\Omega = \mu\dot{\eta}$, $\Theta = \mu\theta$. $v = (u, v, w)$ are the covariant velocities in the two horizontal and vertical directions, respectively, while $\omega = \dot{\eta}$ is the contravariant 'vertical' velocity, u, v, w are the axis components of wind velocity along axis x, y, z, t- is time, θ is the potential temperature, $\varphi = gz$ (geopotential), p (pressure), and $\alpha = 1/p$ (the inverse density). A represents a generic variable. $\gamma = Cp/Cv = 1.4$ is the ratio of the heat capacities for dry air, R_d is the gas constant for dry air, and P_0 is a reference pressure. The right-hand-side terms F_U, F_V, F_W, and F_Θ represent forcing terms arising from model physics, turbulent mixing, spherical projections, and the earth's rotation.

System of equations (4)–(11) is solved by the following boundary conditions (BC): open lateral BC, free-slip bottom BC and gravity wave diffusion absorbing BC at upper level. All the WRF3.6.1 model simulations were initialized at 00:00 UTC on 13th of June 2015, 21th, 23nd of June and 2nd of August 2016 with 0.5° GFS data and then run for 48 h on a coarse 90 × 100 grid point domain with 19.8-km horizontal grid spacing and 54 vertical levels. It should be noted that no cloud information was contained within the WRF initial conditions (i.e. the simulations were run in cold-start mode). The geographical region covered by the nested WRF model and the nested domains configurations are shown in Fig. 1. Prediction of the deep, moist convection processes on the small areas became more vulnerable for

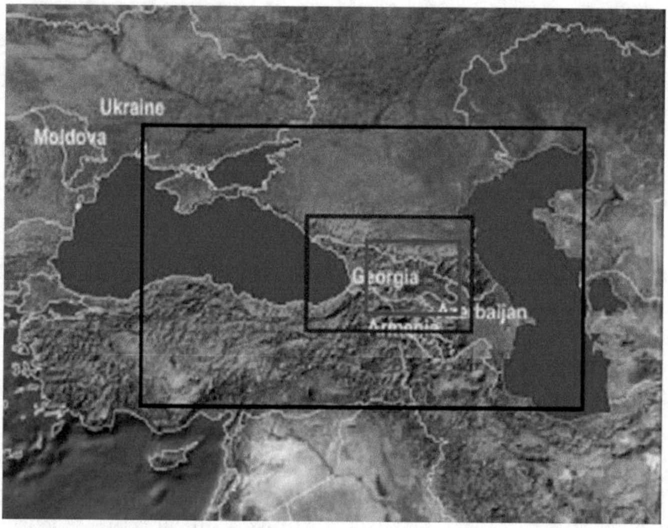

Fig. 1 Model domain used in WRF simulations

Table 1 Five set of the WRF parameterization schemes used in this study

WRF Physics	Set 1	Set 2	Set 3	Set 4	Set 5
Micro physics	WSM6	Thompson	Purdue Lin	Morrison 2-Moment	Goddard
Cumulus parameterization	Kain–Fritsch	Betts–Miller Janjic	Kain–Fritsch	Grell–Devenyi ensemble	Kain–Fritsch
Surface layer	MM5 Simil.	MM5 Simil.	MM5 Simil	(PX) Similarity	MM5 Similarit
Planet. boundary layer	YSU PBL	YSU PBL	YSU PBL	ACM2 PBL	YSU PBL
Land-surface	Noah LSM	Noah LSM	Noah LSM	Noah LSM	Noah LSM
Atmospheric Radiat.	RRTM/Dudhia	RRTM/Dudhia	RRTM/Dudhia	RRTM/Dudhia	RRTM/Dudhia

some regions of Georgia having complex topography [5] and predominantly in the capital city of Georgia, Tbilisi.

In our study we have used one-way nesting of WRF model domains, where an inner nested domain (grid size of 70×70 points, with resolutions 2.2 km) has fully covered the territory of Georgia and it was centered on to the capital city of Georgia, Tbilisi (with the GPS coordinates of $41°43'0.0012''\ N$ and $44°46'59.9988''\ E$). The next outer nested domain (a grid of 94×102 points with resolutions of 6.6km) has fully covered the South Caucasus region. Both of them have used 54 vertical levels including 8 levels below 2 km. A time step of 10 s was used for fulfillment of calculations in the fine mesh resolution grids (see Fig. 1). As known, the WRF 3.6 model contains several physics (micro physics, cumulus parameterization physics, radiation physics, surface layer physics, land surface physics, and planetary boundary layer physics) with a number of different modules and schemes options. In this study, we selected the parameterization schemes have been selected, listed in Table 1.

3 Results and Discussion

In this study all kinds of combinations from five MP and three CPSs have been tested to examine the impact of high-resolution (6.6 and even 2.2km) MP and CPSs schemes on the results of local cumulus processes prediction, but only the results of calculations executed in the frame of Table 1 are presented here.

Convective event on 13 June 2015. According to the weather radar's data of the atmosphere reflecting state on 13 June 2015 from 10:00 during 1.5–2 h there was heavy shower in Tbilisi (see Fig. 2). Due to the heavy rainfall, a big wave (constructed by mass of slush, rocks and trees) run across the Vere river canyon and washed everything away to the square of Heroes in Tbilisi. Unfortunately, at least 20 people (including three zoo workers) and half of the animals from the Tbilisi zoo were killed in this event.

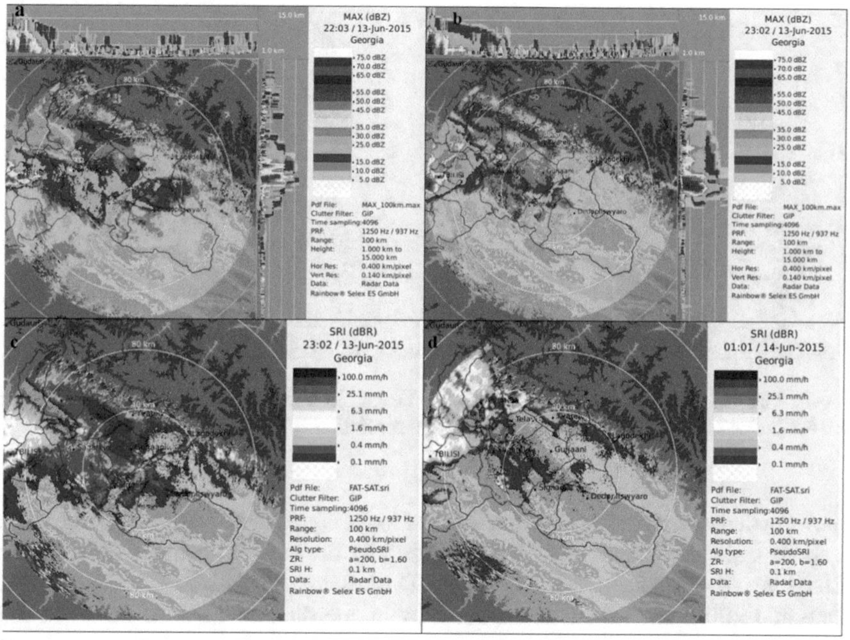

Fig. 2 Weather radar's data of the atmosphere reflecting state (clouds max-intensity dbz) above the territory of eastern Georgia **a** at 22:03, **b** at 23:02 on 13 of June 2015, and values of ATP **c** at 23:02 on 13 of June, **d** at 01:01 on 14 of June

It has been shown that none of the combinations of schemes, listed in Table 1 were able to reproduce occurred deep convection of the 13th of June 2015 in Tbilisi and in its vicinity, accurately. For example on Fig. 3a and b predicted fields of the relative humidity on the 850 hPa for 13 June (21UTC) and 14 June (00UTC) 2015 are presented, respectively, which were simulated by WRF Physics Options set1 (it gave a better result than others). The calculated amounts of water vapour depicted on Fig. 3a and b (nested domain with 6.6 km resolution), presenting the moments when atmospheric event was in full swing, are not in agreement with the real situation (see Fig. 2) which took place in Tbilisi and surroundings on 13 June 2015.

Figure 4 shows Aerological Diagrams (AD) of June 13, 2015 (by Tbilisi time 10:00 a.m.) are shown on the territory of Tbilisi Zoo (with its geographic coordinates $41°71'$ N, $44°77'$ E) by various instants of time (as of June 13, 4a - 06 UTC, 4b - 12 UTC, 4c - 18 UTC, 4d - 14 June 00 UTC). To evaluate the thermal state of the atmosphere, in the diagrams we constructed particle state curves (blue continuous lines). As it is clearly shown on the diagram obtained at 06 UTC (Fig. 4a), instability energy CAPE is at a low level from Level of Free Convection (LFC) to Equilibrium Level (EL). Numerical value of instability energy counted in accordance with READY System of ARL is at a low level, CAPE = 55.6 j/kg that corresponds with weak instability condition (Table 2). As it is shown in this figure, in upper atmospheric layers to the height of 100 mb, curve of particle condition almost follows the stratification curve

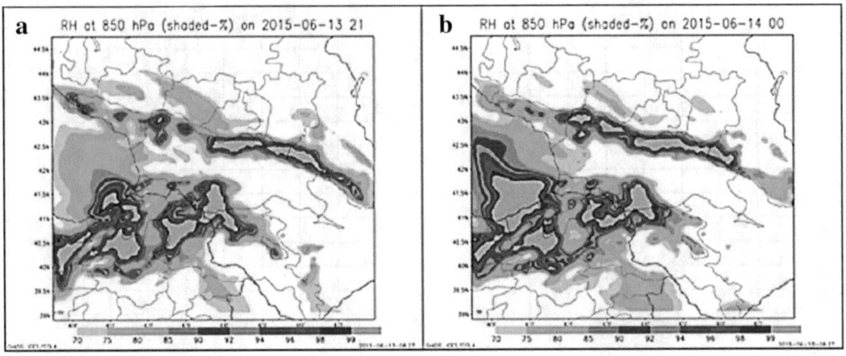

Fig. 3 Maps of the relative humidity (**a** at the 850 hPa at 21:00 13 June 2015 and **b** at 00:00 14 June 2015) obtained by set.1 and simulated with 6.6 km resolution

showing that atmospheric condition is partly indistinguishable, but in the lower layers a stable state is observed, which is indicated by the value of convective inhibition CINH = −9.4 J / kg. Figure 4b shows AD of June 13, at 12:00 UTC (By Tbilisi time at 16:00 p.m.) for the same geographic coordinates. As it is shown in this figure, the curve of condition slightly moved to the right side of stratification curve to certain height above condensation level shown by the increase of instability power CAPE = 109.6 j/kg and decrease of inhibition power CIN = –7.9 j/kg. At 10 o'clock (18 UTC) local time the value of inhibition energy was significantly decreased, by this time CAPE = 48 j/kg (Fig. 4c). If we compare this data with the data as of June 14, in the morning at 4 o'clock (00 UTC) (Fig. 4d), we see that on all levels the curve of condition is located on the left side of stratification curve, meaning that by this time atmospheric condition is absolutely stable. Indeed, by this time, the numerical values of power (calculated by the READY ARL system with a horizontal resolution of 10 latitude (111 km)) CAPE = CINH = 0, which quite well coincides with the data shown in Fig. 4d. Thus, by 12:00 UTC (Tbilisi time 4:00 p.m.), the value of CAPE was maximum, and the value of CINH was minimal (compared with the aerological data calculated at other times of the day). In fact, according to the radar data, later at 22:00 it was found that the clouds were maximally developed and accordingly, we may consider that pursuant to the analysis of AD carried out by morning hours, there may be made preventive forecast on possible complex meteorological processes taken place in atmosphere and in the second half of day about possible heavy atmospheric precipitates.

The another case of local strong convective events was observed on June and August of 2016 in Tbilisi.

Convective event on 21 June 2016. The fair weather suddenly changed by rough weather with a strong wind (wind's velocity reached about 35m/s) at 16:00 on 21 June 2016 in Tbilisi. The cloudy systems gradually grew above the different districts of Tbilisi from 16:15 to 16:25 and after 5 min began heavy unexpected shower, which was accompanied with 2–3 mm diameter hail. Hailing stopped after 8 min and later

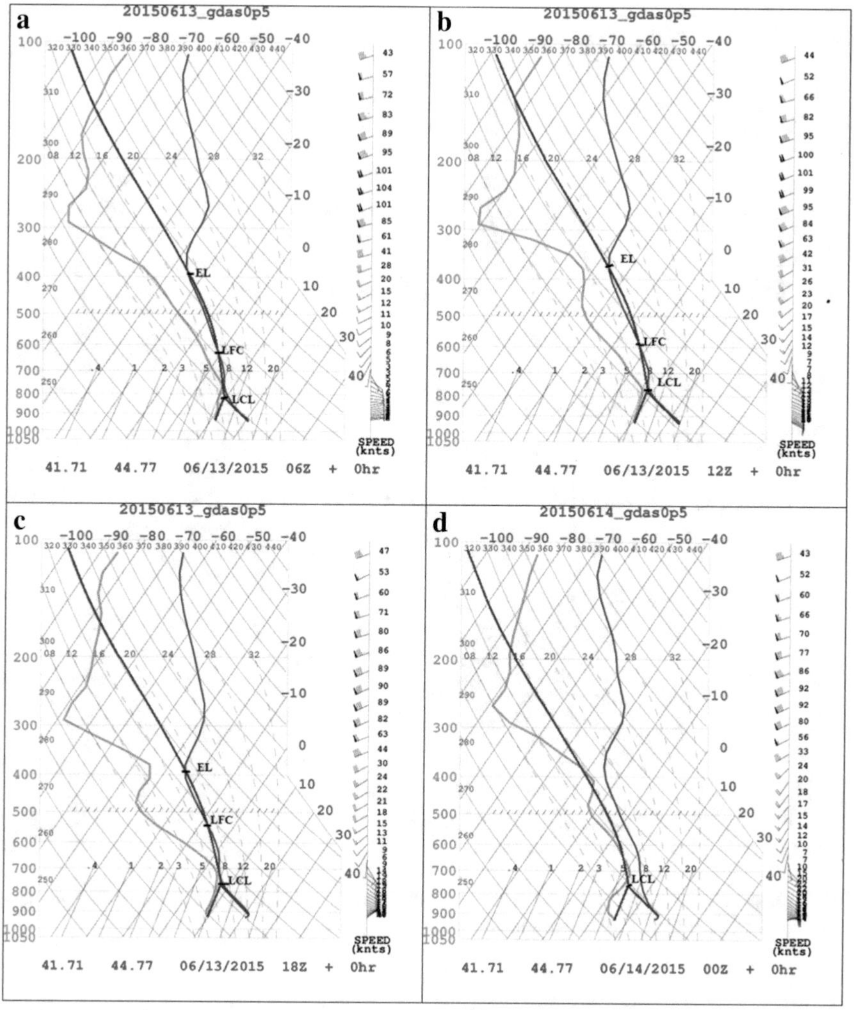

Fig. 4 AD of June 13, 2015 (by Tbilisi time 10:00 a.m.) on the territory of Tbilisi zoo (with its geographic coordinates 41°71′ *N*, 44°77′ *E*) by various instants of time (**a** 06 UTC, **b** 12 UTC, **c** 18 UTC, **d** 14 June 00 UTC)

the shower was accompanied with thunderstorm. At 16:50 the pouring stopped and after 10 min the raining stopped too. Shower, thunderstorm and hails have a very local character and did not extend outside of Tbilisi. This convective event, accompanied by a flood, left Tbilisi with dirty streets, squares and public gardens at 17:00 on June 21, 2016.

The results of calculations executed under WRF model by Set 1 have shown that though the background synoptic processes taken place in Georgia were modelled fairly good, while convective processes that took place above Tbilisi was not mod-

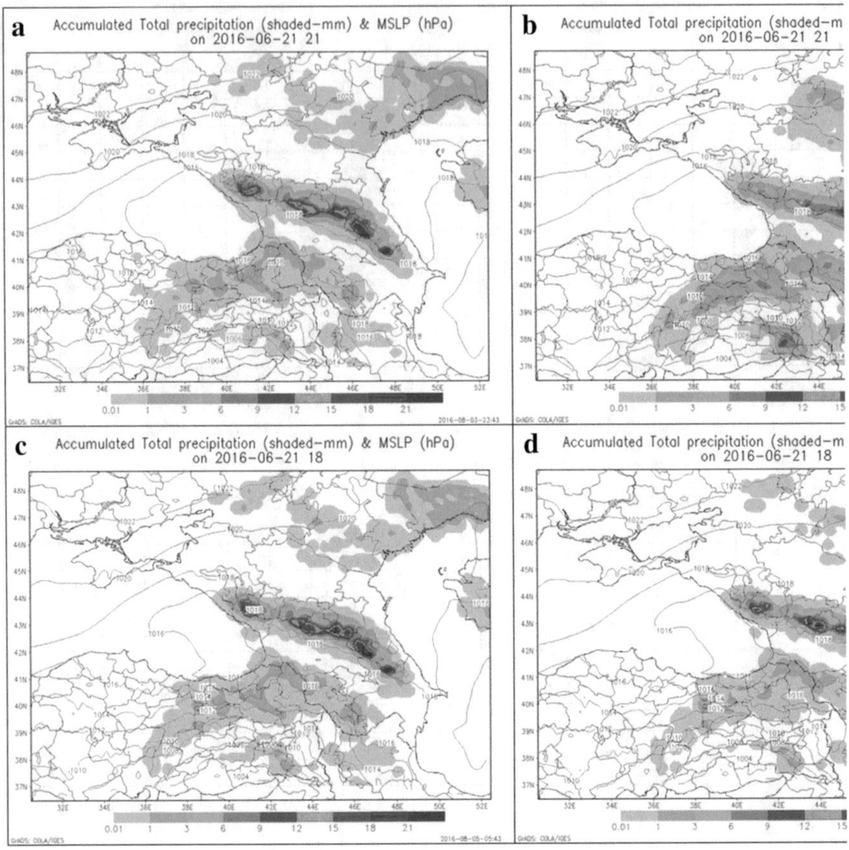

Fig. 5 Maps of ATP. Forecasted **a** at 18:00 21 June 2016 by set. 1,**b** by set. 2, **c** by set. 3 and **d** by set. 5 for the domain with 19.8. km resolution

elled satisfactorily (Fig. 5a). The results of calculations performed by the schemes combination of Set 2 have shown that the cloudy system has invaded on the territory of Georgia from south–west to the north–east direction but from 15 to 18 o'clock there was not raining over the surrounding territory of Tbilisi (Fig. 5b). The results of calculations by the Set 3 have shown that the synoptic processes mentioned above were modelled rather not satisfactorily. Namely, the atmosphere masses were intruded into the territory of Georgia from south–west not to the east direction but rather to the north, north–east direction and taking into consideration placement of Tbilisi it is evident that the Set 3 was not able to model precipitation accumulation surrounding of Tbilisi (Fig. 5c). The results of calculations executed by the schemes combination of Set 4 had evidently showed dry weather surrounding of Tbilisi. The results of calculations performed by the Set 5 have shown that though the background synoptic processes were modelled satisfactorily enough but there was not raining over the surrounding territory of Tbilisi (Fig. 5d).

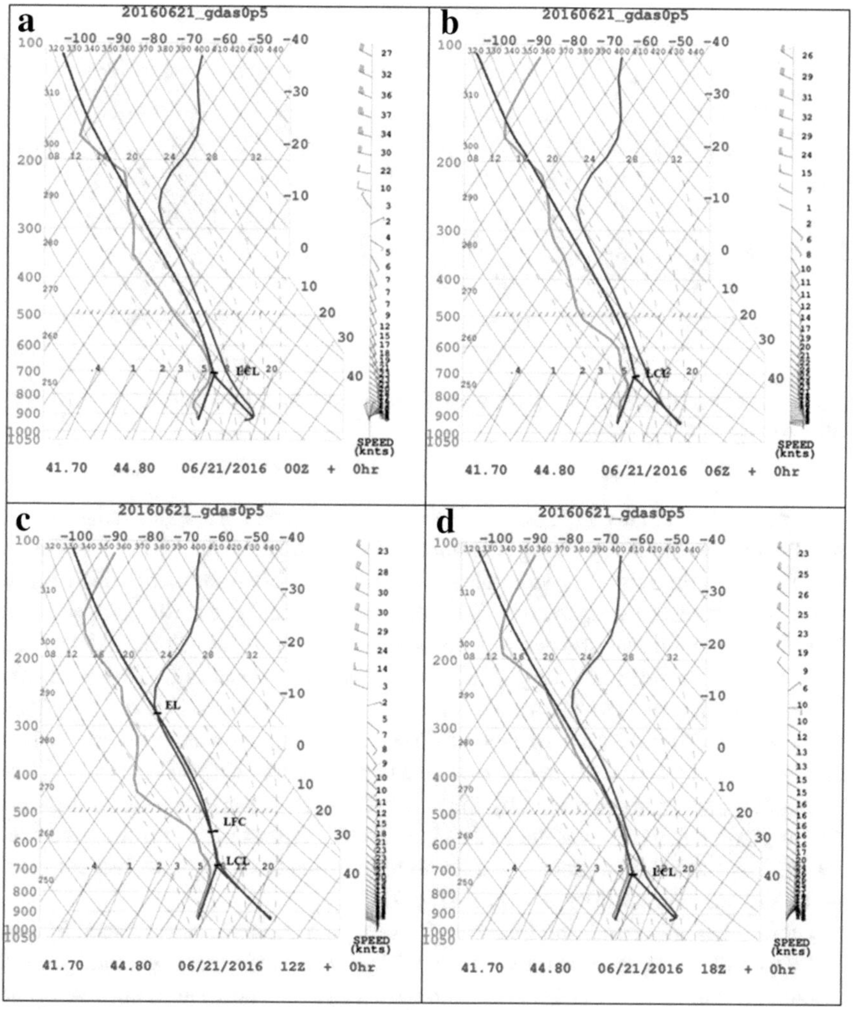

Fig. 6 Aerological diagrams of June 21, 2016 for 41°70′ *N*, 44°80′ *E* geographic coordinates (**a** 00 UTC, **b** 06 UTC, **c** 12 UTC, **d** 18 UTC)

In June 21, 2016 in accordance with the AD, only by 12 UTC was found atmospheric instability, but in other cases atmospheric condition was absolutely stable (Fig. 6). As it is found in Fig. 6a, b, d.) by 00 UTC, 06 UTC, 18 UTC atmospheric condition was absolutely stable and by this time numerical value of energy was CAPE = CINH = 0 j/kg, but by 12 UTC instability power was CAPE = 349 j/kg, but convective inhibition energy was CINH = 32 j/kg. Alike above mentioned cases by June 21, 2016 thermodynamic instability in atmosphere was found by 12 UTC and accordingly, numerical value of instability power was found out to be maximal.

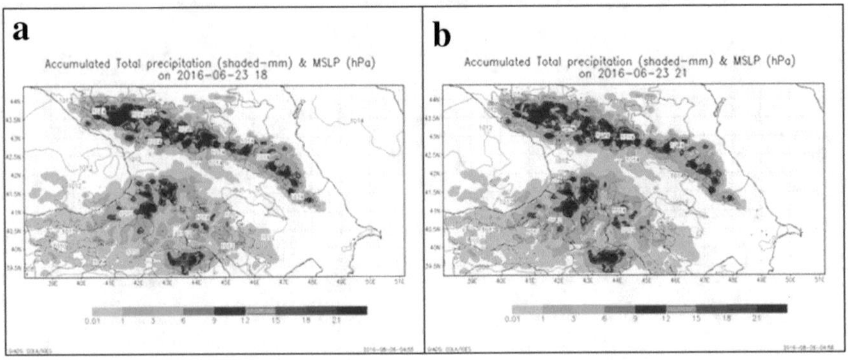

Fig. 7 Maps of ATP forecasted a at 18:00 and b 21:00 on 23 June 2016 by set. 5 for the domain with 6.6 km resolution

Convective event on 23 June 2016. Observations of meteorological radar showed that the weather gradually changed from 18:30 to 18: 40 on June 23, 2016 in Tbilisi. Namely, at this time cloudy system accumulated above Tbilisi and after 5 min it began raining which was accompanied by 3–4 mm diameter hail. The hailing stopped after 5 min but the continual shower was accompanied with thunderstorm. At 19^{10} the down pouring stopped and very soon the raining had stopped too. This event like the severe convective event occurred on the 21 June, 2016 and it had a very local character and did not extend outside of Tbilisi excepting the Kakheti region, where analogous local character convective event had been observed from 16^{15} to 17^{47}. The results of calculations executed by the schemes combination Set 1-Set 5 were not able to predict showers over the territory of Tbilisi. Rather better results have been obtained by the Set 5 but once again it was not raining from 15:00 to 18:00 o'clock on the surrounding territory of Tbilisi (Fig. 7).

Thermodynamic condition of about the same atmosphere was repeated in June 23, 2016. Indeed, in accordance with the AD by 00 UTC atmospheric condition was absolutely stable (Fig. 8a), numerical values of instability energy and convective inhibition energy CAPE = CIN = 0 j/kg that also corresponds to the stable atmospheric condition. By 06:00 UTC instability energy was increased in the atmosphere (Fig. 8b), its numerical value CAPE = 60 j/kg, but CINH = –28 j/kg. During the daytime atmospheric instability was increasing and reached its maximum by 12 UTC, Fig. 8c, accordingly, by this time numerical value of instability energy was also maximal CAPE = 330 j/kg, but CINH = –28 j/kg. By 18 UTC atmospheric condition was absolutely stable CAPE = 0 (Table 2).

Convective events occurred on 2 August 2016 and results of numerical calculations. According to the weather radar's data there was clear (cloudless, unclouded) sky above Tbilisi at 20:49 on 2 August 2016 (Fig. 9a) and due to local inner massive atmosphere processes, during 12 min, there formed deep convective cloudy system (with a height of 15 km and reflecting state 45 db) nearby to Tbilisi (Fig. 9b). Later, strengthened cloudy system (the top level and reflecting state of the cloudy system

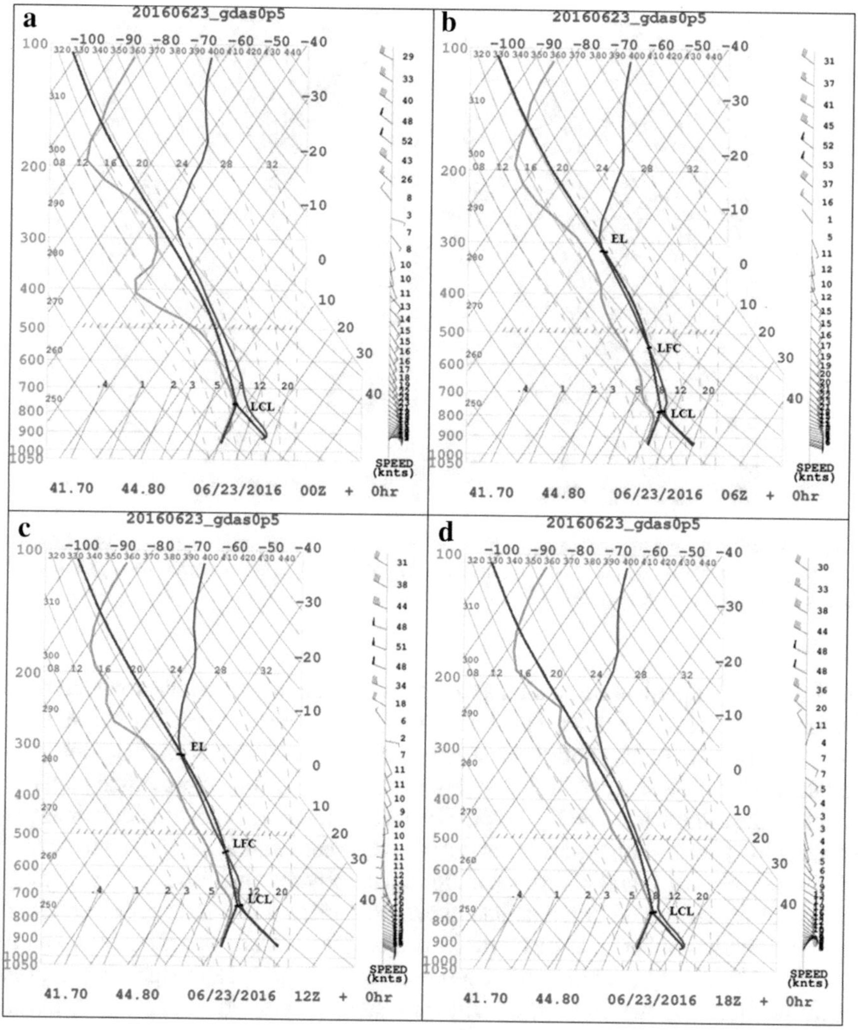

Fig. 8 Aerological diagrams of June 23, 2016 for $41°70'N$, $44°80'E$ geographic coordinates (**a** 00 UTC, **b** 06 UTC, **c** 12 UTC, **d** 18 UTC)

raised to 17 km, and to 60 db, respectively) has invaded on the territory of Tbilisi during 22 min (Fig. 9c). At 21:38 the intensity of precipitations attained to 100 mm/h, an area of the cloudy system increased above Tbilisi, but decreased intensity of raining, as in reality until this moment of time it was continually raining (100 mm/h) and hailing on the territory of Tbilisi (Fig. 9d). Then due to gusty wind (the value of wind velocity reached 25 m/s at 2 m above the surface) the cloudy system was shifted to the north–east direction and the cloudy system disappeared at 21:44.

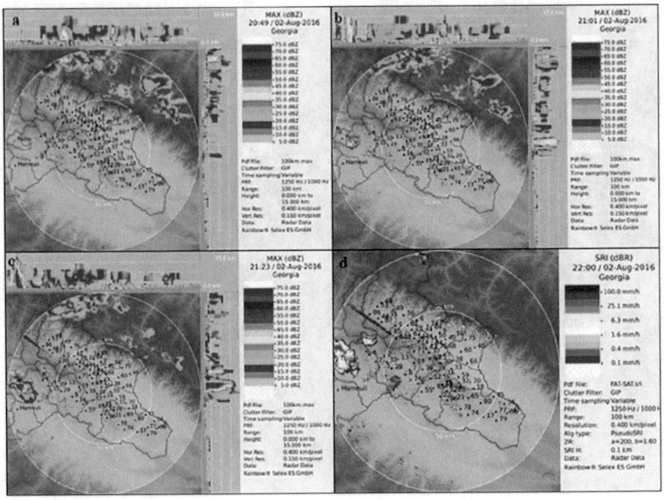

Fig. 9 Weather radar's map of the atmosphere reflecting state (clouds max-intensity dbz) above the territory of eastern Georgia at 20:49 (**a**), at 21:01 (**b**), at 21:23 (**c**), accumulated total precipitation at 22:00 (**d**) on 2nd of August 2016

Results of numerical calculations have shown that 24h predictions executed by all combinations of schemes presented in Table 1, were not able to predict in a satisfactory quality local scale, short term, severe, weather convective event that took place on the 2nd of August, 2016 in Tbilisi. For example in Fig. 10 are presented predicted fields of the accumulated total precipitation (ATP) executed by set1 (calculations performed by other combinations of schemes gave almost similar results) on the nested domains with 6.6 km and 2.2 km resolutions, respectively. The results of calculations, executed on the nested domains had shown that there was not any ATP and there was cloudless weather as at 21:00 on 02 August (Fig. 10a, b) as well at 00:00 on 03 August (Fig. 10c, d) in Tbilisi. Almost the same development atmospheric processes were predicted by the other physical combination sets listed in Table 1. All other predicted maps of ATP demonstrated 24 h WRF-ARW model forecast failure and especially in the investigated region where the both nested domains have predicted absolute dry conditions on the territory of Tbilisi.

In Fig. 11, there is shown AD of Tbilisi are shown (with its geographic coordinates $41 \circ^7 0'N, 44 \circ 80'E$) as of August 2, 2016 by various instants of time (a - 00 UTC, b - 06 UTC, c - 12 UTC, d - 18 UTC). As it is clearly seen on Fig. 11, since 00 UTC instability energy CAPE during the whole day is quickly increasing and reaches its maximum value by 12 UTC (CAPE = 378 j/kg). But during the daytime CINH power changed slowly and minimal value amounts reached at 18 UTC (By Tbilisi time at 22:00 p. m) (Fig. 11). AD as of August 2 at 06 UTC is provided on Fig. 11 b for the same geographic coordinates. As it is found from Fig. 11 there is a stable condition in lower atmospheric layers, but in upper layers we have indistinguishable condition. By this time CAPE = 93 j/kg, CINH = –45 j/kg. AD as of August 2

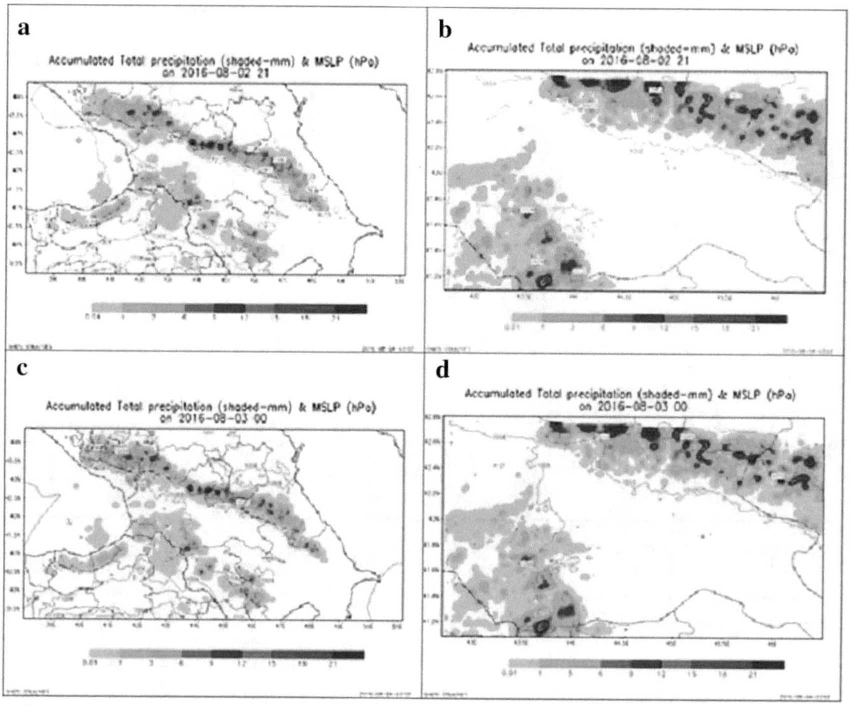

Fig. 10 Map of the ATP simulated by set 1 on the nested domains with 6.6 km resolution **a** with 2.2 km resolution **b** on 02 August 2016 (21:00 local time) and with 6.6 km resolution **c** with 2.2 km resolution **d** on 03 August 2016 (00:00 local time)

by 12:00 UTC (By Tbilisi time at 16:00 p. m) is provided on Fig. 11c for the same geographic coordinates. As it is shown on this figure, curve of condition moved to the right side of stratification curve to certain height above condensation level, meaning that instability energy is increased CAPE = 378 j/kg, but energy CIN = 43 j/kg is decreased. On Fig. 11c by 10 o'clock local time (18 UTC) was decreased the value of instability energy, by this time CAPE = 75 j/kg, at the same time negative energy was even decreased in lower atmospheric layers CINH = 26 j/kg. Pursuant to the aforesaid even in this case by 12 UTC (By Tbilisi time at 16:00 p. m) instability energy CAPE of atmosphere is at maximal level than at other times and accordingly, by 12 UTC instability of atmosphere is at the highest level. In accordance with the data of radar, later by 21:00 UTC was found maximal force of cloud and accordingly, we may consider that pursuant to the analysis of AD carried out by morning hours, there may be made preventive forecast on possible complex meteorological processes taken place in atmosphere and in the second half of day about possible heavy atmospheric precipitates.

In Table 2 numerical values of CAPE, CINH and Updraft velocity by 00, 06, 12, 18 UTC are presented. As it is clearly shown in Table 2, in the cases of convective

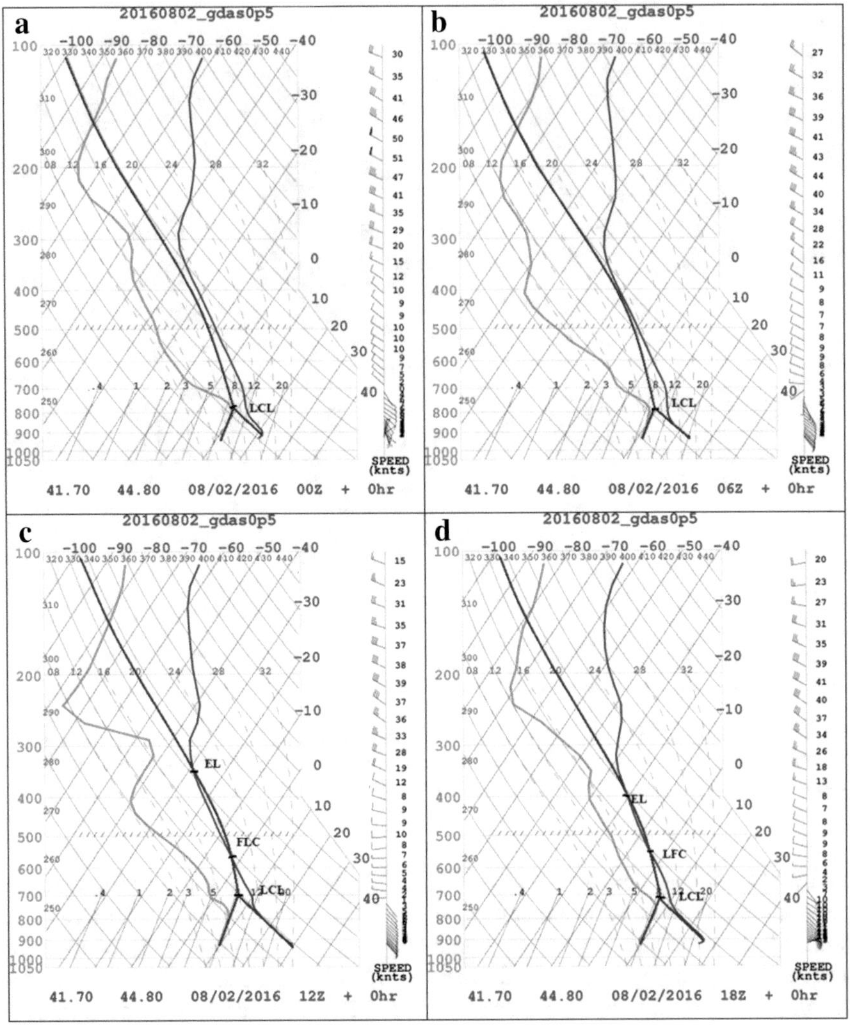

Fig. 11 Aerological diagrams of August 2, 2016 for $41°70'N$, $44°80'E$ geographic coordinates (**a**- 00 UTC, **b** - 06 UTC, **c**- 12 UTC, **d**- 18 UTC)

local processes on local territory (August 2, June 21, 23, 2016, June 13, 2015) numerical value of instability energy does not reach its maximal level, (maximal value of power CAPE = 378 j/kg). According to the data provided by meteorological radar, despite of having short term and local processes, they were still very dangerous and too intensive. That is why even little numerical value of instability power is noteworthy for the purpose of forecasting local atmospheric processes. Herein is to be noted that the best indicator of instability energy (CAPE) for above mentioned local meteorological processes is by 12 UTC (by Tbilisi time 16:00 p. m. when the

Table 2 Numerical values of instability power (CAPE), convective inhibition power (CINH) and Updraft velocity at the time of dangerous meteorological processes developed in some regions of eastern Georgia in 2015–2017

N	Day	Geograf coordinate	Time UTC	CAPE value (j/kg)	CINH value (j/kg)	Updraft velocity (m/s)
1	13.06.15		00	42.9	–9.4	9.3
		41.71 44.77	0.6	55.6	0	10.5
			12	109.6	–7.9	15
			18	48	–48	9.7
2	21.06.16		00	0	0	0
		41.70 44.80	06	0	0	0
			12	349.6	32	26
			18	0	0	0
3	23.06.16		00	0	0	0
		41.70 44.80	06	60	–28	11
			12	330	–55	26
			18	0	–7	0
4	02.08.16		00	50	–47	10
		41.70 44.80	06	93	–45	14
			12	378	–43	27
			18	75	–26	12

value of instability energy reaches its maximal level). in the afternoon temperature of earth surface reaches its maximal level and it is the most advantageous time for the development of convective local processes. So, when we forecast convective local processes under the basis of AD of READY System's ARL, thermodynamic condition is to be studied in accordance with the analysis of forecast AD in the afternoon and accordingly, under the basis of numerical value of forecast instability power calculated during this time. Thus, when it is impossible to forecast dangerous meteorological convective local processes under WRF Model, it is possible to forecast the level of probable dangerous meteorological convective local process on local territory though analyzing numerical value CAPE by READY System of ARL under the basis of forecast aeorological data by the mentioned day in the afternoon.

4 Conclusions

Heavy rainfall, hail and sometimes with subsequently floods not infrequently lead to severe damage, losses for both life and infrastructure. Thus, timely forecast of heavy rainfall and hail is essential issue in meteorology for providing an early warning to the population or minimization aftermath of the disaster (especially in the mountainous regions). Forecasting of local, short term, heavy rainfall based on regional numerical models is considered a difficult task, particularly for the mountainous regions (alike Georgia), as in the majority of cases the stipulating factors of local, short term, heavy rainfalls is not clear. The WRF model is fit out with a wide choice of physical schemes for refinement of mesoscale atmospheric processes prediction. Therefore, in this article we have studied the regional WRF v.3.6.1 model's and READY systems abilities on summer heavy showers prediction in the capital city of Georgia, Tbilisi, characterizing with complex topography. This study also sought after to support the Georgian National Environmental Agency in the performance of everyday operations in the short term weather forecasts. The first step in this process was evaluation numerical predictions' quality obtained by different MP and CPSs schemes of the WRF v.3.6.1 model against the results obtained by READY system and weather radar observations. Analyses of the results of calculations have shown that none of the combinations of MP and CPSs schemes in the WRF v.3.6 model were able to predict those real deep, moist convective atmospheric events which took place on 13 June 2015, 2nd of August, 21 of June and 23 of June 2016 in Tbilisi. While aerological diagrams of READY system precisely showed instability of atmosphere for discussed cases on local territories of Tbilisi. Despite the fact that in all four cases we had different levels of instability, the aerological diagrams of READY system precisely showed these differences which exactly corresponded to real meteorological condition of the specified day (according to the data of weather radar). Based on the obtained results, it can be argued that the analysis of aerological data, obtained for specific regions using the READY system of ARL of NOAA, together with the methods of short-term weather prediction used in operational departments of weather forecasting (alike Georgia), improves a forecast of the thermodynamic state of the atmosphere and an assessment of the level of evolving convective processes over the forecasted local territories. It should be noted that the best instability power indicator (CAPE) for the local meteorological processes, discussed in this article, comes when the power coefficient of instability reaches its maximum. In connection with our specific study (four events in Tbilisi), the calculation results showed that of the four events that took place in Tbilisi, three times the most favorable time for the development of local convective processes was 12 UTC (Tbilisi time is 16:00, when the ratio of the surface temperature of the earth reached its maximum). Accordingly, for predicting local, short-term, convective processes in region with complex orography using the WRF model, it is additionally advisable to study the thermodynamic state of the atmosphere by analyzing upper-air diagrams based on the upper-air diagrams of the READY ARL system.

Acknowledgements This research was supported by the Shota Rustaveli National Science Foundation Grant FR2017/FR17-548. The authors would like to thank Dr. N. Kutaladze and Mr. G. Mikuchadze for their efforts in configuration WRF v.3.6 model for the Caucasus region. The computer time for this study was provided by the Georgian Research and Education National Association GE-01-GRENA under the leadership of Dr. R. Kvatadze.

References

1. Blanchard, D.O.: Assessing the vertical distribution of convective available potential energy. Am. Meteorol. Soc. **l**, 870–877. NOAA/National Severe Storms Laboratory, Boulder, Colorado (1998)
2. Blanchard, D.O.: Notes and correspondence assessing the vertical distribution of convective available potential energy. Weath. Forecast. **13** (1998)
3. Bronnimann, St.: A historical upper air -data set for the 1939-44 period. Int. J. Climatol. **23**, 769–791 (2003). https://doi.org/10.1002/joc.914
4. Brooksa, H.E., Lee, J.W., Cravenc, J.P.: The spatial distribution of severe thunderstorm and tornado environments from global reanalysis data. Atmosph. Res. 73–94 (2003)
5. Davitashvili, T., Kobiashvili, G., Kvatadze, R., Kutaladze, N., Mikuchadze, G.: WRF-ARW Application for Georgia, Report of SEE-GRID-SCI User Forum, 2009, Istanbul. Turkey, pp. 7–10 (2009)
6. Davitashvili, T., Kvatadze, R., Kutaladze, N.: Weather prediction over caucasus region using WRF-ARW model, Proceedings of the 34th International Convection, MIPRO 2011, Print 326–330, ISBN: 978-1-4577-0996-8. Opatija, Croatia (2011)
7. Davitashvili, T., Kutaladze, N. Kvatadze, R., Mikuchadze, G. Modebadze,Z., Samkharadze, I.: Precipitations prediction by different physics of WRF model. Int. J. Environ. Sci. **1**, 294–299 (2016)
8. Davitashvili, T., Kutaladze, N. Kvatadze, R., Mikuchadze, G. Modebadze, Z., Samkharadze, I.: Showers prediction by WRF model above complex terrain. In: Proceedings of the 39th International Convection, MIPRO 2016/DC VIS, Opatija, Croatia, pp. 236–241 (2016)
9. Davis, C.A., Bosart, L.F.: Numerical simulations of the genesis of Hurricane Diana. Part II: Sensitivity of track and intensity prediction. Mon. Weath. Rev. **130**, 1100–1124 (2002)
10. Doswell, ChA, Evans, J.S.: Proximity sounding analysis for derechos and supercells: an assessment of similarities and differences. Atmosph. Res. **67–68**, 117–133 (2003)
11. Fovell, R.,: Impact of microphysics on hurricane track and intensity forecasts. Seventh WRF Users' Workshop, Boulder, CO, Natl. Center Atmosph. Res. **3**(2) (2006)
12. Golden, J.H., Serafin, R., Lally, V., Facundo, J.: Atmospheric sounding systems. In: Ray, P.S. (ed.) Mesoscale Meteorology and Forecasting. American Meteorological Society, Boston (1986)
13. Jankov, I., A.,William, Jr.,Gallus M., Segal, B.,Shaw, Koch, S., E.,: The impact of different wrf model physical parameterizations and their interactions on warm season MCS rainfall, Weath. Forecast. **20**, 1048–1061 (2005)
14. Li, X., Pu, Z.: Sensitivity of numerical simulations of the early rapid intensification of Hurricane Emily to cumulus parameterization schemes in different model horizontal resolutions. J. Meteor. Soc. Jpn. **87**, 403–421 (2009)
15. Lowrey, M., Yang, Z.: Assessing the capability of a regional-scale weather model to simulate extreme precipitation patterns and flooding in central Texas. Weath. Forecast. **23**, 1102–1126 (2008)
16. McFarquhar, G. M., Zhang, H., Heymsfield, G., Hood, R. J. Dudhia, J. B. Halverson, F. Marks Jr.: Factors acting the evolution of Hurricane Erin (2001) and the distributions of hydrometeors: role of microphysical processes. J. Atmos. Sci. **63**, 127–150 (2006)

17. Müller, O.V., Lovino, M.A., Berbery, E.H.: Evaluation of WRF model forecasts and their use for hydroclimate monitoring over Southern South America. Am. Meteorol. Soc. **31**, 1001–1017 (2016)
18. Nasrollahi, N.A., Aghakouchak, J., Li, X., Gao, K., Hsu, Sorooshian, S.: Assessing the impacts of different WRF precipitation physics in hurricane simulations, Weath. Forecast. **27**, 1003–1016 (2012)
19. Nielsen-Gammon. J.: Chapter 5. Weather Observation and Analysis. Soundings. ATMO 251 Home Page. Texas A M University (2007)
20. Ooyama, K.V.: A thermodynamic foundation for modeling the moist atmosphere. J. Atmos. Sci. **47**, 2580–2593 (1990)
21. Parker D.J.: The response of CAPE and CIN to tropospheric thermal variations. Q. J. R. Meteorol. Soc. **128**, 119–130 (2002)
22. Rogers, E., Ek, M., Lin, Y., Mitchell, K., Parrish, D., DiMego, G.: Changes to the NCEP Meso Eta analysis and forecast system: Assimilation of observed precipitation, upgrades to land-surface physics, modified 3DVAR analysis. NWS Tech. Proc. Bull. **488** (2001)
23. Rolph, G., Stein, Ar., Stunder, B.: Real-time environmental applications and display system. READY Environ. Model. Softw. **95**, 210–228 (2017)
24. Second National Communication to the United Nations Framework Convention on Climate Change (SNCUNFCCC); Ministry of Environment Protection and Natural Resources, Republic of Georgia. p. 240 (2009). www.unfccc.int
25. Sikder, S., Hossain, F.: Assessment of the weather research and forecasting model generalized parameterization schemes for advancement of precipitation forecasting in monsoon-driven river basins. J. Adv. Model. Earth Syst. **8**(3), 1210–1228 (2016). https://doi.org/10.1002/2016MS000678
26. Skamarock W.C., Klemp, J.B.: The stability of time-split numerical methods for the hydrostatic and the nonhydrostatic elastic equations, Mon. Weath. Rev. **120**, 2109–2127 (1992)
27. Skamarock, W., Klemp, J., Dudhia, J.,, Gill, D., Barker, D., Wang, W., Powers, J.: A description of the advanced research WRF version 2. NCAR Tech. Note NCAR/TN-4681STR, 88 (2007)
28. Stein, A.F., Draxler, R.R., Rolph, G.D., Stunder, B.J.B., Cohen, M.D., Ngan, F.: NOAA's HYSPLIT atmospheric transport and dispersion modeling system. Bull. Am. Meteorol. Soc. **96**, 2059–2077 (2015). https://doi.org/10.1175/BAMS-D-14-00110.1
29. Sun, B., Reale. A.: Hunt Comparing radiosonde and COSMIC atmospheric profile data to quantify differences among radiosonde types and the effects of imperfect collocation on comparison statistics. J. Geophys. Res. **115**, 1–16 (2010). http://dx.doi.org/10.1029/2010JD014457
30. Tao, W.-K., Simpson, J., McCumber, M.: An ice-water saturation adjustment. Mon. Weath. Rew. **117**, 231–235 (1989)
31. Tao, W.-K., Shi1, J. J., Chen, S.,S., Lang, S., Lin, P.-L., Hong, S.-Y., Peters-Lidard C., Hou, A.: The impact of microphysical schemes on intensity and track of Hurricane. Spec. Issue Asia-Pacific J. Atmosph. Sci. (APJAS), **31**, 1–47 (2010)
32. Wang, W., Seaman, N.L.: A comparison study of convective parameterization schemes in a mesoscale model. Mon. Weath. Rev. **125**, 252–278 (1997)
33. Westphal, M.I., Mehtiyev, M., Shvangiradze, M., Tonoyan, V.: Regional Climate Change Impacts Study for the South Caucasus Region, p. 63 (2011)
34. Wei, L., Jiangping, H.: Study and Development on Fully Automated Sounding System, WMO, IOM-109, TECO-2012 (2012)
35. Weisman, M.L., Klemp, J.B.: The dependence of numerically simulated convective storms on vertical winds hear and buoyancy. Mon. Weather Rev. **110**, 504–520 (1982)
36. Jung, Y., Lin, Y.L.: Assessment of a regional-scale weather model for hydrological applications in South Korea. Environ. Nat. Resour. Res. **6**(2), 28–41 (2016)
37. Yu, E., Wang, H., Gao, Y., Sun, J.: Impacts of cumulus convective parameterization schemes on summer monsoon precipitation simulation over China. Acta Meteorol. Sin. **25**(5), 581–592 (2011)

Extending the ρLog Calculus with Proximity Relations

Besik Dundua, Temur Kutsia, Mircea Marin, and Cleo Pau

Abstract ρLog-prox is a calculus for rule-based programming with strategies, which supports both exact and approximate computations. Rules are represented as conditional transformations of sequences of expressions, which are built from variadic function symbols and four kinds of variables: for terms, hedges, function symbols, and contexts. ρLog-prox extends ρLog by permitting in its programs fuzzy proximity relations, which are reflexive and symmetric, but not transitive. We introduce syntax and operational semantics of ρLog-prox, illustrate its work by examples, and present a terminating, sound, and complete algorithm for the ρLog-prox expression matching problem.

Keywords Rule-based programming · ρLog · fuzzy proximity relations

1 Introduction

ρLog [18] is a calculus for conditional transformation of sequences of expressions, controlled by strategies. It originated from experiments with extending the language of Mathematica [26] by a rule-based programming package [17, 19]. Meanwhile

---

B. Dundua
VIAM, Ivane Javakhishvili Tbilisi State University and International Black Sea University, Tbilisi, Georgia
e-mail: bdundua@egmail.com

T. Kutsia (✉) · C. Pau
RISC, Johannes Kepler University Linz, Linz, Austria
e-mail: kutsia@risc.jku.at

C. Pau
e-mail: ipau@risc.jku.at

M. Marin
West University of Timişoara, Timişoara, Romania
e-mail: mircea.marin@e-uvt.ro

G. Jaiani and D. Natroshvili (eds.), *Applications of Mathematics and Informatics in Natural Sciences and Engineering*, Springer Proceedings in Mathematics & Statistics 334, https://doi.org/10.1007/978-3-030-56356-1_6

there are some tools based on or influenced by ρLog, such as its implementation in Mathematica [16], an extension of Prolog, called PρLog [7], or an extension of Maple, called symbtrans [3].

ρLog objects are logic terms that are built from function symbols without fixed arity and four different kinds of variables: for individual terms, for finite sequences of terms (hedges), for function symbols, and for contexts (special unary higher-order functions). Rules transform finite sequences of terms, when the given conditions are satisfied. They are labeled by strategies, providing a flexible mechanism for combining and controlling their behavior. ρLog programs are sets of rules. The inference system is based on SLDNF-resolution [15]. Program meaning is characterized by logic programming semantics. Rules and strategies are formulated as clauses.

ρLog-based/inspired tools have been used in extraction of frequent patterns from data mining workflows [22], for automatic derivation of multiscale models of arrays of micro- and nanosystems [27], modeling rewriting strategies [6], etc.

The core of ρLog is a powerful pattern matching algorithm [13]. Matching with hedge and context variables is finitary: problems might have finitely many different solutions. In many situations, it can replace recursion, leading to pretty compact and intuitive code. Nondeterministic computations are modeled naturally by backtracking.

The computational mechanism of ρLog is based on the assumption that the provided information is precise and the problems can be solved exactly. However, in many cases, especially in the areas related to applications of artificial intelligence, one has to deal with vague information, which increases demand for the corresponding reasoning and computing techniques. Several approaches to this problem propose methods and tools that integrate fuzzy logic or probabilistic reasoning with declarative programming, see, e.g., [8–11, 14, 20, 21, 23, 24].

ρLog-prox, described in this paper, is an attempt to address this problem by combining approximate reasoning and strategic rule-based programming. It extends ρLog with the capabilities to process imprecise information represented by proximity relations. The latter are binary fuzzy relations, satisfying the properties of reflexivity and symmetry. We develop a matching algorithm that solves the problem of approximate equality between terms that may contain variables for terms, hedges, function symbols and contexts. A particular difficulty is related to the fact that proximity relations are not transitive. We prove that our matching algorithm is terminating, sound, and complete, and integrate it in the ρLog-prox calculus. The integration is transparent: approximate equality is expressed explicitly, no hidden fuzziness is assumed. Multiple solutions to matching problems are explored by nondeterministic computations in the inference mechanism.

The rest of the paper is organized as follows: In Sect. 2 we introduce the terminology, define our language, and discuss proximity relations. Section 3 is about the basics of ρLog-prox: its syntax, semantics, and an illustrative example are presented. In Sect. 4, we develop an algorithm for solving proximity matching problems and prove its properties. Section 5 is the conclusion.

2 Preliminaries

In this section, we introduce the basic notions needed in the rest of the paper.

2.1 Terms, Hedges, Contexts, Substitutions

The alphabet $\mathcal{A}$ consists of the following pairwise disjoint sets of symbols:

- $\mathcal{V}_\mathsf{T}$: term variables, denoted by $x, y, z, \ldots$,
- $\mathcal{V}_\mathsf{S}$: hedge variables, denoted by $\overline{x}, \overline{y}, \overline{z}, \ldots$,
- $\mathcal{V}_\mathsf{F}$: function variables, denoted by $X, Y, Z, \ldots$,
- $\mathcal{V}_\mathsf{C}$: context variables, denoted by $\overline{X}, \overline{Y}, \overline{Y}, \ldots$,
- $\mathcal{F}$: unranked function symbols, denoted by $f, g, h, \ldots$.

Besides, $\mathcal{A}$ contains also auxiliary symbols such as parenthesis and comma, and a special constant $\circ$, called hole. A variable is an element of the set $\mathcal{V} = \mathcal{V}_\mathsf{T} \cup \mathcal{V}_\mathsf{S} \cup \mathcal{V}_\mathsf{F} \cup \mathcal{V}_\mathsf{C}$. A functor, denoted by F, is a common name for a function symbol or a function variable.

We define terms, hedges, contexts, and other syntactic categories over $\mathcal{A}$ as follows:

t	$::= x \mid f(\tilde{s}) \mid X(\tilde{s}) \mid \overline{X}(t)$	Term
$\tilde{t}$	$::= t_1, \ldots, t_n \quad (n \geq 0)$	Term sequence
s	$::= t \mid \overline{x}$	Hedge element
$\tilde{s}$	$::= s_1, \ldots, s_n \quad (n \geq 0)$	Hedge
C	$::= \circ \mid f(\tilde{s}_1, C, \tilde{s}_2) \mid X(\tilde{s}_1, C, \tilde{s}_2) \mid \overline{X}(C)$	Context

Hence, hedges are sequences of hedge elements, hedge variables are not terms, term sequences do not contain hedge variables, contexts (which are not terms either) contain a single occurrence of the hole. We do not distinguish between a singleton hedge and its sole element.

We denote the set of terms by $\mathcal{T}(\mathcal{F}, \mathcal{V})$, hedges by $\mathcal{H}(\mathcal{F}, \mathcal{V})$, and contexts by $\mathcal{C}(\mathcal{F}, \mathcal{V})$. Ground (i.e., variable-free) subsets of these sets are denoted by $\mathcal{T}(\mathcal{F})$, $\mathcal{H}(\mathcal{F})$, and $\mathcal{C}(\mathcal{F})$, respectively.

We make a couple of conventions to improve readability. We put parentheses around hedges, writing, e.g., $(f(a), \overline{x}, b)$ instead of $f(a), \overline{x}, b$. The empty hedge is written as $()$. The terms of the form $a()$ and $X()$ are abbreviated as a and X, respectively, when it is guaranteed that terms and symbols are not confused. For hedges $\tilde{s} = (s_1, \ldots, s_n)$ and $\tilde{s}' = (s'_1, \ldots, s'_m)$, the notation $(\tilde{s}, \tilde{s}')$ stands for the hedge $(s_1, \ldots, s_n, s'_1, \ldots, s'_m)$. We use $\tilde{s}$ and $\tilde{r}$ for arbitrary hedges, while $\tilde{t}$ is reserved for term sequences.

Below we will also need anonymous variables for each variable category. They are variables without name, well-known in declarative programming. We write the single underscore _ for anonymous term and function variables, and the double underscore

__ for anonymous hedge and context variables. The set of anonymous variables is denoted by $\mathcal{V}_{\mathsf{An}}$.

A syntactic expression (or, just an expression) is an element of the set $\mathcal{F} \cup \mathcal{V} \cup \mathcal{T}(\mathcal{F}, \mathcal{V}) \cup \mathcal{H}(\mathcal{F}, \mathcal{V}) \cup \mathcal{C}(\mathcal{F}, \mathcal{V})$. We denote expressions by E.

We also introduce two notations: $\mathcal{V}(E)$ denotes the set of variables occurring in expression E, and $\mathcal{V}(E, \{p_1, ..., p_n\})$, where p_i's are positions in E, is defined as $\mathcal{V}(E, \{p_1, ..., p_n\}) = \cup_{i=1}^{n} \mathcal{V}(E|_{p_i})$, where $E|_{p_i}$ is the standard notation for a subexpression of E at position p_i.

Contexts can apply to contexts or terms. This meta-operation is denoted by $C_1[C_2]$ or $C_1[t]$ and is obtained from C_1 by replacing the hole in it by C_2 or t, respectively. Thus, $C_1[C_2]$ is a context and $C_1[t]$ is a term.

Substitution is a mapping σ from $\mathcal{V}$ to $\mathcal{T}(\mathcal{F}, \mathcal{V}) \cup \mathcal{H}(\mathcal{F}, \mathcal{V}) \cup \mathcal{C}(\mathcal{F}, \mathcal{V}) \cup \mathcal{F} \cup \mathcal{V}_{\mathsf{F}}$, defined as

$$\sigma(x) \in \mathcal{T}(\mathcal{F}, \mathcal{V}), \qquad \sigma(\overline{x}) \in \mathcal{H}(\mathcal{F}, \mathcal{V}),$$
$$\sigma(X) \in \mathcal{F} \cup \mathcal{V}_{\mathsf{F}}, \qquad \sigma(\overline{X}) \in \mathcal{C}(\mathcal{F}, \mathcal{V}),$$

such that $\sigma(v) = v$ for all but finitely many term, hedge, and function variables v, and $\overline{X} = \overline{X}(\circ)$ for all but finitely many context variables $\overline{X}$.

Substitutions are denoted by Greek letters σ, ϑ, φ. The identity substitution is denoted by Id.

A substitution σ may apply to elements of the set $\mathcal{T}(\mathcal{F}, \mathcal{V}) \cup \mathcal{H}(\mathcal{F}, \mathcal{V}) \cup \mathcal{C}(\mathcal{F}, \mathcal{V}) \cup \mathcal{F} \cup \mathcal{V}_{\mathsf{F}}$ in the following way:

$$x\sigma = \sigma(x), \qquad F(\tilde{s})\sigma = (F\sigma)(\tilde{s}\sigma), \qquad \overline{X}(t)\sigma = \sigma(\overline{X})[t\sigma],$$
$$\overline{x}\sigma = \sigma(\overline{x}), \qquad (s_1, \ldots, s_n)\sigma = (s_1\sigma, \ldots, s_n\sigma), \qquad X\sigma = \sigma(X), \quad f\sigma = f,$$
$$\circ\,\sigma = \circ, \qquad F(\tilde{s}_1, C, \tilde{s}_2)\sigma = (F\sigma)(\tilde{s}_1\sigma, C\sigma, \tilde{s}_2\sigma), \qquad \overline{X}(C)\sigma = \sigma(\overline{X})[C\sigma].$$

2.2 Proximity Relations

Basic notions about proximity relations are defined following [11].

A binary fuzzy relation on a set S is a mapping from $S \times S$ to the real interval [0, 1]. If $\mathcal{R}$ is a fuzzy relation on S and λ is a number $0 \le \lambda \le 1$, then the λ-cut of $\mathcal{R}$ on S, denoted $\mathcal{R}_\lambda$, is an ordinary (crisp) relation on S defined as $\mathcal{R}_\lambda := \{(s_1, s_2) \mid \mathcal{R}(s_1, s_2) \ge \lambda\}$.

A fuzzy relation $\mathcal{R}$ on a set S is called a proximity relation, if it reflexive and symmetric:

Reflexivity: $\mathcal{R}(s, s) = 1$ for all $s \in S$;
Symmetry: $\mathcal{R}(s_1, s_2) = \mathcal{R}(s_2, s_1)$ for all $s_1, s_2 \in S$.

In this paper we consider only strict proximity relations:

Strictness: For all $s_1, s_2 \in S$, if $\mathscr{R}(s_1, s_2) = 1$ then $s_1 = s_2$.

A proximity relation is characterized by a set $\Lambda = \{\lambda_1, \ldots, \lambda_n \mid 0 < \lambda_i \leq 1\}$ of approximation levels. They express the degree of relationship of the related elements. We say that a value $\lambda \in \Lambda$ is a cut value. The λ-cut of $\mathscr{R}$, defined as $\mathscr{R}_\lambda = \{(s_1, s_2) \mid R(s_1, s_2) \geq \lambda\}$ is a usual two-valued tolerance (i.e., reflexive and symmetric) relation.

A T-norm $\wedge$ is an associative, commutative, non-decreasing binary operation on $[0, 1]$ with 1 as the unit element. In the rest of the paper, we take minimum in the role of T-norm.

The proximity class of level $\lambda > 0$ of $s \in S$ in a relation $\mathscr{R}$ (a λ-class of s in $\mathscr{R}$) is a set $\mathbf{pc}(s, \mathscr{R}, \lambda) = \{s' \mid \mathscr{R}(s, s') \geq \lambda\}$.

Our proximity relations are defined on the set of function symbols $\mathscr{F}$. We require them to be defined in such a way that the proximity class for each symbol is finite. Given a proximity relation $\mathscr{R}$ defined on $\mathscr{F}$, we extend it to $\mathscr{F} \cup \mathscr{V} \cup \mathscr{T}(\mathscr{F}, \mathscr{V}) \cup \mathscr{H}(\mathscr{F}, \mathscr{V}) \cup \mathscr{C}(\mathscr{F}, \mathscr{V})$:

- For variables, $V \in \mathscr{V}$:
  - $\mathscr{R}(V, V) = 1$.
- For terms, $t, t' \in \mathscr{T}(\mathscr{F}, \mathscr{V})$:
  - If t and t' have the same number of arguments, e.g., $t = F(s_1, \ldots, s_n)$ and $t' = F'(s'_1, \ldots, s'_n)$, then $\mathscr{R}(t, t') = \mathscr{R}(F, F') \wedge \mathscr{R}(s_1, s'_1) \wedge \cdots \wedge \mathscr{R}(s_n, s'_n)$.
- For hedges, $s, s' \in \mathscr{H}(\mathscr{F}, \mathscr{V})$:
  - If $\tilde{s}$ and $\tilde{s}'$ have the same number of elements, e.g., $\tilde{s} = (s_1, \ldots, s_n)$ and $\tilde{s}' = (s'_1, \ldots, s'_n)$, then $\mathscr{R}(\tilde{s}, \tilde{s}') = \mathscr{R}(s_1, s'_1) \wedge \cdots \wedge \mathscr{R}(s_n, s'_n)$.
- For contexts, $C, C' \in \mathscr{C}(\mathscr{F}, \mathscr{V})$:
  - $\mathscr{R}(\circ, \circ) = 1$.
  - If C and C' have the same number of arguments and their context arguments appear in the same position, e.g., $C = F(s_1, \ldots, s_{i-1}, C_1, s_{i+1}, \ldots, s_n)$ and $C' = F'(s'_1, \ldots, s'_{i-1}, C'_1, s'_{i+1} \ldots, s'_n)$, then $\mathscr{R}(C, C') = \mathscr{R}(F, F') \wedge \mathscr{R}(s_1, s'_1) \wedge \cdots \mathscr{R}(s_{i-1}, s'_{i-1}) \wedge \mathscr{R}(C_1, C'_1) \wedge \mathscr{R}(s_{i+1}, s'_{i+1}) \wedge \mathscr{R}(s_n, s'_n)$.
- In all other cases, $\mathscr{R}(E, E') = 0$ for two syntactic expressions $E, E' \in \mathscr{V} \cup \mathscr{T}(\mathscr{F}, \mathscr{V}) \cup \mathscr{H}(\mathscr{F}, \mathscr{V}) \cup \mathscr{C}(\mathscr{F}, \mathscr{V})$.

When $\mathscr{R}$ is strict on $\mathscr{F}$, its extension to $\mathscr{F} \cup \mathscr{V} \cup \mathscr{T}(\mathscr{F}, \mathscr{V}) \cup \mathscr{H}(\mathscr{F}, \mathscr{V}) \cup \mathscr{C}(\mathscr{F}, \mathscr{V})$ is also strict.

The notion of proximity class extends to elements of $\mathscr{F} \cup \mathscr{V} \cup \mathscr{T}(\mathscr{F}, \mathscr{V}) \cup \mathscr{H}(\mathscr{F}, \mathscr{V}) \cup \mathscr{C}(\mathscr{F}, \mathscr{V})$. It is easy to see that each proximity class in this set is also finite.

3 ρLog-prox: ρLog with Proximity Relations

3.1 Syntactic Matching and Proximity Matching Problems

A syntactic matching atom is a formula of the form $E_1 \ll E_2$. It is solved if the expressions E_1 and E_2 are identical, i.e., if $E_1 = E_2$. A substitution σ is a solution (or a matcher) of a matching atom $E_1 \ll E_2$ iff $E_1\sigma = E_2$.

Example 1 The syntactic matching atom

$$(X(a), \overline{x}, \overline{Y}(X(\overline{x}, y)), \overline{z}) \ll (f(a), g(b, f(b), f(a, f(b))), b, c)$$

has two solutions:

$$\sigma_1 = \{X \mapsto f,\ \overline{x} \mapsto (),\ \overline{Y} \mapsto g(b, \circ, f(a, f(b))),\ y \mapsto b,\ \overline{z} \mapsto (b, c)\}$$
$$\sigma_2 = \{X \mapsto f,\ \overline{x} \mapsto (),\ \overline{Y} \mapsto g(b, f(b), f(a, \circ)),\ y \mapsto b,\ \overline{z} \mapsto (b, c)\}$$

A syntactic matching problem is a finite set of syntactic matching atoms. Its solution is a substitution which solves each of the atoms in the problem.

Given a proximity relation $\mathscr{R}$ and a cut value λ, an $(\mathscr{R}, \lambda)$-proximity atom is a formula $E_1 \ll_{\mathscr{R},\lambda} E_2$ for the expressions E_1 and E_2. Its solution is a substitution σ such that $\mathscr{R}(E_1\sigma, E_2) \geq \lambda$. A solution with the proximity degree α is a substitution σ such that $\mathscr{R}(E_1\sigma, E_2) = \alpha \geq \lambda$.

Example 2 Let the proximity relation $\mathscr{R}$ be given by the following:

$$\mathscr{R}(g_1, h_1) = \mathscr{R}(g_2, h_1) = 0.4$$
$$\mathscr{R}(g_1, h_2) = \mathscr{R}(g_2, h_2) = 0.5$$
$$\mathscr{R}(g_2, h_3) = \mathscr{R}(g_3, h_3) = 0.6$$
$$\mathscr{R}(a, b) = 0.7$$

Let the proximity atom be

$$P = f(\overline{x}, x, \overline{Y}(x), \overline{z}) \ll_{\mathscr{R},\lambda} f(g_1(a), g_2(b), f(g_3(a))).$$

Consider the approximation levels $\Lambda = \{0.4, 0.5, 0.6, 0.7\}$. We get the following solutions to P: (In all cases, the proximity degrees of solutions coincide with λ.)

$\lambda = 0.4$:

$$\sigma_1 = \{\overline{x} \mapsto (), x \mapsto h_1(a), \overline{Y} \mapsto \circ, \overline{z} \mapsto f(g_3(a))\}$$
$$\sigma_2 = \{\overline{x} \mapsto (), x \mapsto h_2(a), \overline{Y} \mapsto \circ, \overline{z} \mapsto f(g_3(a))\}$$
$$\sigma_3 = \{\overline{x} \mapsto (), x \mapsto h_1(b), \overline{Y} \mapsto \circ, \overline{z} \mapsto f(g_3(a))\}$$

$$\sigma_4 = \{\overline{x} \mapsto (), x \mapsto h_2(b), \overline{Y} \mapsto \circ, \overline{z} \mapsto f(g_3(a))\}$$
$$\sigma_5 = \{\overline{x} \mapsto (), x \mapsto h_1(a), \overline{Y} \mapsto \circ, \overline{z} \mapsto f(g_3(b))\}$$
$$\sigma_6 = \{\overline{x} \mapsto (), x \mapsto h_2(a), \overline{Y} \mapsto \circ, \overline{z} \mapsto f(g_3(b))\}$$
$$\sigma_7 = \{\overline{x} \mapsto (), x \mapsto h_1(b), \overline{Y} \mapsto \circ, \overline{z} \mapsto f(g_3(b))\}$$
$$\sigma_8 = \{\overline{x} \mapsto (), x \mapsto h_2(b), \overline{Y} \mapsto \circ, \overline{z} \mapsto f(g_3(b))\}$$
$$\sigma_9 = \{\overline{x} \mapsto (), x \mapsto h_1(a), \overline{Y} \mapsto \circ, \overline{z} \mapsto f(h_3(a))\}$$
$$\sigma_{10} = \{\overline{x} \mapsto (), x \mapsto h_2(a), \overline{Y} \mapsto \circ, \overline{z} \mapsto f(h_3(a))\}$$
$$\sigma_{11} = \{\overline{x} \mapsto (), x \mapsto h_1(b), \overline{Y} \mapsto \circ, \overline{z} \mapsto f(h_3(a))\}$$
$$\sigma_{12} = \{\overline{x} \mapsto (), x \mapsto h_2(b), \overline{Y} \mapsto \circ, \overline{z} \mapsto f(h_3(a))\}$$
$$\sigma_{13} = \{\overline{x} \mapsto (), x \mapsto h_1(a), \overline{Y} \mapsto \circ, \overline{z} \mapsto f(h_3(b))\}$$
$$\sigma_{14} = \{\overline{x} \mapsto (), x \mapsto h_2(a), \overline{Y} \mapsto \circ, \overline{z} \mapsto f(h_3(b))\}$$
$$\sigma_{15} = \{\overline{x} \mapsto (), x \mapsto h_1(b), \overline{Y} \mapsto \circ, \overline{z} \mapsto f(h_3(b))\}$$
$$\sigma_{16} = \{\overline{x} \mapsto (), x \mapsto h_2(b), \overline{Y} \mapsto \circ, \overline{z} \mapsto f(h_3(b))\}$$
$$\sigma_{17} = \{\overline{x} \mapsto g_1(a), x \mapsto h_3(a), \overline{Y} \mapsto f(\circ), \overline{z} \mapsto ()\}$$
$$\sigma_{18} = \{\overline{x} \mapsto g_1(a), x \mapsto h_3(b), \overline{Y} \mapsto f(\circ), \overline{z} \mapsto ()\}$$
$$\sigma_{19} = \{\overline{x} \mapsto g_1(b), x \mapsto h_3(a), \overline{Y} \mapsto f(\circ), \overline{z} \mapsto ()\}$$
$$\sigma_{20} = \{\overline{x} \mapsto g_1(b), x \mapsto h_3(b), \overline{Y} \mapsto f(\circ), \overline{z} \mapsto ()\}$$
$$\sigma_{21} = \{\overline{x} \mapsto h_1(a), x \mapsto h_3(a), \overline{Y} \mapsto f(\circ), \overline{z} \mapsto ()\}$$
$$\sigma_{22} = \{\overline{x} \mapsto h_1(a), x \mapsto h_3(b), \overline{Y} \mapsto f(\circ), \overline{z} \mapsto ()\}$$
$$\sigma_{23} = \{\overline{x} \mapsto h_1(b), x \mapsto h_3(a), \overline{Y} \mapsto f(\circ), \overline{z} \mapsto ()\}$$
$$\sigma_{24} = \{\overline{x} \mapsto h_1(b), x \mapsto h_3(b), \overline{Y} \mapsto f(\circ), \overline{z} \mapsto ()\}$$
$$\sigma_{25} = \{\overline{x} \mapsto h_2(a), x \mapsto h_3(a), \overline{Y} \mapsto f(\circ), \overline{z} \mapsto ()\}$$
$$\sigma_{26} = \{\overline{x} \mapsto h_2(a), x \mapsto h_3(b), \overline{Y} \mapsto f(\circ), \overline{z} \mapsto ()\}$$
$$\sigma_{27} = \{\overline{x} \mapsto h_2(b), x \mapsto h_3(a), \overline{Y} \mapsto f(\circ), \overline{z} \mapsto ()\}$$
$$\sigma_{28} = \{\overline{x} \mapsto h_2(b), x \mapsto h_3(b), \overline{Y} \mapsto f(\circ), \overline{z} \mapsto ()\}$$

$\lambda = 0.5:$

$$\sigma_1 = \{\overline{x} \mapsto (), x \mapsto h_2(a), \overline{Y} \mapsto \circ, \overline{z} \mapsto f(g_3(a))\}$$
$$\sigma_2 = \{\overline{x} \mapsto (), x \mapsto h_2(b), \overline{Y} \mapsto \circ, \overline{z} \mapsto f(g_3(a))\}$$
$$\sigma_3 = \{\overline{x} \mapsto (), x \mapsto h_2(a), \overline{Y} \mapsto \circ, \overline{z} \mapsto f(g_3(b))\}$$
$$\sigma_4 = \{\overline{x} \mapsto (), x \mapsto h_2(b), \overline{Y} \mapsto \circ, \overline{z} \mapsto f(g_3(b))\}$$
$$\sigma_5 = \{\overline{x} \mapsto (), x \mapsto h_2(a), \overline{Y} \mapsto \circ, \overline{z} \mapsto f(h_3(a))\}$$
$$\sigma_6 = \{\overline{x} \mapsto (), x \mapsto h_2(b), \overline{Y} \mapsto \circ, \overline{z} \mapsto f(h_3(a))\}$$

$$\sigma_7 = \{\overline{x} \mapsto (), x \mapsto h_2(a), \overline{Y} \mapsto \circ, \overline{z} \mapsto f(h_3(b))\}$$
$$\sigma_8 = \{\overline{x} \mapsto (), x \mapsto h_2(b), \overline{Y} \mapsto \circ, \overline{z} \mapsto f(h_3(b))\}$$
$$\sigma_9 = \{\overline{x} \mapsto g_1(a), x \mapsto h_3(a), \overline{Y} \mapsto f(\circ), \overline{z} \mapsto ()\}$$
$$\sigma_{10} = \{\overline{x} \mapsto g_1(a), x \mapsto h_3(b), \overline{Y} \mapsto f(\circ), \overline{z} \mapsto ()\}$$
$$\sigma_{11} = \{\overline{x} \mapsto g_1(b), x \mapsto h_3(a), \overline{Y} \mapsto f(\circ), \overline{z} \mapsto ()\}$$
$$\sigma_{12} = \{\overline{x} \mapsto g_1(b), x \mapsto h_3(b), \overline{Y} \mapsto f(\circ), \overline{z} \mapsto ()\}$$
$$\sigma_{13} = \{\overline{x} \mapsto h_2(a), x \mapsto h_3(a), \overline{Y} \mapsto f(\circ), \overline{z} \mapsto ()\}$$
$$\sigma_{14} = \{\overline{x} \mapsto h_2(a), x \mapsto h_3(b), \overline{Y} \mapsto f(\circ), \overline{z} \mapsto ()\}$$
$$\sigma_{15} = \{\overline{x} \mapsto h_2(b), x \mapsto h_3(a), \overline{Y} \mapsto f(\circ), \overline{z} \mapsto ()\}$$
$$\sigma_{16} = \{\overline{x} \mapsto h_2(b), x \mapsto h_3(b), \overline{Y} \mapsto f(\circ), \overline{z} \mapsto ()\}$$

$\lambda = 0.6$:

$$\sigma_1 = \{\overline{x} \mapsto g_1(a), x \mapsto h_3(a), \overline{Y} \mapsto f(\circ), \overline{z} \mapsto ()\}$$
$$\sigma_2 = \{\overline{x} \mapsto g_1(a), x \mapsto h_3(b), \overline{Y} \mapsto f(\circ), \overline{z} \mapsto ()\}$$
$$\sigma_3 = \{\overline{x} \mapsto g_1(b), x \mapsto h_3(a), \overline{Y} \mapsto f(\circ), \overline{z} \mapsto ()\}$$
$$\sigma_4 = \{\overline{x} \mapsto g_1(b), x \mapsto h_3(b), \overline{Y} \mapsto f(\circ), \overline{z} \mapsto ()\}$$

$\lambda = 0.7$: No solutions.

A proximity matching problem is a set of proximity atoms. A substitution σ is a solution of a proximity matching problem $\{A_1, \ldots, A_n\}$ (with proximity degree α), if σ is a solution of each atom A_i (with proximity degree α_i and $\alpha = \alpha_1 \wedge \cdots \wedge \alpha_n$).

Note that because of strictness, syntactic matching can be seen as special proximity matching for an arbitrary $\mathscr{R}$ with the lambda-cut equal to 1. Therefore, for simplicity, below we will talk only about proximity matching problems and refer to them briefly as proximity problems.

3.2 *ρLog-prox Programs and Proximity Relations*

ρLog-prox programs consist of conditional rules for hedge transformations. A transformation is an atomic formula (an atom) of the form $\Longrightarrow (t, \langle\tilde{s}_1\rangle, \langle\tilde{s}_2\rangle)$, where $\Longrightarrow$ is a ternary predicate symbol and $\langle\cdot\rangle$ is a function symbol (which appears neither in t nor in $\tilde{s}_1$ and $\tilde{s}_2$). Such an atom is usually written as $t{::}\tilde{s}_1 \Longrightarrow \tilde{s}_2$. Intuitively, it means that the hedge $\tilde{s}_1$ is transformed into the hedge $\tilde{s}_2$ by the strategy t. Atoms are denoted by A and B.

A ρLog-prox query is a conjunction of atoms, written as $B_1, \ldots, B_n$. A ρLog-prox clause has a form $A \leftarrow Q$, where $\leftarrow$ is the inverse implication sign, A is an atom, called the head of the clause, and Q is a query, called the body of the clause. ρLog-prox programs are finite sets of ρLog-prox clauses.

We assume that for each program there is an associated proximity relation defined on the set of function symbols. For such a relation $\mathscr{R}$, the set of (f, g) pairs with $\mathscr{R}(f, g) > 0$ is finite.

A special predefined strategy is **prox**, which takes a single argument, a number from the real interval $(0, 1]$. The atom $\mathbf{prox}(\lambda)::\tilde{s}_1 \Longrightarrow \tilde{s}_2$ is true iff the proximity problem $\tilde{s}_2 \ll_{\mathscr{R},\lambda} \tilde{s}_1$ is solvable for the given $\mathscr{R}$. When $\lambda = 1$, **prox** coincides with the identity strategy **id** of the original ρLog [18] (the strictness assumption is important here).

For the original version of ρLog, semantics of programs can be defined in the same way as it is done for logic programming [1, 15]. Having defined proximity strategies as ρLog-prox atoms, we can do the same for our version of ρLog-prox programs.

Note that the same strategy can be defined by several clauses, which are treated as alternatives.

Now we introduce the inference system of ρLog-prox calculus with proximity relations. It has two rules: resolution and proximity factoring. A program and a proximity relation $\mathscr{R}$ are given.

Resolution takes a query with an atom selected in it and a renamed copy of a program clause and performs the inference step, producing a new query as follows:

$$\frac{str_{\mathrm{q}}::lhs_{\mathrm{q}} \Longrightarrow rhs_{\mathrm{q}}, Q \qquad str_{\mathrm{p}}::lhs_{\mathrm{p}} \Longrightarrow rhs_{\mathrm{p}} \leftarrow Body}{(Body,\ \mathbf{prox}(1)::rhs_{\mathrm{p}} \Longrightarrow rhs_{\mathrm{q}},\ Q)\sigma},$$

where σ is a solution of the proximity problem $\{str_{\mathrm{p}} \ll_{\mathscr{R},1} str_{\mathrm{q}},\ lhs_{\mathrm{p}} \ll_{\mathscr{R},1} lhs_{\mathrm{q}}\}$. The strategy str_{q} does not have the form $\mathbf{prox}(\lambda)$.

Proximity factoring takes a query, in which an atom with the proximity strategy is selected, and produces a new query:

$$\frac{\mathbf{prox}(\lambda)::lhs_{\mathrm{q}} \Longrightarrow rhs_{\mathrm{q}},\ Q}{Q\sigma},$$

where σ is a solution of the proximity problem $\{rhs_{\mathrm{q}} \ll_{\mathscr{R},\lambda} lhs_{\mathrm{q}}\}$.

A derivation of a query Q from a program P (with respect to a proximity relation $\mathscr{R}$) is a sequence of queries $Q_0, Q_1, \ldots$, where $Q_0 = Q$ and Q_i is obtained from Q_{i-1} by resolution or proximity factoring. A derivation is successful if it ends with the empty query. In this case, the union of substitutions computed along the derivation, restricted to variables from Q, is called the answer computed for Q via P. A derivation is failed, if none of the inference rules can apply to the last query, which is nonempty. Like for the original ρLog, the inference system is sound: the computed answers are also correct with respect to the declarative semantics. It is not complete in general due to the leftmost query selection strategy. Completeness is ensured for queries with terminating derivations.

We can allow negations of atoms in queries and clause bodies, as in normal logic programs [1, 2, 15]. Literal is a common name for an atom and its negation. We use the letter L to denote them. To deal with negative literals, the inference system can be extended by the well-known negation-as-failure rule.

In order to guarantee that inference in ρLog-prox is performed by matching and not unification (because the latter problems may have infinitely many solutions [5, 12]), we work with well-moded programs and queries.

Definition 1 (*Well-moded clauses, programs, queries*) Let C be a (normal) clause

$$str_0::\tilde{r}_0 \Longrightarrow \tilde{s}_{n+1} \leftarrow L_1, \ldots, L_n,$$

where for each $1 \leq i \leq n$, the literal L_i is either an atom $str_i::\tilde{s}_i \Longrightarrow \tilde{r}_i$ or a negation of an atom $str_i::\tilde{s}_i \not\Longrightarrow \tilde{r}_i$. C is well-moded if for all $1 \leq i \leq n+1$, we have

- $\mathcal{V}(str_i) \cup \mathcal{V}(\tilde{s}_i) \subseteq \mathcal{V}(str_0) \cup \bigcup_{j=0}^{i-1} \mathcal{V}(\tilde{r}) \setminus \mathcal{V}_{\mathsf{An}}$, and
- if L_i is a negative literal, then $\mathcal{V}(\tilde{r}_i) \subseteq \mathcal{V}(str_0) \cup \bigcup_{j=0}^{i-1} \mathcal{V}(\tilde{r}) \cup \mathcal{V}_{\mathsf{An}}$.

A (normal) ρLog-prox program is well-moded if all clauses in it all well-moded.

A (normal) query $L_1, \ldots, L_n$ is well-moded if the clause $A \leftarrow L_1, \ldots, L_n$ is well-moded, where A is a dummy ground atom.

Example 3 In this rather extended example we illustrate ρLog-prox clauses, strategies, and evaluation mechanism. We borrow the material from [7] and adapt it to ρLog-prox.

An instance of a transformation is finding duplicated elements in a hedge and removing one of them. Let us call this process the merging of duplicates. The following strategy implements the idea:

$$\text{merge\_duplicates}::(\overline{x},\ x,\ \overline{y},\ x,\ \overline{z}) \Longrightarrow (\overline{x},\ x,\ \overline{y},\ \overline{z}).$$

merge_duplicates is the strategy name. The clause is obviously well-moded. It says that if the hedge in *lhs* contains duplicates (expressed by two copies of the variable x) somewhere, then from these two copies only the first one should be kept in *rhs*. That "somewhere" is expressed by three hedge variables, where $\overline{x}$ stands for the subhedge before the first occurrence of x, $\overline{y}$ takes the subhedge between two occurrences of x, and $\overline{z}$ matches the remaining part. These subhedges remain unchanged in the *rhs*.

One does not need to code the actual search process of duplicates explicitly. The matching algorithm is supposed to do the job instead, looking for an appropriate instantiation of the variables. There can be several such instantiations.

Now one can ask, e.g., to merge duplicates in a hedge $(a,\ b,\ c,\ b,\ a)$:

$$\text{merge\_duplicates}::(a,\ b,\ c,\ b,\ a) \Longrightarrow \overline{x}.$$

To this query, ρLog-prox returns two answer substitutions: $\{\overline{x} \mapsto (a,\ b,\ c,\ b)\}$ and $\{\overline{x} \mapsto (a,\ b,\ c,\ a)\}$. Both are obtained from $(a,\ b,\ c,\ b,\ a)$ by merging one pair of duplicates.

Now we generalize merge_duplicates allowing merging of approximate duplicates (we use l as a term variable):

$$\text{merge\_duplicates}(l)::(\overline{x},\ x,\ \overline{y},\ y,\ \overline{z}) \Longrightarrow (\overline{x},\ x,\ \overline{y},\ \overline{z}) \leftarrow \mathbf{prox}(l)::x \Longrightarrow y.$$

This clause (which is well-moded) removes y from the given hedge, if the hedge contains an x such that x and y are close to each other with respect to the given proximity relation with the proximity degree l. The merge_duplicates strategy above is just a special case of merge_duplicates(l) with $l = 1$.

Assume now that in the proximity relation $\mathcal{R}$, we have $\mathcal{R}(a, e) = 0.6$ and $\mathcal{R}(b, d) = 0.7$. Then the query

$$\text{merge\_duplicates}(0.8)::(a,\ b,\ c,\ d,\ e) \Longrightarrow \overline{x}$$

fails, because $(a,\ b,\ c,\ d,\ e)$ does not contain elements which are close to each other with the proximity degree at least 0.8. If we take $l = 0.7$, i.e., the query

$$\text{merge\_duplicates}(0.7)::(a,\ b,\ c,\ d,\ e) \Longrightarrow \overline{x},$$

we get a single answer: $\{\overline{x} \mapsto (a,\ b,\ c,\ e)\}$. Decreasing l further and taking the query

$$\text{merge\_duplicates}(0.6)::(a,\ b,\ c,\ d,\ e) \Longrightarrow \overline{x},$$

we get two answers (via backtracking): $\{\overline{x} \mapsto (a,\ b,\ c,\ d)\}$ and $\{\overline{x} \mapsto (a,\ b,\ c,\ e)\}$.

A hedge without duplicates is a normal form with respect to this single-step merge_duplicates(l) transformation. ρLog-prox has a predefined strategy for computing normal forms, denoted by **nf**, and we can use it to define a new strategy merge_all_duplicates(l) in the following clause:

$$\text{merge\_all\_duplicates}(l)::\overline{x} \Longrightarrow \overline{y}\ \leftarrow\ \mathbf{nf}(\text{merge\_duplicates}(l))::\overline{x} \Longrightarrow \overline{y}.$$

The effect of **nf** is that it applies merge_duplicates to $\overline{x}$, repeating this process iteratively as long as it is possible, i.e., as long as duplicates can be merged in the obtained hedges. When merge_duplicates is no more applicable, it means that the normal form of the transformation is reached. It is returned in $\overline{y}$.

Now, for the query

$$\text{merge\_all\_duplicates}(0.6)::(a,\ b,\ c,\ d,\ e) \Longrightarrow \overline{x}.$$

we get a single answer $\overline{x} \mapsto (a,\ b,\ c)$. However, procedurally, this answer can be computed multiple times (via backtracking). To avoid such multiple computations, we can use another predefined strategy **first_one**:

$$\text{merge\_all\_duplicates}(l)::\overline{x} \Longrightarrow \overline{y} \leftarrow$$
$$\textbf{first\_one}(\textbf{nf}(\text{merge\_duplicates}(l)))::\overline{x} \Longrightarrow \overline{y}.$$

first_one applies to a sequence of strategies, finds the first one among them, which successfully transforms the input hedge, and gives back just one result of the transformation. Here it has a single argument strategy **nf**(merge_duplicates(l)) and returns (by instantiating $\overline{y}$) only one result of its application to $\overline{x}$.

ρLog-prox is good not only in selecting arbitrarily many subexpressions in "horizontal direction" (by hedge variables), but also in working in "vertical direction", selecting subterms at arbitrary depth. Context variables provide this flexibility, by matching the context above the subterm to be selected. With the help of context and function variables, from the merge_duplicates(l) strategy it is pretty easy to define a transformation that merges neighboring branches in a tree, which are approximately the same:

$$\text{merge\_duplicate\_branches}(l)::\overline{X}(Y(\overline{x})) \Longrightarrow \overline{X}(Y(\overline{y})) \leftarrow$$
$$\text{merge\_duplicates}(l)::\overline{x} \Longrightarrow \overline{y}.$$

Now, we can ask to merge neighboring branches in a given tree, which are 0.6-approximate of each other (for the same $\mathcal{R}$ as above):

$$\text{merge\_duplicate\_branches}(0.6)::$$
$$f(g(a, b, e, h(c, c)),\ h(c),\ g(a, e, b, h(c))) \Longrightarrow x.$$

ρLog-prox computes three answers:

$$\{x \mapsto f(g(a, b, h(c, c)),\ h(c),\ g(a, e, b, h(c)))\},$$
$$\{x \mapsto f(g(a, b, e, h(c)),\ h(c),\ g(a, e, b, h(c)))\},$$
$$\{x \mapsto f(g(a, b, e, h(c, c)),\ h(c),\ g(a, b, h(c)))\}.$$

To obtain the first one, ρLog-prox matched the context variable $\overline{X}$ to the context $f(\circ,\ h(c), g(a, a, b, h(c)))$, the function variable Y to the function symbol g, and the hedge variable $\overline{x}$ to the hedge $(a, b, e, h(c, c))$. merge_duplicates(0.6) transformed $(a, b, e, h(c, c))$ to $(a, b, h(c, c))$. The other results have been obtained by taking different contexts and respective subbranches.

The right hand side of transformations in the queries need not be variables. One can have an arbitrary hedge there. For instance, we may be interested in trees that contain $h(c, c)$:

$$\text{merge\_duplicate\_branches}(0.6)::$$
$$f(g(a, b, e, h(c, c)),\ h(c),\ g(a, e, b, h(c))) \Longrightarrow \overline{X}(h(c, c)).$$

We get here two answers, which show instantiations of $\overline{X}$ by the relevant contexts:

$$\{\overline{X} \mapsto f(g(a, b, \circ),\ h(c),\ g(a, e, b, h(c)))\},$$
$$\{\overline{X} \mapsto f(g(a, b, e, \circ),\ h(c),\ g(a, b, h(c)))\}.$$

Similar to merging all duplicates in a hedge above, we can also define a strategy that merges all approximately duplicate branches in a tree repeatedly. Naturally, the built-in strategy for normal forms plays a role also here:

$$\text{merge\_all\_duplicate\_branches}(l){::}x \Longrightarrow y \leftarrow$$
$$\textbf{first\_one}(\textbf{nf}(\text{merge\_duplicate\_branches}(l))){::}x \Longrightarrow y.$$

For the query

$$\text{merge\_all\_duplicate\_branches}(0.6){::}$$
$$f(g(a, b, e, h(c, c)),\ h(c),\ g(a, e, b, h(c))) \Longrightarrow \overline{x}.$$

we get a single answer $\{\overline{x} \mapsto f(g(a, b, h(c)), h(c))\}$.

4 Solving Proximity Problems

As one could see in the previous section, the inference rules of ρLog-prox heavily rely on solving proximity problems. Well-modedness guarantees that only proximity problems with ground right hand side arise during derivations of queries from ρLog-prox programs. Resolving negative literals reduces to the problem of testing whether two ground expressions are in the given proximity relation with respect to the given cut value.

Here we describe an algorithm, which computes not only solutions to proximity problems, but also the degree of proximity for the solutions. They can be used to report the proximity degree of a query instance that is proved from the program.

We say that a set of equations $\{V_1 \approx E_1, \ldots, V_n \approx E_n\}$ is in

- matching pre-solved form, if the E's are ground,
- matching solved form, if it is in matching pre-solved form and each variable V_i appears in the set only once.

If S is a solved form, we define an associated substitution $\sigma_S := \{V_i \mapsto E_i \mid V_i \approx E_i \in S\}$.

The proximity matching algorithm $\mathfrak{P}$ is formulated in a rule-based way. Rules work on configurations, which are either a special symbol $\bot$ or triples of the form $M; S; \alpha$, where M is the proximity matching problem to be solved, S is a set of equations in matching pre-solved form (the candidate set for a solution computed so far), and α is the proximity degree of a solution computed so far. A rule that produces $\bot$ is called a failure rule. We have six success and four failure rules:

RFS: **Removing function symbols**
$\{f(\tilde{s}) \ll_{\mathscr{R},\lambda} g(\tilde{t})\} \uplus M;\ S;\ \alpha \rightsquigarrow M \cup \{\tilde{s} \ll_{\mathscr{R},\lambda} \tilde{t}\};\ S;\ \alpha \wedge \beta$,
where $\mathscr{R}(f, g) = \beta \geq \lambda$.

Dec: **Decomposition**
$\{(t, \tilde{s}) \ll_{\mathscr{R},\lambda} (t', \tilde{t})\} \uplus M;\ S;\ \alpha \rightsquigarrow M \cup \{t \ll_{\mathscr{R},\lambda} t',\ \tilde{s} \ll_{\mathscr{R},\lambda} \tilde{t}\};\ S;\ \alpha$,
where $\tilde{s} \neq ()$ and $\tilde{t} \neq ()$.

FVE: **Function variable elimination**
$\{X(\tilde{s}) \ll_{\mathscr{R},\lambda} g(\tilde{t})\} \uplus M;\ S;\ \alpha \rightsquigarrow M \cup \{\tilde{s} \ll_{\mathscr{R},\lambda} \tilde{t}\};\ S \cup \{X \approx g'\};\ \alpha \wedge \beta$,
where $\mathscr{R}(g', g) = \beta \geq \lambda$.

CVE: **Context variable elimination**
$\{\overline{X}(t_1) \ll_{\mathscr{R},\lambda} C(t_2)\} \uplus M;\ S;\ \alpha \rightsquigarrow M \cup \{t_1 \ll_{\mathscr{R},\lambda} t_2\};\ S \cup \{\overline{X} \approx C'\};\ \alpha \wedge \beta$,
where $\mathscr{R}(C', C) = \beta \geq \lambda$.

TVE: **Term variable elimination**
$\{x \ll_{\mathscr{R},\lambda} t\} \uplus M;\ S;\ \alpha \rightsquigarrow M;\ S \cup \{x \approx t'\};\ \alpha \wedge \beta$,
where $\mathscr{R}(t', t) = \beta \geq \lambda$.

HVE: **Hedge variable elimination**
$\{(\overline{x}, \tilde{s}) \ll_{\mathscr{R},\lambda} (\tilde{t}_1, \tilde{t}_2)\} \uplus M;\ S;\ \alpha \rightsquigarrow M \cup \{\tilde{s} \ll_{\mathscr{R},\lambda} \tilde{t}_2\};\ S \cup \{\overline{x} \approx \tilde{t}_1'\};\ \alpha \wedge \beta$,
where $\mathscr{R}(\tilde{t}_1', \tilde{t}_1) = \beta \geq \lambda$.

Cla1: **Clash 1**
$\{f(\tilde{s}) \ll_{\mathscr{R},\lambda} g(\tilde{t})\} \uplus M;\ S;\ \alpha \rightsquigarrow \bot$, if $\mathscr{R}(f, g) < \lambda$.

Cla2: **Clash 2**
$\{(t, \tilde{s}) \ll_{\mathscr{R},\lambda} ()\} \uplus M;\ S;\ \alpha \rightsquigarrow \bot$.

Cla3: **Clash 3**
$\{() \ll_{\mathscr{R},\lambda} (t, \tilde{t})\} \uplus M;\ S;\ \alpha \rightsquigarrow \bot$.

Inc: **Inconsistency**
$M;\ S;\ \alpha \rightsquigarrow \bot$, if S contains two equations with the same variable in the left hand side.

To solve a proximity matching problem M, we create the initial configuration $M; \emptyset; 1$ and start applying the rules exhaustively. If the same configuration can be transformed by multiple rules, they are applied concurrently except one of the rules is Inc: in this case only Inc applies. Each elimination rule instantiates a variable not exactly with the corresponding expression in the right hand side, but with its approximate expression. Since proximity classes of objects are finite, these choices cause only finite branching. The other source of branching is the choice of a hedge and a context from the right hand side in CVE and HVE rules. Also here, there are finitely many ways to branch. The described process defines the algorithm $\mathfrak{P}$.

Theorem 1 (Termination) *The proximity matching algorithm $\mathfrak{P}$ terminates. Each final configuration has the form either $\bot$ or $\emptyset; S; \alpha$, where S is in matching solved form.*

Proof Let $size(E)$ be the number of symbols in E. By $Msize(M)$ we denote the multiset $\{size(E_2) \mid E_1 \ll_{\mathscr{R},\lambda} E_2 \in M\}$. To each configuration $M; S; \alpha$ we associate the complexity measure, the pair $\langle Msize(M), varocc(M)\rangle$, where $varocc(M)$ is the number of variable occurrences in M. The measures are compared lexicographically, where the used orderings for the components are multiset ordering [4] and the standard ordering on natural numbers. The RFC and Dec rules decrease the first component of the measure. (Note that for Dec it is ensured by the requirement that $\tilde{s}$ and $\tilde{t}$ are not empty hedges.) The elimination rules do not increase the first component and decrease the second one. The failure rules stop immediately, since $\perp$ is not transformed further. Hence, the algorithm terminates.

Since for each possible shape of a proximity problem there is the corresponding rule, the process stops either with $\perp$ or with a configuration of the form $\emptyset; S; \alpha$. In the latter case, S should be in solved form, otherwise Inc would transform it into $\perp$. □

From each final configuration $\emptyset; S; \alpha$, we can extract the corresponding substitution σ_S. These substitutions are called computed answers.

We say that σ is a solution of a (pre-solved) set of equations $\{V_1 \approx E_1, \ldots, V_n \approx E_n\}$ iff $V_i\sigma = E_i$ for each $1 \leq i \leq n$. A solution of a pair $M; S$ of a proximity matching problem M and a set of equations in pre-solved form S is a substitution σ that solves both M and S. The configuration $\perp$ has not solutions.

Theorem 2 (Soundness) *Let M be a proximity problem and σ be its computed answer with the proximity degree α. Then σ is a solution of M with the proximity degree α.*

Proof Let $M_1; S_1; \alpha_1 \rightsquigarrow_{\mathsf{R}} M_2; S_2; \alpha_2$ be the step made by R, where R is one of the rules above. We show that if σ is a solution of M_2 (with the degree α_2) and S_2, then σ is a solution of M_1 (with the same degree α_2) and S_1.

R is RFS. Then $\alpha_2 = \alpha_1 \wedge \beta$, where $\mathscr{R}(f, g) = \beta \geq \lambda$. Obviously, if $\mathscr{R}(\tilde{s}\sigma, \tilde{t}) \geq \alpha_1 \wedge \beta$, then $\mathscr{R}(f(\tilde{s})\sigma, g(\tilde{t})) \geq \alpha_1 \wedge \beta$. Hence, in this case σ is a solution of M_1 with the degree α_2 and S_1 (which is the same as S_2).

R is FVE. Then $\alpha_2 = \alpha_1 \wedge \beta$ where $\mathscr{R}(g', g) = \beta$. Besides, $g' = X\sigma$. Therefore, if $\mathscr{R}(\tilde{s}\sigma, \tilde{t}) \geq \alpha_1 \wedge \beta$, then $\mathscr{R}(X(\tilde{s})\sigma, g(\tilde{t})) \geq \alpha_1 \wedge \beta$, and if σ solves S_2, then it solves also S_1. Hence, also in this case σ is a solution of M_1 with the degree α_2 and S_1.

For the other success rules the proof is similar or easier.

To prove the soundness theorem, we just need to proceed by induction on the length of a successful derivation, using the single-step soundness result we just established. □

Lemma 1 *If $M; S; \alpha \rightsquigarrow \perp$, then $M; S$ has no solution.*

Proof Assume M is a $(\mathscr{R}, \lambda)$-matching problem and analyze the rules that lead to $\perp$. For the Cla1 rule, M is unsolvable, because $\mathscr{R}((f(\tilde{s}))\sigma, g(\tilde{t})) = \mathscr{R}(f(\tilde{s}\sigma), g(\tilde{t})) = \mathscr{R}(f, g) \wedge \mathscr{R}(\tilde{s}\sigma, \tilde{t}) \leq \mathscr{R}(f, g) < \lambda$. In Cla2 and Cla3 rules, unsolvability of M follows from the fact that a nonempty hedge can not be approximated by the empty hedge. In the Inc rule, if we have two equations with the same variable in the left hand side, it means that their right hand sides are different. Since equations in S are solved syntactically, it implies that S has no solution. □

Theorem 3 *rm (Completeness) Let M be a proximity problem and σ be its solution with the proximity degree α. Then there exists a derivation in $\mathfrak{P}$ ending with a configuration $M; \emptyset; 1 \rightsquigarrow^* \emptyset; S; \alpha$, such that $\sigma = \sigma_S$.*

Proof We construct the desired derivation under the guidance of σ. At each variable elimination step, we choose the proximal object of the variable exactly as σ does. This will guarantee that proximity degrees at each such step will be also in accordance to σ. Making RFS and Dec steps will not make the proximity degree differ from α, because σ is a solution. No clashing and inconsistency step will be performed, because by Lemma 1 it would contradict the solvability of M. Hence, if $\beta_1, \ldots, \beta_n$ are all β's in the derivation, then $\beta_1 \wedge \cdots \wedge \beta_n = \alpha$. Since we start from the proximity degree 1, the computed proximity degree will be $1 \wedge \beta_1 \wedge \cdots \wedge \beta_n = \alpha$. By construction, $\sigma_S = \sigma$. □

Example 4 We use the proximity relation and problem from Example 2. The relation $\mathcal{R}$ is

$$\mathcal{R}(g_1, h_1) = \mathcal{R}(g_2, h_1) = 0.4, \qquad \mathcal{R}(g_1, h_2) = \mathcal{R}(g_2, h_2) = 0.5,$$
$$\mathcal{R}(g_2, h_3) = \mathcal{R}(g_3, h_3) = 0.6, \qquad \mathcal{R}(a, b) = 0.7.$$

The proximity problem is

$$f(\overline{x}, x, \overline{Y}(x), \overline{z}) \ll_{\mathcal{R},\lambda} f(g_1(a), g_2(b), f(g_3(a))).$$

We take the cut $\lambda = 0.6$ and show how $\mathfrak{P}$ computes one of the solutions of this problem, the substitution $\sigma_3 = \{\overline{x} \mapsto g_1(b), x \mapsto h_3(b), \overline{Y} \mapsto f(\circ), \overline{z} \mapsto ()\}$:

$$\{f(\overline{x}, x, \overline{Y}(x), \overline{z}) \ll_{\mathcal{R},0.6} f(g_1(a), g_2(b), f(g_3(a)))\}; \emptyset; 1 \rightsquigarrow_{\mathsf{RFS}}$$
$$\{(\overline{x}, x, \overline{Y}(x), \overline{z}) \ll_{\mathcal{R},0.6} (g_1(a), g_2(b), f(g_3(a)))\}; \emptyset; 1 \rightsquigarrow_{\mathsf{HVE}}$$
$$\{(x, \overline{Y}(x), \overline{z}) \ll_{\mathcal{R},0.6} (g_2(b), f(g_3(a)))\}; \{\overline{x} \approx g_1(b)\}; 0.7 \rightsquigarrow_{\mathsf{TVE}}$$
$$\{(\overline{Y}(x), \overline{z}) \ll_{\mathcal{R},0.6} (f(g_3(a)))\}; \{\overline{x} \approx g_1(b),\ x \approx h_3(b)\}; 0.6 \rightsquigarrow_{\mathsf{CVE}}$$
$$\{(x, \overline{z}) \ll_{\mathcal{R},0.6} (g_3(a))\}; \{\overline{x} \approx g_1(b),\ x \approx h_3(b),\ \overline{Y} \approx f(\circ)\}; 0.6 \rightsquigarrow_{\mathsf{TVE}}$$
$$\{\overline{z} \ll_{\mathcal{R},0.6} ()\}; \{\overline{x} \approx g_1(b),\ x \approx h_3(b),\ \overline{Y} \approx f(\circ)\}; 0.6 \rightsquigarrow_{\mathsf{HVE}}$$
$$\emptyset; \{\overline{x} \approx g_1(b),\ x \approx h_3(b),\ \overline{Y} \approx f(\circ),\ \overline{z} \approx ()\}; 0.6.$$

5 Conclusion

We extended the ρLog calculus with the capabilities to work with strict proximity relations. This extension, called ρLog-prox, can process both crisp and fuzzy data. With the help of the corresponding strategies, the user has full control on how fuzzy (proximity) relations are used. There are no hidden assumptions about fuzziness.

We showed that matching modulo proximity can be naturally embedded in the strategy-based transformation rule framework of ρLog-prox. We developed a proximity matching algorithm for expressions involving four different kinds of variables (for terms, for hedges, for function symbols, and for contexts), and proved its termination, soundness, and completeness.

Acknowledgements This research has been partially supported by the Austrian Science Fund (FWF) under the project 28789-N32 and by the Shota Rustaveli National Science Foundation of Georgia (SRNSFG) under the grant YS-18-1480.

References

1. Apt, K.R.: Logic programming. In: van Leeuwen [25], pp. 493–574
2. Apt, K.R., Bol, R.N.: Logic programming and negation: a survey. J. Log. Program. **19**(20), 9–71 (1994)
3. Belkhir, W., Giorgetti, A., Lenczner, M.: A symbolic transformation language and its application to a multiscale method. J. Symb. Comput. **65**, 49–78 (2014)
4. Dershowitz, N., Manna, Z.: Proving termination with multiset orderings. Commun. ACM **22**(8), 465–476 (1979)
5. Dundua, B., Florido, M., Kutsia, T., Marin, M.: CLP(H): constraint logic programming for hedges. TPLP **16**(2), 141–162 (2016)
6. Dundua, B., Kutsia, T., Marin, M.: Strategies in PρLog. In: Fernández, M., (ed.) Proceedings Ninth International Workshop on Reduction Strategies in Rewriting and Programming, WRS 2009, Brasilia, Brazil, 28th June 2009, 15 of EPTCS, pp. 32–43 (2009)
7. Dundua, B., Kutsia, T., Reisenberger-Hagmayer, K.: An overview of PρLog. In: Lierler, Y., Taha, W., (eds.), Practical Aspects of Declarative Languages - 19th International Symposium, PADL 2017, Paris, France, January 16–17, 2017, Proceedings, 10137 of Lecture Notes in Computer Science, pp. 34–49, Springer (2017)
8. Fontana, F.A., Formato, F.: A similarity-based resolution rule. Int. J. Intell. Syst. **17**(9), 853–872 (2002)
9. Guadarrama, S., Muñoz-Hernández, S., Vaucheret, C.: Fuzzy Prolog: a new approach using soft constraints propagation. Fuzzy Sets Syst. **144**(1), 127–150 (2004)
10. Julián-Iranzo, P., Rubio-Manzano, C.: An efficient fuzzy unification method and its implementation into the Bousi-Prolog system. In: FUZZ-IEEE 2010, IEEE International Conference on Fuzzy Systems, Barcelona, Spain, 18–23 July, 2010, Proceedings, pp. 1–8. IEEE (2010)
11. Julián-Iranzo, P., Rubio-Manzano, C.: Proximity-based unification theory. Fuzzy Sets and Syst. **262**, 21–43 (2015)
12. Kutsia, T.: Solving equations with sequence variables and sequence functions. J. Symb. Comput. **42**(3), 352–388 (2007)
13. Kutsia, T., Marin, M.: Matching with regular constraints. In: Sutcliffe, G., Voronkov, A., (eds.), Logic for Programming, Artificial Intelligence, and Reasoning, 12th International Conference, LPAR 2005, Montego Bay, Jamaica, December 2–6, 2005, Proceedings. Lecture Notes in Computer Science, vol. 3835, pp. 215–229. Springer (2005)
14. Lee, R.C.T.: Fuzzy logic and the resolution principle. J. ACM **19**(1), 109–119 (1972)
15. Lloyd, J.W.: Foundations of Logic Programming, 2nd edn. Springer, Berlin (1987)
16. Marin, M.: A System for Rule-Based Programming in Mathematica (2019). http://staff.fmi.uvt.ro/mircea.marin/rholog/
17. Marin, M., Kutsia, T.: On the implementation of a rule-based programming system and some of its applications. In: Konev, B., Schmidt, R., (eds.), Proceedings of the 4th International Workshop on the Implementation of Logics (WIL'03), pp. 55–68. Almaty, Kazakhstan (2003)
18. Marin, M., Kutsia, T.: Foundations of the rule-based system ρLog. J. Appl. Non-Classical Logics **16**(1–2), 151–168 (2006)
19. Marin, M., Piroi, F.: Rule-based programming with mathematica. In: Proceedings of the 6th International Mathematica Symposium, Alberta, Canada (2004)

20. Medina, J., Ojeda-Aciego, M., Vojtás, P.: Similarity-based unification: a multi-adjoint approach. In: Garibaldi, J.M., John, R.I., (eds.), Proceedings of the 2nd International Conference in Fuzzy Logic and Technology, Leicester, United Kingdom, September 5–7, 2001, pp. 273–276. De Montfort University, Leicester (2001)
21. Medina, J., Ojeda-Aciego, M., Vojtás, P.: Similarity-based unification: a multi-adjoint approach. Fuzzy Sets and Syst. **146**(1), 43–62 (2004)
22. Nguyen, P.: Meta-mining: a meta-learning framework to support the recommendation, planning and optimization of data mining workflows. Ph.D. thesis, Department of Computer Science, University of Geneva (2015)
23. Raedt, L.D., Kimmig, A.: Probabilistic (logic) programming concepts. Mach. Learn. **100**(1), 5–47 (2015)
24. Sessa, M.I.: Approximate reasoning by similarity-based SLD resolution. Theor. Comput. Sci. **275**(1–2), 389–426 (2002)
25. van Leeuwen, J. (ed.): Handbook of Theoretical Computer Science, Volume B: Formal Models and Semantics. Elsevier and MIT Press, Amsterdam (1990)
26. Wolfram, S.: The Mathematica Book, 5th Edn. Wolfram-Media (2003)
27. Yang, B., Belkhir, W., Dhara, R.N., Lenczner, M., Giorgetti, A.: Computer-aided multiscale model derivation for MEMS arrays. In Proceedings of the 12th International Conference on Thermal, Mechanical and Multi-Physics Simulation and Experiments in Microelectronics and Microsystems. IEEE Computer Society (2011)

Specification and Analysis of ABAC Policies in a Rule-Based Framework

Besik Dundua, Temur Kutsia, Mircea Marin, and Mikheil Rukhaia

Abstract Attribute-based access control (ABAC) is an access control paradigm whereby access rights to system resources are granted through the use of policies that are evaluated against the attributes of entities (user, subject, and object), operations, and the environment relevant to a request. Many ABAC models, with different variations, have been proposed and formalized. Since the access control policies that can be implemented in ABAC have inherent rule-based specifications, it is natural to adopt a rule-based framework to specify and analyse their properties. We describe the design and implementation of a software tool implemented in Mathematica. Our tool makes use of the rule-based capabilities of a rule-based package developed by us, can be used to specify configurations for the foundational model ABAC_α of ABAC, and to check safety properties.

Keywords Rules-based programming · Access control policies · Safety analysis

B. Dundua · M. Rukhaia
VIAM, Ivane Javakhishvili Tbilisi State University, Tbilisi, Georgia
e-mail: bdundua@gmail.com

M. Rukhaia
e-mail: mrukhaia@yahoo.com

B. Dundua
FBT, International Black Sea University, Tbilisi, Georgia

T. Kutsia
RISC, Johannes Kepler University Linz, Linz, Austria
e-mail: kutsia@risc.jku.at

M. Marin (✉)
West University of Timişoara, Timişoara, Romania
e-mail: mircea.marin@e-uvt.ro

G. Jaiani and D. Natroshvili (eds.), *Applications of Mathematics and Informatics in Natural Sciences and Engineering*, Springer Proceedings in Mathematics & Statistics 334, https://doi.org/10.1007/978-3-030-56356-1_7

1 Introduction

Access (authorization) control is a fundamental security technique concerned with determining the allowed activities of legitimate users, and mediating every attempt by a user to access a resource in a computing environment. Over the years, many access control models have been developed to address various aspects of computer security, including: Mandatory Access Control (MAC) [12], Discretionary Access Control (DAC) [13], and Role-based Access Control (RBAC) [4]. Attribute-Based Access Control (ABAC) has received significant attention recently, although the concept has existed for more than twenty years. According to NIST [5]

> ABAC is an access control method where subject requests to perform operations on objects are granted or denied based on assigned attributes of the subject, assigned attributes of the object, environment conditions, and a set of policies that are specified in terms of those attributes and conditions.

ABAC is considered a next generation authorization paradigm which eliminates many limitations of the previous access control paradigms. It is dynamic: access control permissions are determined when the access control request is made; it is fine-grained: attributes can be added, to form detailed rules for access control policies; it has support for contextual/environmental conditions; and last but not least: it is flexible, and scalable. In fact, the access control policies that can be implemented in ABAC are limited only by the computational language and the richness of the available attributes. In particular, ABAC policies can be easily configured to simulate DAC, MAC and RBAC.

Until recently, there were no widely accepted formal models for ABAC. The foundational operational models $ABAC_\alpha$ and $ABAC_\beta$, and the administrative model GURA were proposed recently [6] as models with "just sufficient" features that can be used to easily and naturally configure the traditional access control models and some advanced features and extensions of RBAC.

The (efficient) implementation and analysis of these formal operational models of ABAC is of great importance. We argue that a rule-based framework is adequate to achieve these goals. For this purpose, we designed and implemented a software tool that allows to specify configurations of $ABAC_\alpha$ policies, and to analyse them. The tool is implemented in Mathematica [15] and is based on the capabilities of ρLog [8, 9], a rule-based system implemented by us on top of the rule-based capabilities of Mathematica. We highlight the main features that make our rule-based system adequate to specify and analyze the configurations of the access control policies of $ABAC_\alpha$.

The rest of this chapter is structured as follows. Section 2 contains a brief description of ρLog and the foundational model $ABAC_\alpha$. In Sect. 3 we describe the rule-based tool designed by us for the specification and analysis of $ABAC_\alpha$. In Sect. 4 we draw some conclusions and directions for future work.

2 Preliminaries

2.1 The ρLog System

ρLog is a system for rule-based programming with strategies and built-in support for constraint logic programming (CLP). This is a programming style similar to Constraint Logic Programming, where programs consist of rules which are used to answer queries using a calculus based on a variation of SLDNF-resolution [2] combined with constraint solving. There are, however, some significant differences.

The specification language has an alphabet $\mathscr{A}$ consisting of the following pairwise disjoint sets of symbols:

- $\mathscr{V}_{\mathsf{T}}$: term variables, denoted by $x, y, z, \ldots$,
- $\mathscr{V}_{\mathsf{S}}$: hedge variables, denoted by $\overline{x}, \overline{y}, \overline{z}, \ldots$,
- $\mathscr{V}_{\mathsf{F}}$: function variables, denoted by $X, Y, Z, \ldots$,
- $\mathscr{V}_{\mathsf{C}}$: context variables, denoted by $\overline{X}, \overline{Y}, \overline{Y}, \ldots$,
- $\mathscr{F}$: unranked function symbols, denoted by $f, g, h, \ldots$.

and distinguishes the following syntactic categories:

$$\begin{array}{ll}
t ::= x \mid f(\tilde{s}) \mid X(\tilde{s}) \mid \overline{X}(t) & \text{Term} \\
\tilde{t} ::= t_1, \ldots, t_n \quad (n \geq 0) & \text{Sequence of terms} \\
s ::= t \mid \overline{x} & \text{Hedge element} \\
\tilde{s} ::= s_1, \ldots, s_n \quad (n \geq 0) & \text{Hedge} \\
C ::= \circ \mid f(\tilde{s}_1, C, \tilde{s}_2) \mid X(\tilde{s}_1, C, \tilde{s}_2) \mid \overline{X}(C) & \text{Context}
\end{array}$$

Hence, hedges are sequences of hedge elements, hedge variables are not terms, term sequences do not contain hedge variables, contexts (which are not terms either) contain a single occurrence of the hole. We do not distinguish between a singleton hedge and its sole element.

We denote the set of terms by $\mathscr{T}(\mathscr{F}, \mathscr{V})$, hedges by $\mathscr{H}(\mathscr{F}, \mathscr{V})$, and contexts by $\mathscr{C}(\mathscr{F}, \mathscr{V})$. Ground (i.e., variable-free) subsets of these sets are denoted by $\mathscr{T}(\mathscr{F})$, $\mathscr{H}(\mathscr{F})$, and $\mathscr{C}(\mathscr{F})$, respectively.

We make a couple of conventions to improve readability. We put parentheses around hedges, writing, e.g., $(f(a), \overline{x}, b)$ instead of $f(a), \overline{x}, b$. The empty hedge is written as $()$. The terms $a()$ and $X()$ are abbreviated as a and X, respectively, when it is guaranteed that terms and symbols are not confused. For hedges $\tilde{s} = (s_1, \ldots, s_n)$ and $\tilde{s}' = (s'_1, \ldots, s'_m)$, the notation $(\tilde{s}, \tilde{s}')$ stands for the hedge $(s_1, \ldots, s_n, s'_1, \ldots, s'_m)$. We use $\tilde{s}$ and $\tilde{r}$ for arbitrary hedges, and $\tilde{t}$ for sequences of terms.

We will also need anonymous variables for each variable category. They are variables without name, well-known in declarative programming. We write just _ for an anonymous term or function variable, and __ for an anonymous hedge or context variable. The set of anonymous variables is denoted by $\mathscr{V}_{\mathsf{An}}$.

A syntactic expression (or, just an expression) is an element of the set $\mathcal{F} \cup \mathcal{V} \cup \mathcal{T}(\mathcal{F}, \mathcal{V}) \cup \mathcal{H}(\mathcal{F}, \mathcal{V}) \cup \mathcal{C}(\mathcal{F}, \mathcal{V})$. We denote expressions by E. Atoms are reducibility formulas $t{::}t_1 \Longrightarrow t_2$ with the intended reading "t_1 reduces to t_2 with strategy t." The negation of this atom is written as $t{::}t_1 \not\Longrightarrow t_2$.

The rules of ρLog are of the form

$$f(\tilde{s}){::}t' \Longrightarrow t'' \leftarrow cond_1, \ldots, cond_n. \tag{1}$$

with the intended reading "$f(\tilde{s}){::}t' \Longrightarrow t''$ holds whenever $cond_1$ and ... and $cond_n$ hold", and provide declarative semantics for reducibility formulas. f is the identifier of the strategy and $\tilde{s}$ is its argument: If $\tilde{s}$ is (), the strategy is atomic, otherwise it is parametric. We view (1) as a partial definition of f.

Some strategies with frequent applications are predefined:

- `id`$::s \Longrightarrow t$ holds if $s = t$.
- `elem`$::l \Longrightarrow e$ holds if e is an element of list l.
- `subset`$::l \Longrightarrow s$ holds if s is subset of set l.
- `fmap`$(t){::}f(s_1, \ldots, s_n) \Longrightarrow f(t_1, \ldots, t_n)$ holds if $t{::}s_i \Longrightarrow t_i$ for $1 \leq i \leq n$.

Another way to specify strategies is by using the predefined combinators:

- $t_1 \circ t_2{::}t' \Longrightarrow t''$ holds if $t_1{::}t' \Longrightarrow u$ and $t_2{::}u \Longrightarrow t''$ hold for some u.
- $t_1|t_2{::}t' \Longrightarrow t''$ holds if either $t_1{::}t' \Longrightarrow t''$ or $t_2{::}t' \Longrightarrow t''$ holds.
- $t^*{::}t' \Longrightarrow t''$ holds if either $t' = t''$ or there exist $u_1, \ldots, u_n$ such that $u_1 = t'$, $u_n = t''$ and $t{::}u_i \Longrightarrow u_{i+1}$ for all $1 \leq i < n$.
- `first_one`$(t_1, \ldots, t_n){::}t' \Longrightarrow t''$ holds if there exists $1 \leq i \leq n$ such that $t_i{::}t' \Longrightarrow t''$ and $t_j{::}t' \not\Longrightarrow t''$ hold for $1 \leq j < i$.
- `nf`$(t){::}t' \Longrightarrow t''$ holds if both $t^*{::}t' \Longrightarrow t''$ and $t{::}t'' \not\Longrightarrow \_$ hold.

ρLog can answer queries of the form $cond_1 \wedge \ldots \wedge cond_m$ where the variables are (implicitly) existentially quantified. The constraints $cond_i$ in queries and programs are of three kinds: reducibility atoms $t{::}t' \Longrightarrow t''$, irreducibility literals $t{::}t' \not\Longrightarrow t''$; and (3) boolean formulas that can be properly interpreted by the constraint solving component of ρLog. To instruct our system to compute one (resp. all) substitution(s) for the variables in the query $cond_1 \wedge \ldots \wedge cond_n$ for which it holds, we can submit requests of the form

Request($cond_1 \wedge \ldots \wedge cond_n$) or RequestAll($cond_1 \wedge \ldots \wedge cond_n$)

Another use of ρLog is to compute one or all reducts of a term with respect to a strategy. The request

ApplyRule(t, t')

instructs ρLog to compute one (if any) reduct of t' with respect to strategy t, that is, a term t'' such that formula $t{::}t' \Longrightarrow t''$ holds. ρLog reports "`no solution found.`" if there is no reduct of t' with t. ρLog can also be instructed to find all reducts of a term with respect to a strategy, with

ApplyRuleList(t, t')

To illustrate, consider the rule-based solutions to the following problems:

1. To eliminate all duplicates of elements in a list L, we submit the request ApplyRule(nf(elim2),L) where strategy elim2 is defined by the rule

   elim2::$\{\overline{x}, x, \overline{y}, x, \overline{z}\} \Longrightarrow \{\overline{x}, x, \overline{y}, \overline{z}\} \leftarrow$.

   For example, ApplyRule(nf(elim2),$\{1, 2, 7, 2, 3, 1\}$) yields answer $\{1, 2, 7, 3\}$.
2. To find out if (or which) e is an element of a list L, we can submit the request Request(elem::$L \Longrightarrow x$) where strategy elim is defined by the rule

   elem::$\{\_\_, x, \_\_\} \Longrightarrow x \leftarrow$.

   For example, Request(elem::$\{1, 2, 3\} \Longrightarrow x$) can return the answer $\{x \mapsto 1\}$, and RequestAll(elem::$\{1, 2, 3\} \Longrightarrow x$) returns $\{\{x \mapsto 1\}, \{x \mapsto 2\}, \{x \mapsto 3\}\}$.
3. To find all function symbols from a list L that occur in an expression E, we can submit the request ApplyRuleList(getF(L), E), where the parametric strategy getF is defined by the rule

   getF(y)::$\_\_(F(\_\_)) \Longrightarrow F \leftarrow$ (elem::$y \Longrightarrow F$).

   For example, {f,g} is the answer to the query

   ApplyRuleList(getF({f,g,u,v,w}),f(g(a(),h(),b())))

Sequence and context variables permit matching to descend to arbitrary depth and width in a tree-like term. The downside of using these kinds of variables in full generality is infinitary unification, and thus the impossibility to find a sound and complete calculus for ρLog. To avoid this problem, we adopted a natural syntactic restriction, called determinism [8], that ensures that all inference steps of our calculus can be performed by computing matchers instead of most general unifiers. The good news is that matching with sequence and context variables is finitary [3].

2.2 *The Operational Model of ABAC$_\alpha$*

ABAC$_\alpha$ is a formal model of ABAC proposed by X. Jin in his Ph.D. thesis [6] with a minimal set of features to configure the well-known access control models DAC, MAC, and RBAC. The core components of this operational model are: : users (U), subjects (S), objects (O), user attributes (UA), subject attributes (SA), object attributes (OA), permissions (P), authorization policy, creation and modification policy, and policy languages (Fig. 1).

Users represent human beings who create and modify subjects, and access resources through subjects. Subjects represent processes created by users to perform some actions in the system. Objects represent system entities that should be

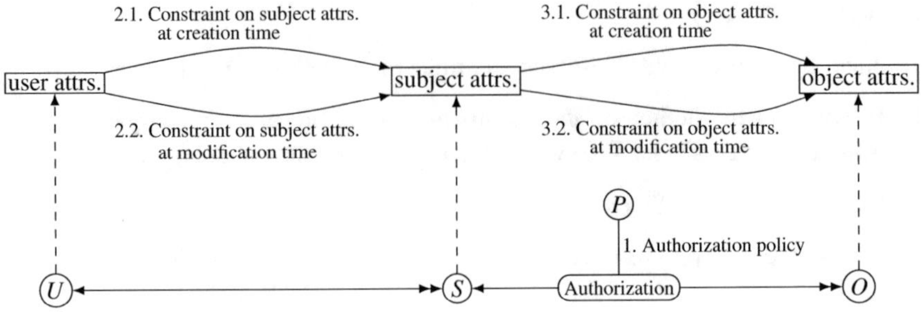

Fig. 1 The structure of ABAC$_\alpha$ model (adapted from [7])

protected. Users, subjects and objects are mutually disjoint in ABAC$_\alpha$, and are collectively called **entities**. Each user, subject, object is associated with a finite set of user attributes (UA), subject attributes (SA) and object attributes (OA) respectively. Every attribute att has a type, scope, and range of possible values. The sets of attributes specific to each kind of entity, together with their corresponding type, scope, and range, are specified in a **configuration type** of ABAC$_\alpha$: there will be one configuration type for DAC, and others for MAC, RBAC, etc.

In ABAC$_\alpha$, the type of an attribute is either atomic or set. The scope of each attribute is a finite set of values SCOPE(at). If at is of atomic type, then at can assume any value from SCOPE(at), otherwise it can assume any subset of values from SCOPE(at). Formally, this means that the range Range(at) of possible values of an attribute at is either SCOPE(at) if at is atomic or $2^{\text{SCOPE}(at)}$ if at is set, where each SCOPE(at) is either an unordered, a totally ordered, or a partially ordered finite set. There are six policies that control the operational behaviour of an ABAC$_\alpha$-based system, and each of them involves the interaction of two entities:

- authorization policies, which control the permissions that a user can hold on objects and exercise through subjects. Every configuration specifies a finite set P of permissions, and an authorization policy for every $p \in P$,
- policies to control the creation of a subject by a user, or of an object by a subject,
- policies for attribute value assignment: to a subject by the user who created it; or to an object by a subject,
- policies to control subject deletion by its creator.

All these policies grant/deny the corresponding operation based on the result of a boolean function which depends on the old and new attribute values of the interacting entities. According to [1, 6], each of these six boolean functions can be specified as a boolean formula in an instance of a language scheme called Common Policy Language (CPL). In CPL, the syntax of any formula ϕ is of the form

$$\begin{aligned}\phi &::= \phi \wedge \phi \mid \phi \vee \phi \mid (\phi) \mid \neg\phi \\ &\quad \mid \exists x \in set.\phi \mid \forall x \in set.\phi \mid set \text{ setcompare } set \\ &\quad \mid atomic \in set \mid atomic \text{ atomiccompare } atomic \\ \text{setcompare} &::= \subset \mid \subseteq \mid \not\subseteq \\ \text{atomiccompare} &::= < \mid = \mid \leq\end{aligned}$$

where set is a finite set of values, and $atomic$ are concrete values.

3 A Rule-Based Framework for ABAC$_\alpha$

Our rule-based tool for the specification and analysis of ABAC$_\alpha$ is built on top of the rule-based programming capabilities of ρLog. The user can specify (1) any particular ABAC$_\alpha$ configuration via the commands DeclareCfgType and DeclareConfiguration, and (2) any specific policies compatible with the operational model of ABAC$_\alpha$ by declaring in ρLog defining rules for the parametric strategies

`ConstrS`(*typeId*) `ConstrO`(*typeId*)
`ConstrModS`(*typeId*) `ConstrModO`(*typeId*) `Auth`(*typeId*, *p*)
`createS`(*cId*) `createO`(*cId*) `modSA`(*cId*) `modOA`(*cId*)

Afterwards, we can check whether it is safe to assume that a subject s can never obtain permission p on an object o in an ABAC$_\alpha$-configuration *cId* with the command `CheckSafety`[*cId*, *s*, *o*, *p*].

The meaning of these commands and parametric strategies is described in the remainder of this section.

Every entity (user, subject, or object) is completely described by its attribute values. Therefore, we chose to represent every entity as a term $K(at_1(v_1), \ldots, at_m(v_m))$ where $K \in \{\texttt{U}, \texttt{S}, \texttt{O}\}$ indicates the kind of entity, and every subterm $at_i(v_i)$ indicates that attribute at_i has value v_i. Every user has a unique identifier given by the value of its attribute `id`. Subjects are created by users and retain the identifier of their creator in the value of subject attribute `id`. From now on, we will assume the existence of a function UId(e) which returns the value of attribute `id` for every entity $e \in U \cup S$. Apart from this, the attribute names, their types and scope are characteristic to a particular configuration of ABAC$_\alpha$.

With our tool we can specify a configuration type for every configuration of interest, with the command

DeclareCfgType(<u>`typeId`</u>,
 $\{\texttt{UA}\to \{uAt_1, \ldots, uAt_m\}, \texttt{SA}\to \{sAt_1, \ldots, sAt_n\}, \texttt{OA}\to \{oAt_1, \ldots, oAt_p\},$
 $\texttt{Scope}\to \{at_1 \to \{sId_1, \tau_1\}, \ldots, at_r \to \{sId_r, \tau_r\}\}\})$

This declaration specifies a <u>configuration type</u> with identifier <u>`typeId`</u>, where

- $\{uAt_1, \ldots, uAt_m\}$ is the set of user attributes; $\{sAt_1, \ldots, sAt_n\}$ is the set of subject attributes, and $\{oAt_1, \ldots, oAt_p\}$ is the set of object attributes;
- the scope of every attribute at_i is the set bound to identifier sId_i in a particular configuration (see below), and its type is $\tau_i \in$ {elem, subset}, where elem stands for atomic and subset for set.

A configuration is an instance of a configuration type, which specifies (1) the configuration type which it instantiates; (2) the sets of values for the identifiers sId_i from the specification of the configuration type, and (3) the initial sets U, S, and O of entities (users, subjects, objects) in the configuration. In our system, the declaration of a concrete configuration of ABAC$_\alpha$ has the syntax

DeclareConfiguration(cId,
 {CfgType→typeId, Users→$\{uId_1 \rightarrow u_1, \ldots, uId_m \rightarrow u_m\}$,
 Range→{UId→$\{uId_1, \ldots, uId_m\}$,
 $sId_2 \rightarrow$ SCOPE$(at_2), \ldots, sId_r \rightarrow$ SCOPE(at_r)},
 Subjects→$\{s_1, \ldots, s_n\}$, Objects→$\{o_1, \ldots, o_q\}$})

Its side effect is to instantiate some globally visible entries:
CfgType(cId) with typeId,
Users(cId) with the set $\{u_1, \ldots, u_m\}$ of terms for users,
every User(cId,uId_i) with the term u_i,
Subjects(cId) with the set $\{s_1, \ldots, s_n\}$ of terms for subjects, and
Objects(cId) with the set $\{o_1, \ldots, o_q\}$ of terms for objects.

To illustrate, consider the mandatory access control model (MAC). Users and subjects have a clearance attribute of type elem, whose value is a number from a finite set of integers $L = \{1, 2, \ldots, N\}$, which indicates the security level of the corresponding entity. Objects have a sensitivity attribute of type elem whose value is also from L, and represents the sensitivity degree of the information in that object. When read and write are the only permissions on objects, we can assume the set of permissions P to be {read, write}.

A configuration type for MAC can be defined as follows:

DeclareCfgType(MAC,
 {UA→ {id, clearance}, SA→ {id, clearance}, OA→ {sensitivity},
 Scope→{id →{uId, elem}, clearance →{level, elem},
 sensitivity →{level, elem}}})

A particular MAC configuration can be defined by

DeclareConfiguration(MAC-Cfg01,
 {CfgType→MAC,
 Users→ {u1 → U(id(u), clearance(3)),
 u2 → U(id(u2), clearance(4))},
 Range→{uId→ {u1, u2}, level→ {1, 2, 3, 4, 5}},
 Subjects→ {S(id(u1), clearance(3)),
 S(id(u2), clearance(2))},
 Objects→ {O(sensitivity(1)), O(sensitivity(4))}})

3.1 Rules for the Policies of the Configuration Points

The constraint solving component of ρLog allows to specify and interpret correctly all formulas written in instances of the CPL scheme. Therefore, for every configuration type typeId, we can use ρLog to define parametric strategies for the policies of interaction between system entities:

- A user u can create a subject s if `ConstrS(typeId)::`$\{u, s\} \Longrightarrow$ `true` holds, where the defining rule of strategy `ConstrS` is of the form

  `ConstrS(typeId)::{`$\texttt{U}(\tilde{s}_1), \texttt{S}(\tilde{s}_2)$`}`$\Longrightarrow$ `true` $\leftarrow \phi_1$.

- A subject s can create an object o if `ConstrO(typeId)::`$\{s, o\} \Longrightarrow$ `true` holds, where the defining rule of strategy `ConstrO` is of the form

  `ConstrO(typeId)::{`$\texttt{S}(\tilde{s}_1), \texttt{O}(\tilde{s}_2)$`}`$\Longrightarrow$ `true` $\leftarrow \phi_2$.

- A user u can modify a subject s to become a subject s' if the reducibility formula `ConstrModS(typeId)::`$\{u, s, s'\} \Longrightarrow$ `true` holds, where the defining rule of strategy `ConstrModS` is of the form

  `ConstrModS(typeId)::{`$\texttt{U}(\tilde{s}_1), \texttt{S}(\tilde{s}_2), \texttt{S}(\tilde{s}_3)$`}`$\Longrightarrow$ `true` $\leftarrow \phi_3$.

- A subject s can modify an object o to become an object o' if the reducibility formula `ConstrModO(typeId)::`$\{s, o, o'\} \Longrightarrow$ `true` holds, where the defining rule of strategy `ConstrModO` is of the form

  `ConstrModO(typeId)::{`$\texttt{S}(\tilde{s}_1), \texttt{O}(\tilde{s}_2), \texttt{O}(\tilde{s}_3)$`}`$\Longrightarrow$ `true` $\leftarrow \phi_4$.

- A subject s is authorized to hold permission p $\in P$ on an object o if the reducibility formula `Auth(typeId,p)::`$\{s, o\} \Longrightarrow$ `true` holds, where the defining rule of strategy `Auth` is of the form

  `Auth(`x, z`)::{`$\texttt{S}(\tilde{s}_1), \texttt{O}(\tilde{s}_2)$`}`$\Longrightarrow$ `true` $\leftarrow \phi_{5,p}$.

In these rule-based specifications, ϕ_i and $\phi_{5,p}$ are formulas written in the instance of the CPL scheme for the values of the attributes of the interacting entities mentioned in the left-hand side of the corresponding rule.

For example, the mandatory access control (MAC) configuration type with read and write permissions can have the following rule-based specifications

```
ConstrS(MAC)::{U(x, clearance(y)), S(x, clearance(z))}
                                        ⟹ true ← (z ≤ y).
ConstrO(MAC)::{S(_, clearance(x)), O(sensitivity(y))}
                                        ⟹ true ← (x ≤ y).
ConstrModS(MAC)::{_,_,_}⟹ false ←.
ConstrModO(MAC)::{_,_,_}⟹ false ←.
Auth(MAC,read)::{S(_, clearance(x)), O(sensitivity(y))}
                                        ⟹ true ← (y ≤ x).
Auth(MAC,write)::{S(_, clearance(x)), O(sensitivity(y))}
                                        ⟹ true ← (x ≤ y).
```

These policies do not allow to modify the attribute values of subjects and objects.

3.2 Rules for the Operational Model

3.2.1 Subject and Object Creation

These are nondeterministic operations: at any time, a user can create any subject whose attribute values satisfy the CPL-formula for the subject creation policy; similarly, a subject can create any object whose attribute values satisfy the CPL-formula for the object creation policy. These operations are implemented in two steps:

1. We use the auxiliary functions sSeed($\underline{\texttt{cId}}$) to compute the term
   $\texttt{S}(sAt_1(\text{SCOPE}(sAt_1), \tau_1), \ldots, sAt_n(\text{SCOPE}(sAt_n), \tau_n))$
   and oSeed($\underline{\texttt{cId}}$) which computes the term
   $\texttt{O}(oAt_1(\text{SCOPE}(oAt_1), \tau_1), \ldots, oAt_p(\text{SCOPE}(oAt_p), \tau_p))$,
   where τ_i is the corresponding attribute type.
   For example, for the MAC configuration `MAC-Cfg01` illustrated before, the terms computed by sSeed(`MAC-Cfg01`) and oSeed(`MAC-Cfg01`) are
   $\texttt{S(id(\{u1, u2\}, elem), clearance(\{1, 2, 3, 4, 5\}, elem))}$ and
   $\texttt{O(sensitivity(\{1, 2, 3, 4, 5\}, elem))}$.
2. We use the terms computed by sSeed($\underline{\texttt{cId}}$) and oSeed($\underline{\texttt{cId}}$) as "seeds" to create any entity allowed by the creation policies. In rule-based thinking, an entity (subject or object) $K(att_1(v_1), \ldots, att_k(v_k))$ can be generated from the "seed" term $K(att_1(scope_1, \tau_1), \ldots, att_k(scope_k, \tau_k))$ if and only if the reducibility formulas $scope_i \rightarrow_{\tau_i} v_i$ hold. If we define the auxiliary strategy

   $$\texttt{setAt}::F_{at}(y_{scope}, x_{type}) \Longrightarrow F_{at}(x) \leftarrow (x_{type}::y_{scope} \Longrightarrow x).$$

   then the set of entities that can be generated from a seed term st is the set of all e for which the reducibility formula $\texttt{fmap(setAt)}::st \Longrightarrow e$ holds. Therefore, for a given ABAC$_\alpha$ configuration $\underline{\texttt{cId}}$:

   (1) a user u can create a subject s if $\texttt{createS(}\underline{\texttt{cId}}\texttt{)}::u \Longrightarrow s$ holds, where the defining rule of the parametric strategy `createS` is

   $$\begin{aligned}&\texttt{createS}(x_{cId})::x_u \Longrightarrow x_s \leftarrow \texttt{(fmap(setAt)}::\text{sSeed}(x_{cId}) \Longrightarrow x_s\texttt{)},\\ &\qquad \texttt{(id}::\text{UId}(x_u) \Longrightarrow \text{UId}(x_s)\texttt{)},\\ &\qquad \texttt{(ConstrS(}\text{CfgType}(x_{cId})\texttt{)}::\{x_u, x_s\} \Longrightarrow \texttt{true)}.\end{aligned}$$

   (2) a subject s can create an object o if $\texttt{createO(}\underline{\texttt{cId}}\texttt{)}::s \Longrightarrow o$ holds, where the defining rule of the parametric strategy `createO` is

   $$\begin{aligned}&\texttt{createO}(x_{cId})::x_s \Longrightarrow x_o \leftarrow \texttt{(fmap(setAt)}::\text{oSeed}(x_{cId}) \Longrightarrow x_o\texttt{)},\\ &\qquad \texttt{(id}::\text{UId}(x_u) \Longrightarrow \text{UId}(x_s)\texttt{)},\\ &\qquad \texttt{(ConstrO(}\text{CfgType}(x_{cId})\texttt{)}::\{x_s, x_o\} \Longrightarrow \texttt{true)}.\end{aligned}$$

3.2.2 Modification of Entity Attributes

Users can try to modify the attributes of subjects created by them, and subjects can try to modify the attributes of objects. A simple way to model these operations for an ABAC$_\alpha$ configuration cId of type typeId is as follows:

(1) Modification of the attribute values of a subject s by a user u can be viewed as generating a subject s' for which ConstrModS(typeId)::$\{u, s, s'\} \Longrightarrow$true holds. The outcome of changing the attribute values of s is s'. We define

$$\begin{aligned}\texttt{modSA}(x_{cId})::\{x_u, x_s\} \Longrightarrow x_s' \leftarrow\ & (\texttt{fmap}(\texttt{setAt}::\text{sSeed}(x_{cId}) \Longrightarrow x_s'),\\ & (\texttt{id}::\text{UId}(x_u) \Longrightarrow \text{UId}(x_s)),\ (\texttt{id}::\text{UId}(x_s) \Longrightarrow \text{UId}(x_s')),\\ & (\texttt{ConstrModS}(\text{CfgType}(x_{cId})::\{x_u, x_s, x_s'\} \Longrightarrow \texttt{true}).\end{aligned}$$

and note that modSA(cId)::$s \Longrightarrow s'$ holds if and only if the user u who created subject s is allowed to modify the attribute values of s to become s'.

(2) Modification of the attribute values of an object o by a subject s can be viewed as generating an object o' for which ConstrModO(typeId)::$\{s, o, o'\} \Longrightarrow$ true holds. The outcome of changing the attribute values of o is o'. We define

$$\begin{aligned}\texttt{modOA}(x_{cId})::\{x_s, x_o\} \Longrightarrow x_o' \leftarrow\ & (\texttt{fmap}(\texttt{setAt}::\text{oSeed}(x_{cId}) \Longrightarrow x_o'),\\ & (\texttt{ConstrModO}(\text{CfgType}(x_{cId})::\{x_s, x_o, x_o'\} \Longrightarrow \texttt{true}).\end{aligned}$$

3.2.3 State Transitions

A system with an ABAC$_\alpha$ access control model can be viewed as a state transition system whose states are triples $\{U, S, O\}$ consisting of the existing users (U), subjects (S), and objects (O), and whose transitions correspond to the six operations controlled by the policies of ABAC$_\alpha$.

Except for authorized access, the other five operations from the functional specification of ABAC$_\alpha$ determine state transitions. Their rule-based specifications are:

$$\begin{aligned}
&\texttt{createSubj}(x_{cId})::\{\{\overline{x}, x_u, \overline{y}\}, x_S, x_O\} \Longrightarrow\\
&\qquad \{\{\overline{x}, x_u, \overline{y}\}, x_S \cup \{x_s\}, x_O\} \leftarrow (\texttt{createS}(x_{cId})::x_u \Longrightarrow x_s), x_s \notin x_S.\\
&\texttt{deleteSubj}(_)::\{\{\overline{x}_1, x_u, \overline{x}_2\}, \{\overline{y}_1, x_s, \overline{y}_2\}, x_O\} \Longrightarrow\\
&\qquad \{\{\overline{x}_1, x_u, \overline{x}_2\}, \{\overline{y}_1, \overline{y}_2\}, x_O\} \leftarrow (\texttt{id}::\text{UId}(x_u) \Longrightarrow \text{UId}(x_s)).\\
&\texttt{createObj}(x_{cId})::\{x_U, \{\overline{x}, x_s, \overline{y}\}, x_O\} \Longrightarrow\\
&\qquad \{x_U, \{\overline{x}, x_s, \overline{y}\}, x_O \cup \{x_o\}\} \leftarrow (\texttt{createO}(x_{cId})::x_s \Longrightarrow x_o), x_o \notin x_O.\\
&\texttt{modifySubj}(x_{cId})::\{x_U, \{\overline{x}, x_s, \overline{y}\}, x_O\} \Longrightarrow\\
&\qquad \{x_U, \{\overline{x}, x_s', \overline{y}\}, x_O\} \leftarrow (\texttt{modSA}(x_{cId})::\{x_U, x_s\} \Longrightarrow x_s').\\
&\texttt{modifyObj}(x_{cId})::\{x_U, \{\overline{x}_1, x_s, \overline{x}_2\}, \{\overline{y}_1, x_o, \overline{y}_2\}\} \Longrightarrow\\
&\qquad \{x_U, \{\overline{x}_1, x_s, \overline{x}_2\}, \{\overline{y}_1, x_o', \overline{y}_2\}\} \leftarrow (\texttt{modOA}(x_{cId})::\{x_s, x_o\} \Longrightarrow x_o').
\end{aligned}$$

In the state transitions defined by these rules, the entities matched by x_u, x_s, x_o are those who interact during rule application.

3.3 Safety Analysis

Safety is a fundamental problem for any protection system. The safety problem for ABAC_α asks whether a subject s can obtain permission p for an object o. Recently, it has been shown that this problem is decidable [1], by identifying a state-matching reduction from ABAC_α to the pre-authorization usage control model with finite attribute domains ($\mathrm{UCON}_{\mathrm{preA}}^{\mathrm{finite}}$). The result follows from the facts that (1) the safety problem of $\mathrm{UCON}_{\mathrm{preA}}^{\mathrm{finite}}$ is decidable [11], and (2) state-matching reductions, like the one defined in [1], preserve security properties including safety. It provides an indirect way to implement an algorithm to decide the safety problem of ABAC_α. In [10] we noticed that this indirection can be avoided: a direct analysis of the operational model of ABAC_α revealed the main reasons when a configuration is unsafe. In this section we recall the theoretical results reported in [10], and illustrate how to use ρLog to turn our theoretical findings into rule-based specifications that can be directly executed. We claim that our approach is a natural and effective way to solve the safety problem for any configuration of ABAC_α.

3.3.1 Properties of ABAC_α Derivations

We start from the state transition view of the operational model described in Sect. 3.2.3. If $e \in S \cup O$ then a derivation $\Pi : St = \{U, S, O\} \Longrightarrow \ldots \Longrightarrow \{U, S', O'\}$ whose transition steps do not delete e may modify the attributes values of e. To analyze the possible changes of the attribute values of e in ABAC_α, we introduce the auxiliary notion of descendant of e in Π: $desc_\Pi(e)$ is the entity $e' \in S' \cup O'$ which represents e after performing the operations $op_1, \ldots, op_n$ in this order. Another useful auxiliary notion is $Desc^{St}(e) = \{desc_\Pi(e) \mid \Pi : St \Longrightarrow^* \{U, S', O'\}\}$.

With these preparations, the safety problem for ABAC_α is

Given an ABAC_α configuration `cId` with initial state $St = \{U, S, O\}$, a subject $s \in S$, an object $o \in O$, and a permission $p \in P$,

Decide if there is a derivation $\Pi : St \Longrightarrow \ldots \Longrightarrow \{U, S', O'\}$ whose transitions steps do not delete the descendants of s, such that subject $desc_\Pi(s)$ can be authorized to obtain permission p on object $desc_\Pi(o)$. Formally, this means that the formula `Auth(typeId,` p`)::{`$desc_\Pi(s), desc_\Pi(o)$`}` $\Longrightarrow$ `true` holds, where `typeId` is the configuration type of `cId`.

In this state transition system, objects can only participate at changing their own attributes. Therefore, objects from $O - \{o\}$ do not affect the truth value of the formula `Auth(typeId,` p`)::{`$desc_\Pi(s), desc_\Pi(o)$`}` $\Longrightarrow$ `true`. Hence it is harmless to assume that the initial state is $\{U, S, \{o\}\}$) and Π has no object creation steps. Also, if $\{U, S, O\} \Longrightarrow \{U, S', O'\}$ then $\{U, S \cup S'', O'\} \Longrightarrow \{U, S \cup S'', O'\}$ holds too, because we can choose the same participating entities to perform the transition. Therefore, we can assume that Π has no subject deletion steps.

Thus, we can assume without loss of generality that the safety problem is

Given an ABAC$_\alpha$ configuration cId with initial state $St_0 = \{U, S, \{o\}\}$ with $s \in S$, and a permission $p \in P$,

Decide UNSAFE if there is a derivation $\Pi : St \to^* (U, S', \{o'\})$ without subject deletion and object creation steps, such that the reducibility formula

$$\texttt{Auth}(\text{CfgType}(\underline{\texttt{cId}}), p)::\{desc_\Pi(s), o'\} \Longrightarrow \texttt{true}$$

holds, and SAFE otherwise.

By [10, Theorem 1], the answer is UNSAFE if and only if there exist $s' \in Desc^{St}(s)$ and $o' \in Desc^{St}(o)$ such that $\texttt{Auth}(\underline{\texttt{typeId}}, p)::\{s', o'\} \Longrightarrow \texttt{true}$ holds. In ABAC$_\alpha$, all attributes assume values from finite sets specified for cId, therefore $Desc^{St}(s)$ and $Desc^{St}(o)$ are finite sets that can be computed. Based on this observation, we designed a safety decision algorithm that computes incrementally the finite sets $Desc^{St}(s)$ and $Desc^{St}(o)$, and interleaves their computation with testing if $\texttt{Auth}(\underline{\texttt{typeId}}, p)::\{s', o'\} \Longrightarrow \texttt{true}$ holds for some $s' \in Desc^{St}(s)$ and $o' \in Desc^{St}(o)$.

3.3.2 A Rule-Based Safety Decision Algorithm

Suppose u is the creator of s. If $u \notin U$ then $Desc^{St}(s) = \{s\}$, otherwise $Desc^{St}(s) = \bigcup_{k=1}^{\infty} S_k$ where $S_1 = \{s\}$ and

$$S_{n+1} = \left\{ s'' \notin \bigcup_{k=1}^{n} S_k \mid \exists s' \in \bigcup_{k=1}^{n} S_k.(\texttt{ModSA}(\texttt{cId})::\{u, s'\} \Longrightarrow s'') \right\} \quad \text{if } n \geq 1.$$

Because $Desc^{St}(s)$ is finite, $Desc^{St}(s) = \bigcup_{k=1}^{n_0} S_k$ where $n_0 = \min\{n \in \mathbb{N} \mid S_n = \emptyset\}$. The partition $\{S_k \mid 1 \leq k \leq n_0\}$ of $Desc^{St}(s)$ can be computed iteratively: $S_1 = \{s\}$, and $S_{k+1} = \text{ApplyRuleList}(\texttt{nextS}(\underline{\texttt{cId}}, \bigcup_{i=1}^{k} S_i), \{U, S_k\})$ where the parametric strategy nextS is defined by the rule

$$\texttt{nextS}(x_{cId}, xs)::\{\{\_, x_u, \_\}, \{\_, x_s, \_\}\} \Longrightarrow$$
$$x_s' \leftarrow (\texttt{modSA}(x_{cId})::\{x_u, x_s\} \Longrightarrow x_s'), x_s' \notin xs.$$

We can speed up the safety decision algorithm by interleaving the computation of every S_k with testing if $\texttt{Auth}(\text{CfgType}(\underline{\texttt{cId}}), p)::\{s', o\} \Longrightarrow \texttt{true}$ holds for some $s' \in S_k$. We can do this test by checking if $\text{ApplyRule}(\texttt{auth?}(p, \underline{\texttt{cId}}), \{S_k, \{o\}\})$ yields true, where the parametric strategy auth? is defined by the rule

$$\texttt{auth?}(x_p, x_{cId})::\{\{\_, x_s, \_\}, \{\_, x_o, \_\}\} \Longrightarrow \texttt{true} \leftarrow$$
$$\texttt{Auth}(\text{CfgType}(x_{cId}, x_p))::\{x_s, x_o\} \Longrightarrow \texttt{true}).$$

As soon as any of these tests yields true, the decision algorithm stops by returning UNSAFE. Otherwise, we end up computing the set $Desc^{St}(s)$ and will start computing $Desc^{St}(o)$. The computation of this set can proceed in two steps:

1. First, we compute the set S_{all} of all subjects that can show up in the system: $S_{all} = \bigcup_{k=1}^{\infty} S_k$ where $S_1 = S$, S_2 is the set of all subjects that can be created by users in U, and

$$S_{n+1} = \left\{ s'' \notin \bigcup_{k=1}^{n} S_k \mid \exists u \in U.\exists s' \in \bigcup_{k=1}^{n} S_k.(\texttt{ModSA}(\underline{\texttt{cId}})::\{u, s'\} \Longrightarrow s'') \right\}$$

If $n \geq 2$. Because S_{all} is finite, $S_{all} = \bigcup_{k=1}^{n_1} S_k$ where $n_1 = \min\{n \geq 2 \mid \wedge S_n = \emptyset\}$. The partition $\{S_k \mid 1 \leq k \leq n_1\}$ of S_{all} can be computed incrementally:

$$S_2 = \bigcup_{u \in U} \text{ApplyRuleList}(\texttt{createS}(\underline{\texttt{cId}}), u)$$

$$S_{n+1} = \text{ApplyRuleList}(\texttt{nextS}(\underline{\texttt{cId}}, \bigcup_{k=1}^{n} S_k), \{U, S_k\}) \qquad \text{if } n \geq 2.$$

2. $Desc^{st}(o) = \bigcup_{k=1}^{\infty} O_k$ where $O_1 = \{o\}$ and

$$O_{n+1} = \left\{ o'' \notin \bigcup_{k=1}^{n} o_k \mid \exists s' \in S_{all}.\exists o' \in \bigcup_{k=1}^{n} O_k.(\texttt{ModOA}(\underline{\texttt{cId}})::\{s', o'\} \Longrightarrow o'') \right\}$$

if $n \geq 1$. Since $Desc^{st}(o)$ is finite, $Desc^{st}(o) = \bigcup_{k=1}^{n_2} O_k$ where $n_2 = \min\{n \in \mathbb{N} \mid O_n = \emptyset\}$.

With ρLog, it is easy to compute incrementally the partition $\{O_k \mid 1 \leq k \leq n_2\}$ of $Desc^{st}(o)$: for every $k \geq 1$ we have

$$O_{k+1} = \text{ApplyRuleList}(\texttt{nextO}(\underline{\texttt{cId}}), \bigcup_{i=1}^{k} O_i), \{S_{all}, O_k\})$$

where the parametric strategy `nextO` is defined by the rule

$$\texttt{nextS}(x_{cId}, x_O)::\{\{\_\_, x_s, \_\_\}, \{\_\_, x_o, \_\_\}\} \Longrightarrow$$
$$x_o' \leftarrow (\texttt{modOA}(x_{cId})::\{x_s, x_o\} \Longrightarrow x_o'), x_o' \notin x_O.$$

Here, again, we can speed up the safety decision algorithm by interleaving the computation of every O_k with testing if `Auth`(CfgType($\underline{\texttt{cId}}$), p)::$\{s', o'\} \Longrightarrow$ `true` holds for some $s' \in S_{all}$ and $o' \in O_k$. We can do this test by checking if the request ApplyRule(`auth?`(p, $\underline{\texttt{cId}}$), S_{all}, O_k) yields `true`. As soon as this happens, the algorithm stops by returning UNSAFE. Otherwise, we stop and return SAFE.

This decision algorithm is implemented in the method `CheckSafety` $[cId, s, o, p]$, which returns SAFE if, in configuration cIt, subject s can not get permission p on object o, and UNSAFE otherwise.

For example, the command

```
CheckSafety(MAC-Cfg01,S(id(u1),clearance(3)),
    O(sensitivity(1)),write)
```

returns SAFE because the clearance of subject `S(id(u1),clearance(3))` is too high to grant `write` permission to object `O(sensitivity(1))`.

4 Conclusion

State-matching reduction [14] is a powerful technique to prove security properties (including safety) of state transition systems. This indirect way to define an algorithm for the safety problem of ABAC_α configurations makes hard to observe some important properties that can be used to improve its performance. The direct rule-based analysis performed by us has the following advantages:

1. It provides a unified framework to specify policies for ABAC_α configurations, the operational model, execute them, and verify some security properties, including safety.
2. It allowed us to detect some useful properties of the transition model, that simplified significantly the design of our decision algorithm for safety. In particular, it allowed us to reduce the safety problem of to a simpler one: check if `Auth`(CfgType(`cId`), p)::$\{s', o'\} \Longrightarrow$ `true` holds for some $s' \in Desc^{St}(s)$ and $o' \in Desc^{St}(o)$. We solved it by identifying rule-based algorithms that interleave detection of unsafety with the incremental computation of $Desc^{St}(s)$ and $Desc^{St}(o)$.
3. With ρLog, we turned such a rule-based specification into executable code and obtained a practical tool to check the safety of any configuration of interest. The rule-based specification is parametric with respect to the configuration types of ABAC_α. Therefore, whenever we want to check that, for a given configuration, a subject s never gets permission p on an object o, it is enough to do the following:
   a. specify the configuration and its type, as indicated in Sect. 3.
   b. call the method CheckSafety(`cfgId`, s, o, p) which runs our safety-check algorithm. It returns SAFE if s never gets permission p on o, and UNSAFE otherwise.

There are many other rule-based systems with support for strategic programming, that can be used to formalize state transition systems and study their properties. But ρLog has some outstanding capabilities for this purpose:

1. It has four kinds of variables which give the user flexible control to select the components of the term which is transformed. The code is usually quite short and declaratively clear, as witnessed by the rule-based specification of ABAC_α.
2. It inherits from the Wolfram language of Mathematica a rich variety of constraints that can be used in requests and the conditional parts of rules. In particular, the

boolean formulas that constrain the operations of ABAC_α have direct translations as constraints in the CLP component of ρLog.

3. It can generate human-readable traces of the reductions that yield an answer. For the safety problems of ABAC_α, this capability could be used to produce scenarios that indicate the sequence of transitions that yield a state where a subject s can exercise a permission p on an object o. This capability could become a useful tool to detect security holes of ABAC_α configurations, and to fix them. We leave the extension of our a tool with this capability as a direction of future work.

Acknowledgements This work was supported by Shota Rustaveli National Science Foundation of Georgia under the grant no. FR17_439 and by the Austrian Science Fund (FWF) under the project P 28789-N32.

References

1. Ahmed, T., Sandhu, R.: Safety of ABAC_α is decidable. In: Yan, Z., Molva, R., Mazurczyk, W., Kantola, R. (eds.) Network and System Security, pp. 257–272. Springer International Publishing, New York (2017)
2. Apt, K.R., van Emden, M.H.: Contributions to the theory of logic programming. JACM **29**(3), 841–862 (1982)
3. Buchberger, B., Campbell, J.A. (eds.): Proceedings of Artificial Intelligence and Symbolic Computation (AISC 2004). LNCS, vol. 3249. Springer, Berlin (2004)
4. Ferraiolo, D.F., Sandhu, R., Gavrila, S., Kuhn, D.R., Chandramouli, R.: Proposed NIST standard for role-based access control. ACM Trans. Inf. Syst. Secur. (TISSEC) **4**(3), 224–274 (2001)
5. Hu, V., Ferraiolo, D., Kuhn, R., Schnitzer, A., Sandlin, K., Miller, R., Scarfone, K.: Guide to attribute based access control (ABAC) definition and considerations. NIST Special Publication 800-162 (2014)
6. Jin, X.: Attribute-based access control models and implementation in cloud infrastructure as a service. Ph.D. thesis, University of Texas at San Antonio (2014)
7. Jin, X., Krishnan, R., Sandhu, R.: A unified attribute-based access control model covering DAC, MAC and RBAC. In: Cuppens-Boulahia, N., Cuppens, F., Garcia-Alfaro, J. (eds.) Data and Applications Security and Privacy XXVI. LNCS, vol. 7371, pp. 41–55. Springer, Berlin (2012)
8. Marin, M., Kutsia, T.: Foundations of the rule-based system ρLog. J. Appl. Non-Class. Logics **16**(1–2), 151–168 (2006)
9. Marin, M., Piroi, F.: Rule-based programming with mathematica. In: Proceedings of International Mathematica Symposium (IMS 2004), Banff, Canada (2004)
10. Marin, M., Kutsia, T., Dundua, B.: A rule-based approach to the decidability of ABAC_α. In: Proceedings of the 24th ACM Symposium on Access Control Models and Technologies, SACMAT 2019, New York, NY, USA, pp. 173–178. Association for Computing Machinery (2019)
11. Rajkumar, P.V., Sandhu, R.: Safety decidability for pre-authorization usage control with identifier attribute domains. IEEE Trans. Dependable Secur. Comput. 1 (2018)
12. Sandhu, R.S.: Lattice-based access control models. Computer **26**(11), 9–19 (1993)
13. Sandhu, R.S., Samarati, P.: Access control: principle and practice. IEEE Commun. Mag. **32**(9), 40–48 (1994)
14. Tripunitara, M.V., Li, N.: A theory for comparing the expressive power of access control models. J. Comput. Secur. **15**(2), 231–272 (2007)
15. Wolfram, S.: The Mathematica Book, 5th edn. Wolfram Media, Champaign (2003)

A Strategic Graph Rewriting Model of Rational Negligence in Financial Markets

Nneka Ene, Maribel Fernández, and Bruno Pinaud

Abstract We propose to use strategic port graph rewriting as a visual modelling tool to analyse financial market processes. We illustrate the approach by specifying a basic "rational negligence" model in which investors may choose to trade securities without performing independent evaluations of the underlying assets. We show that our model is correct with respect to the equational model and can be used to simulate simple market behaviours. The model has been implemented within PORGY, a graph-based specification and simulation environment.

Keywords Graph rewriting · Strategies · PORGY · Rational negligence · Financial modelling

1 Introduction

Rational negligence [1] has been identified as a behavioural pattern in financial tradings, where transactions are performed without proper checks in order to maximise benefits and reduce operational costs. For example, in 2008 ratings from credit agencies (later found to be inaccurate) were used to replace costly checks, leading to a financial crisis that the DSGE (Dynamic Stochastic General Equilibrium) models [23] were unable to anticipate. This motivated a quest for more effective and transparent tools in the modelling of capital markets [26].

As an alternative to traditional top-down macro equilibrium models, Agent-Based Models (ABM) have been proposed, which examine behaviour at a micro-level [13]. In this paper we explore an alternative approach: we seek to formalise the rational negligence theory using graph rewriting. We provide an example to illustrate the

N. Ene · M. Fernández (✉)
King's College London, London, UK
e-mail: maribel.fernandez@kcl.ac.uk

B. Pinaud
University of Bordeaux, Bordeaux, France

G. Jaiani and D. Natroshvili (eds.), *Applications of Mathematics and Informatics in Natural Sciences and Engineering*, Springer Proceedings in Mathematics & Statistics 334, https://doi.org/10.1007/978-3-030-56356-1_8

ideas, as a step towards the development of alternative tools for the analysis of markets to complement the current agent-based implementations.

Rewrite rules are an intuitive and natural way of expressing dynamic, structural changes which are generally more difficult to model in traditional simulation approaches where the structure of the model is usually fixed [8]. Graph rewriting languages are well-suited to the study of the dynamic behaviour of complex systems: their declarative nature and visual aspects facilitate the analysis of the processes of interest producing a shorter distance between mental picture and implementation; they can be used for rapid prototyping, to run system simulations, and, thanks to their formal semantics, also to reason about system properties.

We use attributed port graphs, that is, graphs where edges are connected to nodes at specific points called ports, and where attributes are attached to ports, nodes and edges. Attributed port graphs are useful in the development of graph models, due to their support of both topology (via ports and edges) and data (via attributes). To control the rewriting process, we use strategies that permit to select which rules to apply and where, including probabilistic rule applications. We present first a basic model of asset trading following a discretised equational model presented in [1], where the probability of asset toxicity, due diligence analysis cost and asset cost are fixed. We then briefly discuss a more general version of the model where stochasticity is introduced by using a probabilistic choice model of logit type [13].

Summary of Contributions

We provide port graph rewrite rules and strategies that specify basic asset-trading transactions, starting with an auction to select a potential buyer. These rules and strategies model the rational negligence phenomenon [1, 20], whereby investors may choose to trade securities without performing independent evaluations of the underlying assets. The model has been implemented in PORGY,[1] an interactive, visual port graph rewriting tool. The graph rewriting approach we advocate produces flexible models that are easy to validate, experiment with and reason about. We illustrate it by showing the correctness of our graph rewrite rules and strategies with respect to the equations defining the rational negligence phenomenon, and using the implemented model to analyse simple market behaviours.

Related Work

Graph Transformation Systems (GTSs) have been used as a modelling framework in many areas: for example, RuleBENDER[2] is a simulation tool that supports rule-based modelling of biochemical systems [30], Kappa [22] is a rule-based language for modelling protein interaction networks, graph transformation has also been used to outline the semantics of domain specific modelling languages [8].

A basic set of port graph rewrite rules to model rational negligence was presented by Ene [11], focusing on implementation aspects. Here we extend the rules to include an abstract representation of an auction process and we analyse the properties of the

[1] http://porgy.labri.fr.

[2] http://www.rulebender.org.

model: we prove that the rewrite rules and the strategies we provide correctly simulate the equational model of rational negligence [1].

Previous rational negligence models followed an agent-based approach (see, for example [1, 26]). Test results for our model line up with results form traditional agent-based models (see Sect. 4 and [11] for a discussion of experimental results). General purpose agent-based simulation tools (see [21] for a survey) support an imperative object-oriented approach to model development. The graph rewriting approach used in this paper is declarative: the program consists of graph transformation rules and a strategy. Languages like Stratego [6, 36], Maude [10, 27] and ELAN [5] support a term rewriting approach with user-defined strategies to control the application of rules. Rascal [32] (and its predecessor ASF+DSF [33]) are closely related, using algebraic specifications as a basis to define programs, with traversal functions to control the application of rules. Tom [3] is an extension of Java with algebraic terms, rule definitions and a strategy language, thus allowing programmers to combine imperative object-oriented programming and strategic term rewriting. The symbolic transformation language symbtrans designed in the context of MEMSALab [4] (where models are defined using partial differential equations) extends Maple™ with conditional rewriting, strategies and pattern-matching modulo associativity and commutativity.

An alternative rule-based approach uses rules to define predicates, as in the logic programming language Prolog and its variants, including in some cases domain-specific constraint solvers or special-purpose languages to handle constraints [17]. The multi-paradigm language Claire [7] combines the imperative, functional and object-oriented styles with rule processing capabilities, including constructs to create new branches in the search-tree and to backtrack if the current branch fails. The language Prholog [9] extends logic programming with strategic conditional transformation rules, combining Prolog with the ρLog calculus [25] to enable strategic programming.

We have chosen to develop our models using port graph rewriting in PORGY [15], since it provides a visual rule-based programming-style, including user-defined strategies. The visual, declarative nature of GTS tools such as PORGY is welcome in the cases where users seek to primarily focus on describing what the system should accomplish, and is especially useful for the analysis of complex systems in interactive environments.

A benchmark analysing the differences between several GTS tools has been developed by Varró et al. [35]. A variety of GTS tools are available: among others we can cite GROOVE [19], a graph-based model checker for object oriented systems; AGG (the Attributed Graph Grammar System) [31], a graph-based language for the transformation of attributed graphs that comes with a visual programming environment; PROGRES (Programmed Graph Rewriting Systems) [29] that offers backtracking and nondeterministic constructs; GrGen (Graph Rewrite Generator) [18] that uses attributed typed multigraphs and includes features such as Java/C code generation, and GP [28], a graph programming language, where users can define rules and strategy expressions, with support for conditional rewriting. PORGY [15] has been used to model social networks [14] and database design [16, 34], as well as biochemical processes [2], where non-determinism, backtracking, positioning constructs, and probabilistic rule application are key features. A distinctive feature of PORGY is that rewriting derivations are directly available to users via the so-called derivation tree, which provides a visual representation of the dynamics of the system modelled and can be used to plot parameters and generate charts as illustrated in Sect. 4.

Overview

We first recall key notions on securitisation and graph rewriting in Sect. 2. Section 3 describes the proposed approach to the modelling of securitisation, including a short description of rules and associated strategies. Section 4 examines key properties of the model. We finally conclude and briefly outline future plans in Sect. 5.

2 Background

In this section we recall the main notions of asset trading and port graph rewriting that are needed in the rest of the paper.

2.1 Asset-Backed Securities

Assets [20] represent loans to clients or obligors who make regular installment payments to the originator to clear their debts. In a securitisation, assets are selected, pooled and transferred to a special purpose vehicle (SPV), who funds them by issuing securities. In general, an ABS (asset-backed security), or simply asset if there is no ambiguity, is any securitisation issue backed by consumer loans, car loans, etc.

In the core rational negligence model [1], the profit $\mathcal{U}_w$ expected by an agent (e.g., a bank) w from trading an asset depends on whether or not w follows the negligence rule, i.e., the rule of not performing independent risk assessment. Let z be a binary variable indicating whether or not the agent is following the negligence rule, then $\mathcal{U}_w$ is a function of z. According to [1], $\mathcal{U}_w(z)$ can be characterised by the following equations, where p is the probability of asset toxicity, Z is the average of all z's in the domain, c is the cost of purchasing an asset (note that the payoff from successfully reselling the asset is normalised to unity), x_w is the cost of performing a complete risk analysis, k is the number of trading partners of the seller bank and $\mathcal{N}_i$ is the set of agents.

- Expected profit for w when following the negligence rule, i.e., when $z(w) = 1$, if w buys an asset and then tries to sell it to w':

$$\mathcal{U}_w(1) =^{def} -p(1 - z(w'))c + [1 - p(1 - z(w'))](1 - c) \approx 1 - p(1 - Z) - c$$

This is because if the asset is toxic then w will loose c if w' checks, and will have a profit of $1 - c$ if w' does not check. Of course w does not know a priori whether

w' will or not follow the rule, but it can estimate $z(w')$ as the average of all the values of z in the system, Z. Note that when $p = 0$ the profit is $1 - c$ as expected.

- Similarly, the expected profit for w when the rule is not followed, i.e., $z(w) = 0$, is defined by:

$$\mathcal{U}_w(0) =^{def} (1-p)(1-c) - x_w$$

This is because if the asset is toxic, then w will not buy it (losing only x_w), but if it is not toxic then it will resell it with a profit of $1 - c - x_w$. Note that when $p = 1$ the loss is x_w as expected.

So the best response of agent w to a buying request is determined by the value of $\mathcal{U}(1) - \mathcal{U}(0)$. If it is positive, then negligence is better, otherwise diligence is better. Note that

$$\mathcal{U}(1) - \mathcal{U}(0) = p(Z-c) + x_w = p\left(\frac{1}{k}\sum_{j\in\mathcal{N}i} z_j - c\right) + x_w$$

Following [1], in this paper we study the behaviour produced by the trading of one asset since this is sufficient to perform validations against equivalent DSGE analyses. The goal is to study the evolution of the system till fixed point (that is, a stable state) is reached i.e., in this case, a state such that all potential buyers in the universe of discourse no longer alternate between diligent and negligent behaviour in their handling of the purchase of a particular asset.

2.2 Port Graph Rewriting

A port graph is a graph where nodes have explicit connection points, called ports, and edges are attached to ports. Nodes, ports and edges are labelled by a set of attributes, including a mandatory attribute Name that characterises the type of the node, port or edge. Attributes describe properties such as colour, size, etc. In PORGY [15] labels are records, i.e., lists of attribute-value pairs. The values can be concrete (numbers, Booleans, etc.) or abstract (expressions in a term algebra, which may contain variables). For example, the port graph in Fig. 1 depicts a toy ABS market universe represented by a community of banks (B nodes), one of which owns a tradeable asset (A), together with a global environment represented by the nodes Z, Change and Auction. The edge between A and B represents ownership.

Transactions between banks are specified by means of rewrite rules. A port graph rewrite rule $L \Rightarrow_C R$ is itself a port graph consisting of two port graphs L and R together with an "arrow" node. Intuitively, the pattern, L, is used to identify subgraphs (redexes) in a given graph which should be replaced by an instance of the right-hand side, R, provided the condition C holds. The arrow node may have ports and edges that connect it to L and R; these edges specify a partial morphism between

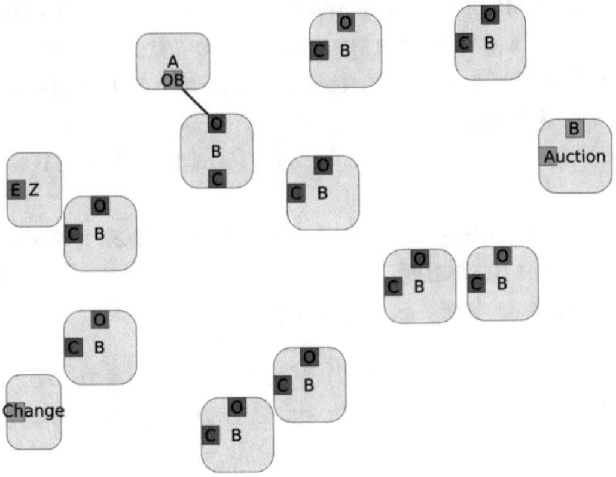

Fig. 1 Sample port-graph: model's starting graph

the ports in L and R, following the single push-out approach [24] to graph rewriting (see [15] for more details). Operationally, the arrow-node edges are used during rewriting to redirect edges that arrive to ports in the redex from outside, ensuring that no edges are left dangling. Table 3 shows the rules used in our model (these will be discussed in the next sections). The arrow-node edges can be optionally displayed in PORGY; when displayed, they are shown in red. In PORGY attribute values can be updated in the right-hand side of a rule by means of an "algorithm tab" (see Table 3).

For a given graph, several different rewriting steps may be possible (due to the intrinsic non-determinism of rewriting). Strategies in rewriting systems are a means of controlling the creation of rewriting steps. A sequence of rewriting steps is called a derivation. A derivation tree is a collection of derivations with a common root. Intuitively, the derivation tree is a representation of the possible evolutions of the system starting from a given initial state (each derivation provides a trace, which can be used to analyse and reason about the behaviour of system).

PORGY's strategy language allows us to specify not only the rule to be used in a rewriting step, but also the position where the rule should (or should not) be applied. Formally, the rewriting relation is defined on located graphs, which are port-graphs with two distinguished subgraphs P (Position subgraph, the focus of rewriting) and Q (Banned subgraph, where rewriting steps are forbidden). The keywords `crtGraph`, `crtPos`, `crtBan` in the strategy language denote, respectively the current graph being rewritten and its Position and Banned subgraphs. For example, the strategy expression `setPos(crtGraph)` sets the position graph as the full current graph. If T is a rule, then the strategy $one(T)$ randomly selects one possible occurrence of a match of rule T in the current graph G, which should superpose the position subgraph P but not superpose the banned subgraph Q. This strategy fails if the rule cannot be applied. Id and $Fail$ denote success and failure, respectively. The strategy expression $match(T)$ is used to check if the rule T can be applied but does not apply the rule. $(S)orelse(S')$ tries strategy S and if it fails then tries to apply S'. If both strategies fail then the whole statement fails. The strategy $ppick(T_1, \ldots, T_n, \Pi)$ selects one of the transformations $T_1, \ldots T_n$ according

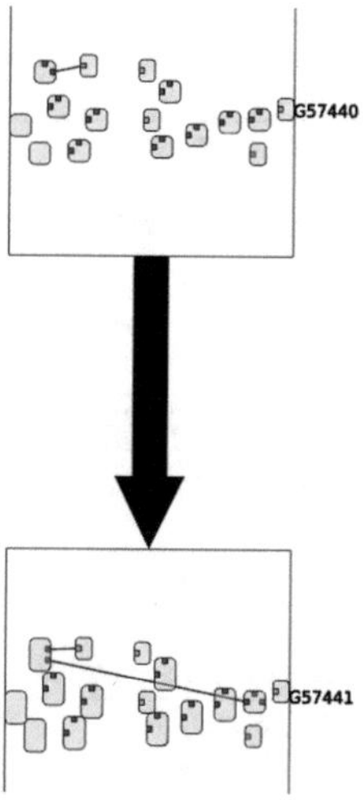

Fig. 2 A portion of the derivation tree in PORGY. The square boxes are nodes in the derivation tree: they contain graphs, and the black arrow represents the application of a rewrite rule

to the given probability distribution Π. The strategy $while(S)[(n)]do(S')$ executes strategy S' (not exceeding n iterations if the optional parameter n is specified) while S succeeds. $repeat(S)[max\ n]$ repeatedly executes a strategy S, not exceeding n times; it can never fail (when S fails, it returns Id). We refer the reader to [15] for the full definition of PORGY's strategy language.

PORGY [15] offers an in-built strategy editor, a navigable derivation tree widget, and widgets for the creation of rules and graphs. By navigating on the derivation tree and zooming on different nodes, we can see the various stages in the simulation (see Fig. 2); if we click on the black arrows in the derivation tree we can see which rule has been applied and identify the cause of the change in the model state.

3 The ABS-GTS Model

In this section we provide a graph-based model of the ABS process as specified by the equations given in Sect. 2.1. The ABS trading process is modelled hierarchically. The asset trading model sits at the top level of the model hierarchy. It is non-deterministic in nature. Below this system lie several subsystems that model origination, structuring of the deal, SPV transfers and profitability of the sale. In the rest of the paper we focus on the top tier level, which is where the 'rational negligence' phenomenon can be observed.

Asset-transfer transactions are modelled using a combination of global and local data: the global state includes Z (an indicator of market behaviour obtained as the

Table 1 Nodes and attributes

Entity name	Attribute	Description
Bank/Bidder/Potential Buyer (B/BD/PB)	Payoff (payoff)	Returns from re-selling an asset
	z	Indicates whether or not, as a rule, the bank performs independent risk analyses
	Bank ID (b_id)	Bank identifier
Asset (A)	Current value (c_val)	Cost of purchasing an asset
	Probability of toxicity (p_tox)	An asset is toxic if the borrowers of the underlying loans are likely to default or are in default
	Actualised toxicity (a_tox)	Current toxicity level
	Perception (pe)	External rating of the asset by rating agencies
	Due diligence cost (ddcost)	Full cost of an independent risk assessment
Change	Change	Change in bank approach
	Sum of change (sumofchange)	Sums all changes in a current cycle
Z	z	Represents the global average z
	Number of iterations (numofiterations)	Counter that keeps track of AllTrade iterations
	Number of agents (numofagents)	Variable that keeps track of number of banks
Theta	U1	Profitability of being negligent
	U0	Profitability of being diligent
	DeltaU1U0	Difference between U1 and U0
Auction	Kind	Abstract single-sided auction

average value of each individual bank's approach, represented by the bank's attribute *z*) and a *Change* indicator, to detect whether the market has reached a stable state. See Tables 1 and 2 for a description of the nodes used. Similar nodes were used in the model implemented in [11]; here we have additional nodes to represent the Auction and Bidders.

Model execution begins with a parameterised initialisation phase that produces a sample universe with one asset, linked to the owner bank (see Fig. 1). Colour attributes in nodes and ports are used to distinguish between classes of objects and to aid in the identification of states of interest (such as negligent behaviour, as explained below).

Tables 3 and 4 describe the rewrite rules handling asset transfer in our model. As in the foundational paper [1], our current implementation has been limited to the trading of one asset among *k* banks. The starting state of the model is the graph shown in Fig. 1

Table 2 Ports in each kind of node

Entity	Ports	Description
Bank, bidder	O (Owns)	Edges attached to this port link to assets owned by the bank
	C (Contacts)	Communication channel with another bank
Asset	OB (Owned_by)	Connects the asset to its current owner
Z	E (Environment)	Global entity that tracks current average sentiment
PotentialBuyer	O (Owns)	Links to assets owned by the bank
	C (Contacts)	Communication channel with another bank
	GE (Generates)	Declares a relationship with an analysis node
Change	CH (change)	Keeps track of behaviour changes
Theta	PB (Produced_by)	Links to entity that produces this node
Auction	B (Buyers)	Links to bidders
	S (Seller)	Links to seller

and it is from this point that the derivation tree begins to undergo construction as the execution strategy calls on rules that create step-wise transformations. Specifically, the asset transfer processes are governed by the strategies Auction, AllTrade and FixedPointSearch (see Strategies 1, 2 and 3 below).

Auction (Strategy 1) starts by specifying that rules will apply anywhere in the current graph (line 1). Line 2 applies rule sellorder, to represent a sell request from the asset owner. After a number of buy orders are received (specified in line 3 by repeated applications of the rule buyorder), an auction takes place and one of the bidders is selected (line 4, rule matchorders). The auction is then closed (line 5).

A basic description of the strategy AllTrade (Strategy 2) is as follows: Line 1 starts a trading cycle (the number of iterations is bound by the number of banks, k). Each iteration corresponds to one transaction: First an auction takes place (the Strategy Auction is called in line 1). After the auction, the potential buyer then begins the analysis in line 2 to decide whether or not to follow the negligence rule. It does this by computing the profitability of choices as described in Sect. 2.1 using rule beginanalysis. If diligence is more profitable the deviation rules will apply, otherwise the bank follows the negligence rule (see the `orelse` in lines 3 and 4). The rule *updatez* used in line 5 updates the global Z. We repeat k times in order to give all banks an opportunity to trade.

Strategy 3 controls the full execution: AllTrade is iterated until there are no changes in the agent behaviours (i.e., as long as the change rule can be applied).

Table 3 Rewrite rules

Name of rule	Description
sellorder	Initiates sell-order communication with Auction 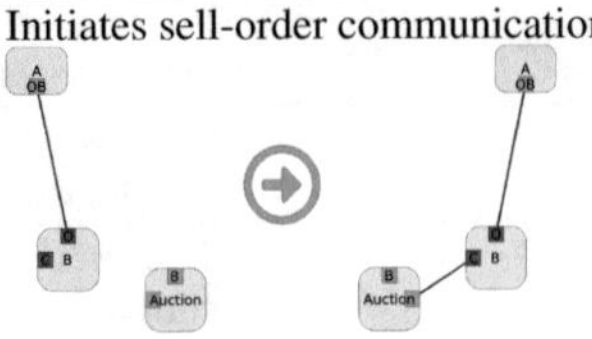
buyorder	Initiates buy-order communication with Auction 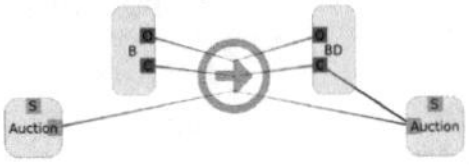
matchorders	Handles the match of sell-buy orders 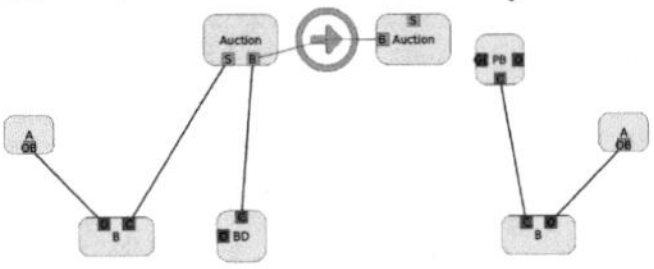
close	Closes the auction 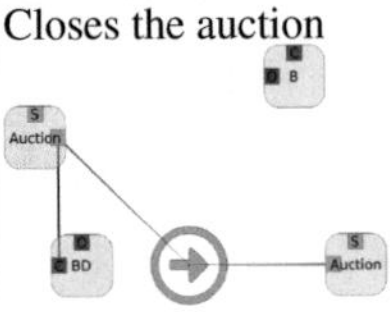
beginanalysis	Computes profitability $\mathscr{U}(1)$, $\mathscr{U}(0)$ of PB, generating a node <u>Theta</u> with attribute <u>DeltaU1U0</u> $= \mathscr{U}(1) - \mathscr{U}(0)$ 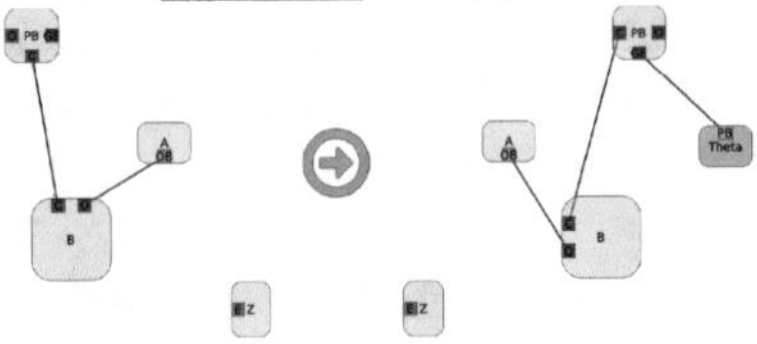 Algorithm tab $Theta.U1 = 1 - A.p\_tox(1 - Z.z) - A.c\_val$ $Theta.U0 = (1 - A.p\_tox)(1 - A.c\_val) - A.ddcost$ $Theta.DeltaU1U0 = Theta.U1 - Theta.U0$
updatez	Updates the attribute z in node Z  Algorithm tab: $Z.z = ((Z.z * (Z.numofagents - 1)) + B.z)/Z.numofagents$

Table 4 Rewrite rules

Name of rule	Description
followresult	Applies if DeltaU1U0 ≥ 0. It generates a follow node if more profitable to not do a full risk analysis 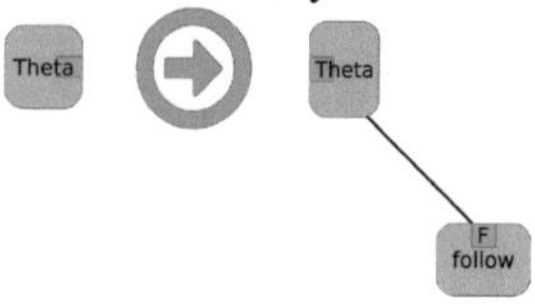 Arrow-node condition If $Theta.DeltaU1U0 \geq 0$
deviationresult	Applies if DeltaU1U0 < 0. It generates a deviation node if more profitable to do a full risk analysis (Similar to followresult)
followdecision	Transfers asset and prepares for a new transaction (i.e. cleans up after the decision negligence rule), updating bank's attribute z, updating the Change counter if necessary 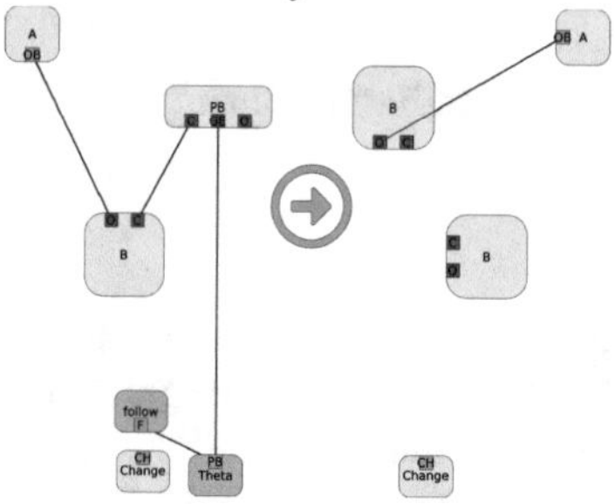
deviationdecision	Transfers asset and prepares for a new transaction (i.e. cleans up after the decision to deviate from the negligence rule), updating bank's attribute z, updating the Change counter if necessary (Similar to followdecision)
change	Sets the Change counter back to 0 if greater than 0  Algorithm tab $Change.change = 0$ $Change.sumofchange = 0$

A variant of strategy AllTrade replaces the `orelse` operator (lines 3 and 4) by a `ppick` operator, to model probabilistic choice of logit type between following or deviating from the negligence rule. The probability distribution used in this case implements the stochastic "trembles" described in [13] and can be written within our strategy environment as follows:

```
ppick((one(followresult);one(followdecision)),
     (one(deviationresult);one(deviationdecision)),
```

```
setPos(crtGraph);
one(sellorder);
repeat(one(buyorder))(n);
one(matchorders);
repeat(one(close))
```

Strategy 1: Auction

```
repeat(#Auction#;
  one(beginanalysis);
  (one(deviationresult);one(deviationdecision)) orelse
  (one(followresult);one(followdecision));
  one(updatez))(k)
```

Strategy 2: AllTrade

```
#AllTrade#;
while(match(change))do(
  one(change);
  #AllTrade#)
```

Strategy 3: FixedPointSearch

`udfLogitModel)`

where `udfLogitModel` is a function that reads the profitability of being negligent or diligent (attributes U1 and U0 in the node Theta of the graph produced by the rule beginanalysis) and returns the following values as a list:

$$\frac{\exp^{\mathscr{B}U_i(z=1)}}{\exp^{\mathscr{B}U_i(z=1)}+\exp^{\mathscr{B}U_i(z=0)}} \quad and \quad 1-\left(\frac{\exp^{\mathscr{B}U_i(z=1)}}{\exp^{\mathscr{B}U_i(z=1)}+\exp^{\mathscr{B}U_i(z=0)}}\right)$$

where i is the current agent number and $\mathscr{B}$ is the intensity of choice parameter that controls the ease at which fixed point is reached (as specified in [13]).

Levels of toxicity, asset value and due diligence cost are parameters of the simulation, which can be changed in our model by updating values of bank and asset node attributes.

4 Model Properties

First, we show that our model specification is correct with respect to the equational semantics (Sect. 2.1). This ensures that our model captures the ABS process of interest, and predictions from the ABS models under the same conditions coincide with the predictions produced by our system.

Lemma 1 *Starting from an initial graph that contains at least two bank nodes, one of which owns an asset, and an Auction node, Strategies 1 (Auction), 2 (AllTrade), and 3 (FixedPointSearch) never fail.*[3]

Proof Strategy 1 starts with a *set Pos* command, which cannot fail, and then executes a sell order (which cannot fail if the graph has at least two banks, one of which owns an asset, and an auction node), followed by a repeat command, which according to PORGY's semantics [15] can never fail, then matches the buying and selling orders (this rule cannot fail since the previous repeat command generates a redex) and finally the strategy executes another repeat command which cannot fail. Strategy 2 (AllTrade) executes a command of the form $repeat(S)(k)$, which can never fail. Since AllTrade cannot fail, strategy FixedPointSearch can only fail if the rule *change* in the body of the while loop fails, which is impossible due to the condition in the "while" (there is at least one match for change).

Theorem 1 (Correctness) *The graph-based model defined by the initial state, rewrite rules and strategies defined above is correct with respect to the equationally defined ABS process (see Sect. 2.1). More precisely, the graphs generated by the application of the rewrite rules with the given strategy represent states reached by the system governed by the equational ABS model.*

Proof We show that one trading transaction in our system corresponds to one trading transaction in the equational model. Let w be the bank that owns the asset (i.e., the bank linked by an edge to the asset), and let w' be the potential buyer (selected by auction). Rule beginanalysis computes the value of the projected profitability made by w' following and not following the negligence rule using the attributes p-tox, c-val and dd-cost (i.e., probability of toxicity, current value and due diligence cost) in the asset, which correspond to the values of p, c and x in the equational model. It computes the difference between $\mathcal{U}_w(1)$ and $\mathcal{U}_w(0)$ using the equations given in Sect. 2.1 and stores it in the attribute $DeltaU1U0$, as indicated in its algorithm tab. The result of this computation is the value specified by the equational model. The strategy ensures that the potential buyer selects the most profitable choice (lines 3–4 of Strategy 2), and the rule updatez recomputes the global Z value as outlined in Table 3, as follows:

$$Z_{i+1} = \frac{Z_i * (k-1) + z(w')}{k}$$

which gives the average value specified in the equational model.

Theorem 2 (Completeness) *The graph-based model defined by the initial state, rewrite rules and strategies defined above is complete with respect to stability as specified by the equational ABS model (see Sect. 2.1). More precisely, if the equational model reaches a stable state, so does our model.*

[3] A strategy fails if it attempts to apply a rule that is not applicable.

Proof (*Sketch*) The transactions of the equational ABS model are mimicked by the iterations in our strategy. A stable state in the equational model is reached when banks do not change their approach to negligence, which corresponds to absence of "Change" in our model: the *Change* flag is updated as required when rules *followdecision* and *deviationdecision* are applied (see Table 4).

Theorems 1 and 2 ensure that our model reaches a stable state if and only if the ABS equational model (see Sect. 2.1 and [1]) reaches the same stable state.

Theorem 3 (Termination) *The graph program consisting of the initial graph, rewrite rules and strategy described above terminates.*

More generally, if the rule updatez also changes the values of the asset attributes (reflecting external changes in risk analysis cost, toxicity and asset value) then the graph program terminates if and only if stable state is reached.

Proof A state is stable if no bank has changed its mind regarding its negligence choice when given an opportunity to trade. If stable state has been reached, there is no change after executing AllTrade hence the while loop found in line 2 of Strategy 3 stops. Conversely if our strategy terminates then the change rule does not apply since this is the condition to exit the while, hence no bank has changed its behaviour in AllTrade (stability has been reached). Thus, the graph program terminates if and only if the initial graph reaches a stable state.

Moreover, if the parameters of the asset do not change during the simulation then the program is guaranteed to terminate, because in this case Z is monotonic (once a bank decides towards diligence or negligence, the rest follow the trend). Thus, in this simple case, the program terminates.

Experiments and Analysis

A base case validation of the model is described in [11], in which test results line up with results from a traditional ABM simulation given in [1]. In Fig. 3 we recall some experimental results, where average Z value is plotted versus depth of the simulation. A natural question arises: What events could have mitigated or further instigated a negligent behaviour? By increasing toxicity values for example we can take into account the increase in interest rates that led to increased default rates and the 2008 crisis. Our experiments show that when toxicity is increased (attribute p in node A) the system reaches a stable state where all banks perform independent risk analysis, as expected. In particular, for high values of p (that is, high probability of toxicity), we observe the expected result when the initial state contains a mixture of negligent and diligent agents: a sharp drop in Z, corresponding to a sharp switch towards diligence which in turn will generate stability. An illustration of this can be seen in Fig. 3c and notice that given high due diligence costs Fig. 3b, d highlight a negligent approach whereas Fig. 3c, e reflect the favouring of a diligent approach. However, even for high toxicity, if the initial state is a set of negligent agents, the model reaches equilibrium without switching approach as seen in Fig. 3f.

We observe the following behaviours:

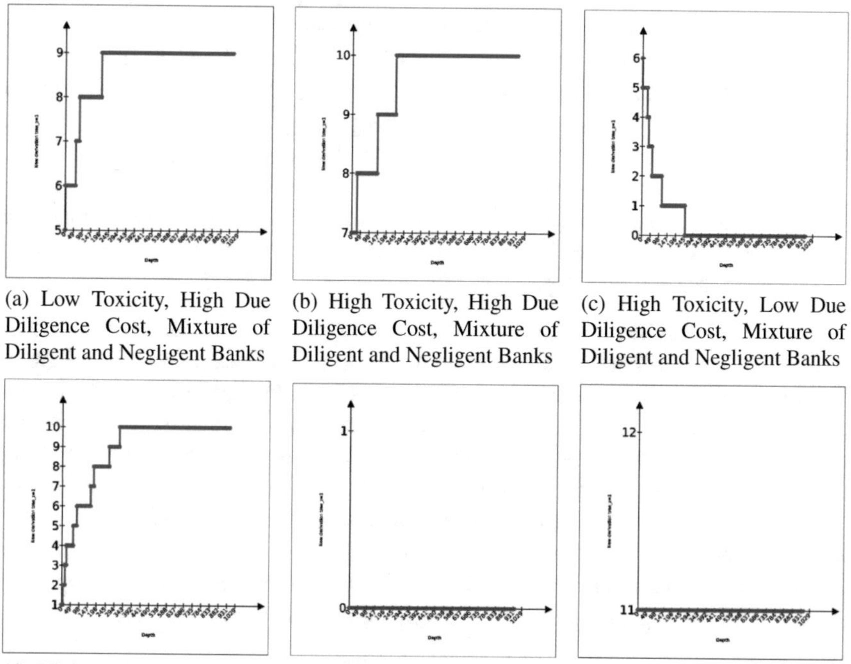

(a) Low Toxicity, High Due Diligence Cost, Mixture of Diligent and Negligent Banks

(b) High Toxicity, High Due Diligence Cost, Mixture of Diligent and Negligent Banks

(c) High Toxicity, Low Due Diligence Cost, Mixture of Diligent and Negligent Banks

(d) High Toxicity, High Due Diligence Cost, Diligent Banks

(e) High Toxicity, Low Due Diligence Cost, Diligent Banks

(f) High Toxicity, Low Due Diligence Cost, Negligent Banks

Fig. 3 Experiment results. y-axis: count of the number of negligent banks. The intersection of x and y axes in the case of a starting universe of purely diligent banks corresponds to the co-ordinates (0, 0) as opposed to (11, 0) in the case where we begin with negligent banks. Curves tending upwards reflect a negligent equilibrium result

1. Negligent equilibrium: If in the initial graph $Z \approx 1$, then the system arrives at negligent equilibrium (i.e., a result that reflects a community decision to no longer perform due diligence on a particular asset) even when the asset has high probability of toxicity.
   Explanation: For $Z \approx 1$ the profitability equation outlined in Sect. 2.1 reduces to: $\mathscr{U}(1) - \mathscr{U}(0) \approx 1 - c + x_w$ given that the difference between $\mathscr{U}(1)$ and expected profit when the rule is not followed (i.e. $\mathscr{U}(0)$) is $p(Z - c) + x_w$. This linear equation is computed at each iteration of the repeat loop. The result is positive given that c and x_w are both positive constants smaller than 1.
   Similarly, we observe that if p is high but in the initial graph the majority of banks are deviating from the negligence rule, then the system reaches a due diligence stable state.
2. Indefinite propagation: A continuous increase in the number of negligent bank agents means that a market crash can be postponed. This condition, although not feasible in a real market, is valid in equational models.

Explanation: A continuous increase in the number of agents used in calculating the average current sentiment, Z, as outlined in Sect. 2.1 and as computed by PORGY, means that the value of Z used in deciding whether or not to perform an independent risk analysis can remain unchanged.

3. Dangerous Equilibrium: A negligent stable state can be reached despite high toxicity under certain circumstances (high due diligence costs).
   Explanation: A sensitivity analysis shows that for a certain range of high due diligence cost values, negligent equilibrium can be obtained despite high toxicity values, even if initially negligence is not the norm, see Fig. 3b, d.

The results obtained with the basic experiments performed so far suggest that the graph rewriting approach, and in particular the derivation tree provided by PORGY, could be used to get insights beyond simulation runs. For example, the derivation tree could be used to search for states with specific properties, or to identify the occurrence of specific events (e.g., the first application of a specific rule). More meaningful analyses could be carried out, such as calculating propagation speeds (i.e., number of steps it takes for rule sentiment to be adopted by all agents relative to the size of network or the rate of change of average sentiment within different environments), taking into the account the pay-down factor of the loans supporting the asset and the expected contractual degradation of the asset itself, etc.

5 Conclusions

We have shown that strategic port-graph rewriting provides a basis for the design and implementation of graph models of the rational negligence phenomenon. Whilst ABMs rely on the internal processing of its agents, GTSs provide at each point in time a holistic view of the system state and a visual trace of the specific rules that trigger specific behaviours. In future, we will further develop the model using hierarchical graphs [12] to capture all tiers of the model, and also generalise the rules to permit dynamic changes in key attributes such as asset toxicity and costs.

References

1. Anand, K., Kirman, A., Marsili, M.: Epidemics of rules, rational negligence and market crashes. Eur. J. Financ. **19**(5), 438–447 (2013)
2. Andrei, O., Fernández, M., Kirchner, H., Pinaud, B.: Strategy-driven exploration for rule-based models of biochemical systems with Porgy. In: Hlavacek, W.S. (ed.) Modeling Biomolecular Site Dynamics: Methods and Protocols. Springer, Berlin (2018)
3. Balland, E., Brauner, P., Kopetz, R., Moreau, P., Reilles, A.: Tom: piggybacking rewriting on Java. In: Baader, F. (ed.) Term Rewriting and Applications, 18th International Conference, RTA 2007, Paris, France, 26–28 June 2007, Proceedings. Lecture Notes in Computer Science, vol. 4533, pp. 36–47. Springer (2007)

4. Belkhir, W., Giorgetti, A., Lenczner, M.: A symbolic transformation language and its application to a multiscale method. J. Symb. Comput. **65**, 49–78 (2014)
5. Borovanský, P., Kirchner, C., Kirchner, H., Ringeissen, C.: Rewriting with strategies in ELAN: a functional semantics. Int. J. Found. Comput. Sci. **12**(1), 69–95 (2001)
6. Bravenboer, M., Kalleberg, K.T., Vermaas, R., Visser, E.: Stratego/XT 0.17. A language and toolset for program transformation. Sci. Comput. Program. **72**(1–2), 52–70 (2008)
7. Caseau, Y., Josset, F.-X., Laburthe, F.: CLAIRE: combining sets, search and rules to better express algorithms (2004) arXiv
8. De Lara, J., Guerra, E., Boronat, A., Heckel, R., Torrini, P.: Domain-specific discrete event modelling and simulation using graph transformation. Softw. Syst. Model. **13**(1), 209–238 (2014)
9. Dundua, B., Kutsia, T., Reisenberger-Hagmayer, K.: An overview of pρlog. In: Lierler, Y., Taha, W. (eds.) Practical Aspects of Declarative Languages - 19th International Symposium, PADL 2017, Paris, France, 16–17 January 2017, Proceedings. Lecture Notes in Computer Science, vol. 10137, pp. 34–49. Springer (2017)
10. Durán, F., Eker, S., Escobar, S., Martí-Oliet, N., Meseguer, J., Rubio, R., Talcott, C.L.: Programming and symbolic computation in Maude. J. Log. Algebr. Methods Program. **110** (2020)
11. Ene, N.: Implementation of a port-graph model for finance. In: Proceedings of TERMGRAPH 2018: Computing with Terms and Graphs, pp. 14–25 (2019)
12. Ene, N., Fernández, M., Pinaud, B.: Attributed hierarchical port graphs and applications. In: Proceedings Fourth International Workshop on Rewriting Techniques for Program Transformations and Evaluation, WPTE@FSCD 2017, Oxford, UK, 8th September 2017, pp. 2–19 (2017)
13. Farmer, J., Gallegati, M., Hommes, C., Kirman, A., Ormerod, P., Cincotti, S., Sanchez, A., Helbing, D.: A complex systems approach to constructing better models for managing financial markets and the economy. Eur. Phys. J. Spec. Top. **214**(1), 295–324 (2012)
14. Fernández, M., Kirchner, H., Pinaud, B., Vallet, J.: Labelled graph strategic rewriting for social networks. J. Log. Algebr. Methods Program. **96**, 12–40 (2018)
15. Fernández, M., Kirchner, H., Pinaud, B.: Strategic port graph rewriting: an interactive modelling framework. Math. Struct. Comput. Sci. **29**(5), 615–662 (2019)
16. Fernández, M., Pinaud, B., Varga, J.: A port graph rewriting approach to relational database modelling. In: Proceedings 29th International Conference on Logic-Based Program Synthesis and Transformation, LOPSTR 2019, Porto, October 2019. Lecture Notes in Computer Science. Springer (2020)
17. Frühwirth, T.W.: Parallelism, concurrency and distribution in constraint handling rules: a survey. Theory Pract. Log. Program. **18**(5–6), 759–805 (2018)
18. Geiß, R., Kroll, M.: GrGen.NET: a fast, expressive, and general purpose graph rewrite tool. In: Applications of Graph Transformations with Industrial Relevance, Third International Symposium, AGTIVE 2007, Kassel, Germany, 10–12 October 2007, Revised Selected and Invited Papers, pp. 568–569 (2007)
19. Ghamarian, A.H., de Mol, M., Rensink, A., Zambon, E., Zimakova, M.: Modelling and analysis using GROOVE. STTT **14**(1), 15–40 (2012)
20. Gorton, G., Metrick, A.: Securitization. Working Paper 18611, National Bureau of Economic Research (2012)
21. Kravari, K., Bassiliades, N.: A survey of agent platforms. J. Artif. Soc. Soc. Simul. **18**(1), 11 (2015)
22. Krivine, J., Danos, V., Benecke, A.: Modelling epigenetic information maintenance: a Kappa tutorial. In: Computer Aided Verification, 21st International Conference, CAV 2009, Grenoble, France, June 26–July 2, 2009. Proceedings, pp. 17–32 (2009)
23. Leung, C.K.Y., Lubik, T.A.: Introduction: dynamic stochastic general equilibrium modelling and the study of Asia-Pacific economies. Pac. Econ. Rev. **17**(2), 204–207 (2012)
24. Löwe, M.: Algebraic approach to single-pushout graph transformation. Theor. Comput. Sci. **109**(1&2), 181–224 (1993)

25. Marin, M., Kutsia, T.: Foundations of the rule-based system ρLog. J. Appl. Non Class. Logics **16**(1–2), 151–168 (2006)
26. Markose, S., Dong, Y., Oluwasegun, B.: An multi-agent model of RMBS, credit risk transfer in banks and financial stability: implications of the subprime crisis (2008)
27. Martí-Oliet, N., Meseguer, J., Verdejo, A.: Towards a strategy language for Maude. Electron. Notes Theor. Comput. Sci. **117**, 417–441 (2005). Proceedings of the Fifth International Workshop on Rewriting Logic and Its Applications (WRLA 2004)
28. Plump, D.: The graph programming language GP. In: Bozapalidis, S., Rahonis, G. (eds.) Algebraic Informatics: Third International Conference, CAI 2009, Thessaloniki, Greece, 19–22 May 2009, Proceedings, pp. 99–122. Springer (2009)
29. Schürr, A., Winter, A.J., Zündorf, A.: The PROGRES approach: language and environment. Handbook of Graph Grammars and Computing by Graph Transformation, pp. 487–550. World Scientific Publishing Co., Inc., Singapore (1999)
30. Smith, A.M., Xu, W., Sun, Y., Faeder, J.R., Marai, G.: Rulebender: integrated modeling, simulation and visualization for rule-based intracellular biochemistry. BMC Bioinform. **13**(8) (2012)
31. Taentzer, G.: AGG: a graph transformation environment for modeling and validation of software. Applications of Graph Transformations with Industrial Relevance. Lecture Notes in Computer Science, vol. 3062, pp. 446–453. Springer, Berlin (2004)
32. Van den Bos, J., Hills, M., Klint, P., Van der Storm, T., Vinju, J.J.: Rascal: from algebraic specification to meta-programming. Electron. Proc. Theor. Comput. Sci. **56**, 15–32 (2011)
33. van den Brand, M.G.J., van Deursen, A., Heering, J., de Jong, H.A., de Jonge, M., Kuipers, T., Klint, P., Moonen, L., Olivier, P.A., Scheerder, J., Vinju, J.J., Visser, E., Visser, J.: The ASF+SDF meta-environment: a component-based language development environment. In: Wilhelm, R. (ed.) Compiler Construction, pp. 365–370. Springer, Berlin (2001)
34. Varga, J.: Finding the transitive closure of functional dependencies using strategic port graph rewriting. In: Proceedings Tenth International Workshop on Computing with Terms and Graphs, TERMGRAPH@FSCD 2018, Oxford, UK, 7th July 2018, pp. 50–62 (2018)
35. Varró, G., Schürr, A., Varró, D.: Benchmarking for graph transformation. In: 2005 IEEE Symposium on Visual Languages and Human-Centric Computing (VL/HCC 2005), 21–24 September 2005, Dallas, TX, USA, pp. 79–88. IEEE Computer Society (2005)
36. Visser, E.: Stratego: a language for program transformation based on rewriting strategies. In: Middeldorp, A. (ed.) Rewriting Techniques and Applications, 12th International Conference, RTA 2001, Utrecht, The Netherlands, 22–24 May 2001, Proceedings. Lecture Notes in Computer Science, vol. 2051, pp. 357–362. Springer (2001)

On Lie Algebras with an Invariant Inner Product

Alice Fialowski

Abstract In this note we consider low dimensional metric Lie algebras with an invariant inner product over the complex and real numbers up to dimension 5. We study their metric deformations with the help of cyclic cohomology, and give explicit formulas for the cocycles and deformations.

Keywords Lie algebra · Invariant bilinear form · Cohomology · Deformation

1 Introduction

Lie algebras with an invariant inner product became a current topic of research in Lie theory. Any reductive Lie algebra has such an invariant inner product, and they are related to the Killing form, which is an invariant inner product on a semisimple Lie algebra. Usually, one considers such algebras over the real numbers, but complex forms also play a role. One advantage in considering the complex case is that in that case, all inner products are equivalent, whereas over the real numbers, the signature of the form also plays a role.

Lie algebras with an invariant inner product are also referred to as metric Lie algebras or quadratic Lie algebras in the literature (see [3, 8, 14]), and include the diamond and oscillator algebras as examples of solvable algebras with an invariant inner product.

These Lie algebras are interesting by several reasons. Let G be the corresponding connected and simply connected Lie group of the Lie algebra $\mathfrak{g}$. The inner product on $\mathfrak{g}$ induces a left-invariant Riemannian metric on G in a natural way [10]. Such Lie algebras correspond to special pseudo-Riemannian symmetric spaces, see [16]. They also show up also in the formulation of some physical problems, like in the

A. Fialowski (✉)
Eötvös Loránd University, Budapest, Hungary

University Pécs, Pécs, Hungary
e-mail: fialowsk@cs.elte.hu; fialowsk@ttk.pte.hu

G. Jaiani and D. Natroshvili (eds.), *Applications of Mathematics and Informatics in Natural Sciences and Engineering*, Springer Proceedings in Mathematics & Statistics 334, https://doi.org/10.1007/978-3-030-56356-1_9

Adler–Kostant–Symes scheme [16] or in conformal field theory, as precisely the Lie algebras for which a Sugawara construction exists [5].

Not much is known about these Lie algebras. The classification is known up to dimension 7, see [2, 9, 11], but in higher dimension it is a hard unsolved problem. Representations of the diamond Lie algebra have been studied [4, 13, 14] but in general, the invariants of metric Lie algebras and the cohomology and deformation theory of such algebras has not been studied. A good survey about the known facts see in [15].

There is a notion of cyclic cohomology, which plays a role in the study of deformations preserving an invariant inner product. We examine this notion in some depth in this paper, and apply a variant of this type of cyclic cohomology to study metric deformations of Lie algebras preserving an invariant inner product.

2 Cohomology and Cyclic Cohomology

Let us consider the Lie algebra $\mathfrak{g}$ on a vector space V. Recall that the space of cochains $C(\mathfrak{g}, M)$ of a Lie algebra $\mathfrak{g}$ with coefficients in a module M is given by $C(\mathfrak{g}, M) = \mathrm{Hom}(\bigwedge, M)$. If $C^n(\mathfrak{g}, M) = \mathrm{Hom}(\bigwedge^n \mathfrak{g}, M)$, then $C(\mathfrak{g}, M) = \Pi_{n=0}^{\infty} C^n(\mathfrak{g}, M)$ or write $C(\mathfrak{g}, M) = \bigoplus_{n=0}^{\infty} C^n(\mathfrak{g}, M)$.

Consider the coboundary operator $D : C(\mathfrak{g}, M) \to C(\mathfrak{g}, M)$, which is determined by the maps $D : C^n(\mathfrak{g}, M) \to C^{n+1}(\mathfrak{g}, M)$,

$$
\begin{aligned}
D(\varphi)(v_1, \cdots, v_{n+1}) = &\sum_{\sigma \in \mathrm{Sh}(1,n)} (-1)^{\sigma} v_{\sigma(1)}.\varphi(v_{\sigma(2)}, \cdots, v_{\sigma(n+1)}) \\
&+ \sum_{\sigma \in \mathrm{Sh}(2,n-1)} (-1)^{\sigma} \varphi([v_{\sigma(1)}, v_{\sigma(2)}], v_{\sigma(3)}, , \cdots, v_{\sigma_{n+1}}),
\end{aligned}
$$

Here $\mathrm{Sh}(k, \ell)$ is the set of (un)shuffles of type (k, ℓ), that is those permutations of $k + \ell$ which are increasing on $1 \cdots k$ and $k + 1 \cdots k + \ell$, $(-1)^{\sigma}$ is the sign of the permutation σ and $[\cdot, \cdot]$ represents the Lie bracket on V.

A standard fact from Lie theory is that the map D satisfies $D^2 = 0$ and so we can define the (Eilenberg–Chevalley) cohomology

$$
H^n(\mathfrak{g}, M) = \ker(D : C^n(\mathfrak{g}, M) \to C^{n+1}(\mathfrak{g}, M)) / \mathrm{Im}(D : C^{n-1}(\mathfrak{g}, M) \to C^n(\mathfrak{g}, M)).
$$

Denote the cochains $C(\mathfrak{g}, \mathfrak{g})$ with coefficients in the adjoint representation by $C(\mathfrak{g})$, so that $C^n(\mathfrak{g}) = C^n(\mathfrak{g}, \mathfrak{g})$ are the n-cochains of the Lie algebra, and $H^n(\mathfrak{g}) = H^n(\mathfrak{g}, \mathfrak{g})$ is called the cohomology of $\mathfrak{g}$.

An inner product $\langle \cdot, \cdot \rangle$ on V is invariant with respect to a Lie algebra structure if

$$
\langle [u, v], w \rangle = \langle u, [v, w] \rangle.
$$

Given any inner product, define a cochain $\varphi \in C^n(V)$ to be <u>cyclic</u> if the associated element $\tilde{:} \bigwedge V \otimes V \to \mathbb{K}$, given by

$$\tilde{\varphi}(v_1, \cdots, v_{n+1}) = \langle(\varphi(v_1, \cdots, v_n), v_{n+1})\rangle,$$

satisfies

$$\tilde{\varphi}(v_{n+1}, v_1, \cdots, v_n) = (-1)^n \tilde{\varphi}(v_1, \cdots, v_{n+1}).$$

Denote a Lie algebra structure on the vector space V by $d \in C^2(V)$. An inner product is invariant with respect to a Lie algebra structure $d \in C^2(V)$ precisely when $\tilde{d}$ is cyclic with respect to the inner product. Also, φ is cyclic precisely when $\tilde{\varphi}$ is antisymmetric, that is $\tilde{\varphi} \in C^{n+1}(V, \mathbb{K})$, where $\mathbb{K}$ is the trivial module structure.

If φ and ψ are cyclic cocycles, then $[\varphi, \psi]$ is also cyclic. This allows us to extend the bracket to $C(V, \mathbb{K})$ by defining

$$[\tilde{\varphi}, \tilde{\psi}] = \widetilde{[\varphi, \psi]},$$

which equips $C^k(V, \mathbb{K})$ with a graded Lie algebra structure. In fact, we can compute that if $\varphi \in C^k(V)$, $\psi \in C^\ell(V)$, then

$$[\tilde{\varphi}, \tilde{\psi}] = \sum_{\sigma \in \mathrm{Sh}(l,k)} \tilde{\varphi}(\psi(v_{\sigma(1)}, \cdots, v_{\sigma(\ell)}), v_{\sigma(\ell+1)}, \cdots, v_{\sigma(k+\ell)}).$$

It is not obvious why this formula is correct, since it seems unbalanced in terms of φ and ψ, but it can be checked that under this bracket

$$[\tilde{\psi}, \tilde{\varphi}] = (-1)^{(k+1)(\ell+1)+1}[\tilde{\varphi}, \tilde{\psi}],$$

which is the graded antisymmetry corresponding to the antisymmetry of the bracket of φ and ψ. Note that the coboundary operator on an element $\tilde{\varphi} \in C^n(V, \mathbb{K})$ is just given by

$$\begin{aligned}(D\varphi)(v_1, \cdots, v_{n+2}) &= \sum_{\sigma \in \mathrm{Sh}(2,n)} (-1)^\sigma \tilde{\varphi}([v_{\sigma(1)}, v_{\sigma(2)}], v_{\sigma(3)}, \cdots, v_{\sigma(n+2)}) \\ &= [\tilde{\varphi}, \tilde{d}](v_1, \cdots, v_{n+2}),\end{aligned}$$

which means that if we define the coboundary operator D on $CC^n(\mathfrak{g})$ by $D(\tilde{\varphi}) = [\tilde{\varphi}, \tilde{d}]$, then the associated cohomology $HC(\mathfrak{g})$, called the <u>cyclic cohomology</u> of $\mathfrak{g}$, coincides with the shifted cohomology $H(\mathfrak{g}, \mathbb{K})$ of $\mathfrak{g}$ with coefficients in $\mathbb{K}$. That is, we have $HC^n(\mathfrak{g}) = H^{n+1}(\mathfrak{g}, \mathbb{K})$.

When there is no invariant inner product, the cohomology $H(\mathfrak{g}, \mathbb{K})$ is still well defined, and it is customary to define the cyclic cohomology of $\mathfrak{g}$ by $HC^n(\mathfrak{g}) = H^{n+1}(\mathfrak{g}, \mathbb{K})$, so that the definition of cyclic cohomology can be given independently of any invariant inner product, but the bracket of cyclic cochains makes sense only in the presence of an invariant inner product.

In the presence of an invariant inner product, there is a connection between cyclic cohomology and deformations of such an algebra. However, the connection is not as straightforward as one might think.

Example Let us consider $\mathfrak{g} = \mathfrak{sl}(2, \mathbb{C})$ with the invariant inner product Id. Then $H^1(\mathfrak{g}, \mathbb{C})$ and $H^2(\mathfrak{g}, \mathbb{C})$ both vanish and $H^3(\mathfrak{g}, \mathbb{C})$ is 1-dimensional. But then $HC^2(\mathfrak{g}) = \langle d \rangle$ is generated by the cochain represented by the algebra itself, so this cochain is nontrivial. As it turnes out, d is not the coboundary of a cyclic 1-cochain, which means that it really is true that the cyclic cohomology $HC^2(\mathfrak{g})$ has an extra basis element.

This means that in some sense, the algebra deforms along itself, which would not make any sense in the usual notion of deformation of an algebra. In fact, we will explain later that the two algebras d and $(1+t)d$, while isomorphic, are not formally isomorphic in the metric sense, and that is the source of the problem, because deformation theory considers a deformation to be trivial only when it is generated by a formal isomorphism. The problem turns out to be related to the fact that the identity matrix is never a cyclic 1-cochain, and for reductive algebras, a multiple of the identity matrix determines the only trivial deformation taking d to $(1+t)d$.

The problem can be expressed as follows:

$$\ker(D : CC^n \to CC^{n+1}) = \ker(D : C^n \to C^{n+1}) \bigcap CC^n$$

$$\mathrm{Im}(D : CC^{n-1} \to CC^n) \subseteq \mathrm{Im}(D : C^{n-1} \to C^n) \bigcap CC^n$$

The problem is that the inclusion on the bottom may be strict. One idea is to replace the left hand side with the right hand side, and define a new type of cohomology, which we call reduced cyclic cohomology

$$HRC^n = \ker(D : C^n \to C^{n+1}) \bigcap CC^n / (\mathrm{Im}(D : C^{n-1} \to C^n) \bigcap CC^n).$$

If we do this, then we obtain that $HRC^2(\mathfrak{g})$ vanishes, since d is a coboundary of an ordinary 1-cochain. In fact, it is always true that $d = [d, I]$, where $I \in Hom(V, V)$ is the identity map, but the identity map is never a cyclic 1-cochain. Note that the invariant inner product was used in the definition of reduced cyclic cohomology. In fact, if the algebra is reductive (or more generally, contains a simple direct summand), then HC^2 does not vanish. However, for the solvable algebras that we have studied, we have noticed that the reduced and ordinary cyclic cohomology coincide, which is a function of the type of invariant inner products that arise in those cases.

Let me explain the relation between cyclic and reduced cyclic cohomology in words. In his paper [17], Michael Penkava explained that, for a Lie algebra $\mathfrak{g}$ with invariant inner product, the second Lie algebra cyclic cohomology classifies infinitesimal deformations of $\mathfrak{g}$ preserving the inner product. One can distinguish between two types of deformations. The first type deforms the metric Lie algebra, but does not change the isomorphism class of the underlying Lie algebra. The second type deforms also the underlying Lie algebra structure. Since cyclic cohomology does not distinguish between the two types, the author introduced the reduced cyclic cohomology HRC^*. The reduced cyclic cohomology describes deformations of the second type.

I should mention that even though the reduced cyclic cocycles give nontrivial metric deformations of the metric Lie algebra, but some of the deformations may lead to isomorphic objects (see in the case of the Lie algebra W_3).

The reduced cyclic cohomology expresses deformations of metric Lie algebras if we allow formal deformations to be induced by any 1-cochain, instead of restricting to only cyclic cochains. This arises because we allow cyclic coboundaries of non-cyclic cochains in the reduced cyclic cohomology, so we allow formal deformations which take a metric algebra to another metric formal algebra, rather then just metric formal deformations.

It is well known that the Killing form gives an invariant inner product for a semisimple Lie algebra, and thus reductive Lie algebras have an invariant inner product. However, these are not the only types of Lie algebras with invariant inner product. Examples of real 4 dimensional solvable Lie algebras with an invariant inner product are the diamond and oscillator algebras, while in dimension 5, the nilpotent Lie algebra W_3 also has an invariant inner product. These cases have been well studied [1, 3, 11, 12]. We will study low dimensional examples of real and complex metric Lie algebras and their deformations preserving the invariant inner product.

3 Deformations

Recall that a 1-parameter deformation of a Lie algebra structure d on a vector space V is a formal power series of the form

$$d_t = d + t\psi_1 + t^2\psi_2 + \ldots$$

where $\psi_k \in C^k(V) = \mathrm{Hom}(\wedge^k V, V)$ are 2-cochains. The connection with cohomology is given by the fact that ψ_1 is a 2-cocycle, and we have

$$D(\psi_n) = -1/2 \sum_{k+l=n} [\psi_k, \psi_l],$$

where $[\psi_k, \psi_l]$ is the bracket of the cochains ψ_k and ψ_l. If d_t is isomorphic to some algebra structure d', then we say that d_t and d' are equivalent and we write $d_t \sim d'$.

If $d_t \sim d'$ for all t in some neighborhood of the origin, then the deformation is called a jump deformation from d to d'. If, on the other hand, $d_t \nsim d_{t'}$ for $t' \neq t$ in some neightborhood of the origin, then the deformation is called a smooth deformation. In this case, the set of algebras d_t form a family of nonisomorphic algebras.

Multiparameter deformations are also possible, and there is a special type of such deformation, the so called versal deformation, which is of the form

$$d^\infty = d + t_i\delta^i + \text{higher order terms},$$

where the expression $t_i\delta^i$ represents the Einstein summation notation for the sum of basis elements for the cohomology H^2. There are also relations of the form

$$r_i = t_k t_l r_i^{kl} + \text{higher order terms},$$

which are formal power series of order at least 2, with the number of relations being equal to the dimension of H^3. The base A of the deformation is the formal algebra $A = \mathbb{K}[[t_1, t_2, ..., t_n]]/(r_i)$, that is, the quotient of the ring of formal power series over the field $\mathbb{K}$ by the ideal generated by the relations. (See more about versal deformations of Lie algebras in [6, 7].)

For cyclic cohomology, we obtain a similar picture for deformations, including versal deformation, where this time the δ^i-s come from a basis of the cyclic cohomology HC^2 (or HRC^2), and the relations are determined by the dimension of HC^3. The source of the problem with deformations arises because of the notion of formal (or infinitesimal) equivalence of algebras. Two infinitesimal (or first order) deformations d and d' are formally equivalent if there is a linear map $\beta : g \to g$ such that if $g = \exp(t\beta)$ then

$$d' = g^*(d) = d + t[d, \beta].$$

We want d' to be a metric algebra, so there are two ways to guarantee this. We can require β to be cyclic with respect to d, in which case $[d, \beta]$ is automatically cyclic, since the bracket of two cyclic cochains is cyclic; or we can just require that $[d, \beta]$ be cyclic. The latter case is potentially problematic for formal deformations, because in that case, higher order terms arise, which may not be cyclic. Thus, the restriction that β is cyclic is natural, and gives rise to a consistent deformation picture. However, this is precisely what we obtain from the cyclic, rather than the reduced cyclic, cohomology. This means that the algebra d and $(1 + t)d$ may not be formally equivalent, and this is exactly what happens for any simple algebra, because the identity map is a never a cyclic cochain.

4 Dimension 3

We have already studied the complex 3-dimensional Lie algebra $\mathfrak{sl}(2, \mathbb{C})$. This is the only nontrivial 3-dimensional complex Lie algebra with an invariant inner product.

There are two real forms for this complex algebra, $\mathfrak{sl}(2, \mathbb{R})$ and $\mathfrak{so}(3, \mathbb{R})$. The first can be given by $\mathfrak{sl}(2, \mathbb{C})$. It has an invariant inner product given by the matrix

$$\begin{bmatrix} 0 & 1 & 0 \\ 1 & 0 & 0 \\ 0 & 0 & 2 \end{bmatrix}.$$

The signature of this matrix is (2, 1), but the signature of a metric form is only determined up to the transposition of the signature, so that there is an invariant inner product with signature (1, 2) as well. However, we can give it uniquely by requiring the first number to be greater that or equal to the second. In fact, the signature of a metric bilinar form of a metric Lie algebra is not always determinate. For example, for the 2-dimensional trivial algebra, any invertible symmetric matrix will serve as a metric, so the signature is not well defined by the algebra.

As this is a real form for the complex simple Lie algebra, it has vanishing reduced cyclic cohomology.

The second real Lie algebra is $\mathfrak{so}(3, \mathbb{R})$, which can be given by the nontrivial brackets $[e_1, e_2] = e_3$, $[e_1, e_2] = -e_2$, $[e_2, e_3] = e_1$, with invariant inner product given by the identity matrix, so its signature is $(3, 0)$.

Note that is it not surprising that the invariant form is not the same as for $\mathfrak{sl}(2, \mathbb{R})$, because the form of the matrix depends on the choice of basis. However, here, since we are over $\mathbb{R}$, the signature of the form comes into play, so in fact, the two invariant inner products are not of the same type, since they have different signatures.

In the Table below, we give the dimensions of the cohomology for the simple 3-dimensional Lie algebras for HC^n, HRC^n, and the standard cohomology H^n, with coefficients in the adjoint representation (Table 1).

This Table applies for the complex simple Lie algebra and both of its real forms. The only variations will be for the basis elements of the cohomology.

While there are no deformations of $\mathfrak{sl}(2, \mathbb{C})$, or any of its real forms, because the cohomology H^2 vanishes, something interesting occurs with cyclic cohomology. With respect to HC^2, we have a nontrivial cocycle given by d itself, and therefore, the

Table 1 Cohomology of the 3-dimensional simple algebra

n	HC^n	HRC^n	H^n
0	0	0	0
1	0	0	0
2	1	0	0
3	0	0	0

1-parameter deformation $d_t = (1+t)d$ is not a trivial deformation. This is unusual, since this type of deformation would be trivial with respect to the usual notion of deformation. Note that this deformation preserves the metric.

5 4-Dimensional Lie Algebras with Invariant Inner Product

5.1 *The Direct Sum $\mathfrak{sl}(2,\mathbb{C}) \oplus \mathbb{C}$ and Its Real Forms*

This algebra can be given by the nontrivial brackets $[e_1, e_2] = e_3$, $[e_2, e_3] = 2e_2$, $[e_1, e_3] = -2e_1$. In this form, an invariant inner product is given by the matrix

$$B = \begin{bmatrix} 0\,1\,0\,0 \\ 1\,0\,0\,0 \\ 0\,0\,2\,0 \\ 0\,0\,0\,1 \end{bmatrix}.$$

Its signature is (3, 1).

Theorem 1 $\mathfrak{sl}(2,\mathbb{C}) \oplus \mathbb{C}$ *does not have (metric) deformations, neither do its real forms.*

Proof We could calculate the cyclic cochains, coboundaries and cocycles by our computer technology, but here we think it is interesting and technically easier to use standard results on the computation of cohomology of Lie algebras. First, we recall that

$$H^n(\mathfrak{g}, M \oplus N) = H^n(\mathfrak{g}, M) \oplus H^n(\mathfrak{g}, N)$$

if $\mathfrak{g}$ is a Lie algebra and M and N are $\mathfrak{g}$-modules. Secondly, we recall the Künneth formula:

$$H^n(\mathfrak{g} \oplus \langle,, M\rangle) = \bigoplus_{k+\ell=n} H^k(\mathfrak{g}, M) \otimes H^\ell(\langle,, M\rangle),$$

Finally, we use the fact that if $\mathfrak{g}$ is simple (or semisimple), then $H^n(\mathfrak{g}, \mathfrak{g}) = 0$ for all n. Let let $\mathfrak{g} = s$ be $\mathfrak{sl}(2,\mathbb{C})$ and $\langle =, \mathbb{C}\rangle$ in the above. We obtain that

$$\begin{aligned} H^n(s \oplus \mathbb{C}, s \oplus \mathbb{C}) &= H^n(s \oplus \mathbb{C}, s) \oplus H^n(s \oplus \mathbb{C}, \mathbb{C}) \\ &= \bigoplus_{k+\ell=n} H^k(s, s) \otimes H^\ell(\mathbb{C}, s) \oplus H^k(s, \mathbb{C}) \otimes H^\ell(\mathbb{C}, \mathbb{C}) \\ &= \bigoplus_{k+\ell=n} H^k(s, \mathbb{C}) \otimes H^\ell(\mathbb{C}, \mathbb{C}). \end{aligned}$$

We need some elementary facts about the dimensions of $H^k(s, \mathbb{C})$ and $H^\ell(\mathbb{C}, \mathbb{C})$.

$$h^0(s,\mathbb{C}) = 1, \quad h^1(s,\mathbb{C}) = 0, \quad h^2(s,\mathbb{C}) = 0, \quad h^3(s,\mathbb{C}) = 1$$
$$h^0(\mathbb{C},\mathbb{C}) = 1, \quad h^1(\mathbb{C},\mathbb{C}) = 1, \quad h^2(\mathbb{C},\mathbb{C}) = 0, \quad h^3(\mathbb{C},C) = 0.$$

where $h^k = \dim(H^k)$. Thus we obtain that

$$\begin{aligned}
h^0(s\oplus\mathbb{C}, s\oplus\mathbb{C}) =& h^0(s,\mathbb{C})\cdot h^0(\mathbb{C},\mathbb{C}) = 1\cdot 1 = 1\\
h^1(s\oplus\mathbb{C}, s\oplus\mathbb{C}) =& h^1(s,\mathbb{C})\cdot h^0(\mathbb{C},\mathbb{C}) + h^0(s,\mathbb{C})\cdot h^1(\mathbb{C},\mathbb{C}) = 0\cdot 1 + 1\cdot 1 = 1\\
h^2(s\oplus\mathbb{C}, s\oplus\mathbb{C}) =& h^2(s,\mathbb{C})\cdot h^0(\mathbb{C},C) + h^1(s,\mathbb{C})\cdot h^1(\mathbb{C},\mathbb{C}) + h^0(g,\mathbb{C})\cdot h^2(\mathbb{C},\mathbb{C})\\
=& 0\cdot 1 + 0\cdot 1 + 1\cdot 0 = 0
\end{aligned}$$

$$\begin{aligned}
h^3(s\oplus\mathbb{C}, s\oplus\mathbb{C}) = h^3(s,\mathbb{C})\cdot h^0(\mathbb{C},\mathbb{C}) + h^2(s,\mathbb{C})\cdot h^1(\mathbb{C},\mathbb{C})\\
+h^1(s,\mathbb{C})\cdot h^2(\mathbb{C},\mathbb{C}) + h^0(s,\mathbb{C})\cdot h^3(\mathbb{C},\mathbb{C}) = 1\cdot 1 + 0\cdot 1 + 0\cdot 0 + 1\cdot 1 = 1
\end{aligned}$$

To compute hc^n, we obtain

$$hc^0 = z^1(s\oplus\mathbb{C},\mathbb{C}) = 1, \quad hc^1 = h^2(s\oplus\mathbb{C},\mathbb{C}) = 0$$
$$hc^2 = h^3(s\oplus\mathbb{C},\mathbb{C}) = 1, \quad hc^3 = h^4(g\oplus\mathbb{C},\mathbb{C}) = 1.$$

Let us summarize these results in the following Table (Table 2).

The real form $\mathfrak{sl}(2,\mathbb{R})\oplus\mathbb{R}$ has an invariant inner product which can be given by the same matrix as for the complex case, which has signature (3, 1). However, there is another real form with signature (2, 2). For $\mathfrak{so}(3,\mathbb{R})\oplus\mathbb{R}$ with nonzero brackets $[e_1, e_2] = e_3$, $[e_1, e_3] = -e_2$, $[e_2, e_3] = e_1$, we have an invariant inner product given by the matrix

$$B = \begin{bmatrix} 1\,0\,0\,0\\ 0\,1\,0\,0\\ 0\,0\,1\,0\\ 0\,0\,0\,1 \end{bmatrix},$$

whose signature is (4, 0). On the other hand, there is also a real form with signature (3, 1).

Table 2 Cohomology of the 4-dimensional Lie algebra $\mathfrak{sl}(2,\mathbb{C})\oplus\mathbb{C}$

n	HC^n	HRC^n	H^n
0	1	0	1
1	0	0	1
2	1	0	0
3	1	0	1

The reason that there are real forms with multiple signatures is that these algebras are direct sums of a simple and trivial algebra, and when combining the signatures, there are some variants possible. For example, with $\mathfrak{sl}(2, \mathbb{R})$, the forms of signature (2, 1) can combine with either the form of signature $(1, 0)$ on $\mathbb{R}$ to give a form of signature (3, 1), or it can combine with the form of signature (0, 1) on $\mathbb{R}$ to give a form of signature (2, 2).

The fact that there is an overlap in the possible signatures of these two real forms turns out to be important when we study deformations of the diamond and oscillator algebra.

The bases of the cohomology change for the $\mathfrak{sl}(2, \mathbb{R}) \oplus \mathbb{R}$ case, but the dimensions of the cohomology are the same, so are also given by the table above.

5.2 The Complex Diamond Algebra and Its Two Real Forms

The complex diamond algebra can be given by $[e_1, e_2] = e_3$, $[e_1, e_3] = -e_2$, $[e_2, e_3] = e_4$. It has an invariant inner product given by the matrix

$$B = \begin{bmatrix} 1 & 0 & 0 & 1 \\ 0 & 1 & 0 & 0 \\ 0 & 0 & 1 & 0 \\ 1 & 0 & 0 & 0 \end{bmatrix}$$

with signature $(2, 2)$.

This is the first algebra we have encountered which has deformations, thus $H^2(\mathfrak{g})$ does not vanish. The computation of the cohomology of the diamond algebra is not too difficult, and for brevity, we omit it. Just summarize the results in the Table below (Table 3).

Theorem 2 *The complex diamond algebra has a metric deformation to* $\mathfrak{sl}(2, \mathbb{C}) \oplus \mathbb{C}$.

Proof Notice that since the dimension of HC^2 is 1, we expect to have the diamond algebra deform to exactly one other metric algebra.

Table 3 Cohomology of the 4-dimensional complex diamond Lie algebra

n	HC^n	HRC^n	H^n
0	1	1	1
1	0	0	2
2	1	1	2
3	1	1	2

A miniversal deformation of the complex diamond algebra depends on two parameters, say t_1 and t_2. It turns out that there are only first order terms of t_1 and t_2, which means that the versal deformation is infinitesimal. From the relations on the parameters that must be satisfied, called relations on the base, one of the parameters must vanish, and we get two different 1-parameter infinitesimal deformations:

$$[e_1, e_2]_{t_1} = (1 + t_1)e_2, [e_1, e_3]_{t_1} = e_3, [e_2, e_3]_{t_1} = e_4, [e_1, e_3]_{t_1} = t_1 e_4,$$
$$[e_1, e_2]_{t_2} = e_2, [e_1, e_3]_{t_2} = -e_3, [e_2, e_3]_{t_2} = e_4 + t_2 e_1.$$

The first deformation, d_{t_1}, is a smooth deformation along a family of algebras to which the diamond algebra belongs, and the algebras to which it deforms do not have an invariant inner product.

The second deformation d_{t_2} is a jump deformation to the complex algebra $\mathfrak{sl}(2, \mathbb{C}) \oplus \mathbb{C}$, and a matrix of an invariant inner product which is well defined for small values of t_2 is

$$B = \begin{bmatrix} 1 & 0 & 0 & 1 \\ 0 & t_2+1 & 0 & 0 \\ 0 & 0 & t_2+1 & 0 \\ 1 & 0 & 0 & -t_2 \end{bmatrix},$$

which has determinant $-(1 + t_2)^3$, which does not vanish for $t_2 \neq -1$.

For the complex case, this just means it deforms to $\mathfrak{sl}(2, \mathbb{C}) \oplus \mathbb{C}$. The versal (and infinitesimal) cyclic deformation, given by d_{t_2} is isomorphic with $\mathfrak{sl}(2, \mathbb{C}) \oplus \mathbb{C}$, with the same metric as for the diamond algebra. It turns out that $d \sim d' = \mathfrak{sl}(2, \mathbb{C}) \oplus \mathbb{C}$ with the same metric as for the diamond algebra. In fact, the matrix

$$G = \begin{bmatrix} 0 & 0 & 1 & 1/2 \\ 1 & 0 & 0 & 0 \\ 0 & t^{-1} & 0 & 0 \\ 0 & 0 & t^{-1} & -1/2\,t^{-1} \end{bmatrix}$$

satisfies the property that $G(d_t(e_i, e_j)) = d'(G(e_i), G(e_j))$, which shows that G gives an isomorphism between the algebras d_t and d'.

5.3 *The Real Diamond Algebra*

The structure we gave for the complex diamond Lie algebra coincides with the structure for the real diamond Lie algebra, so all of the above information on the cohomology and cyclic cohomology is the same. However, there is an important difference related to the fact that there are two real forms for the complex algebra.

In this case, we see that the signature of the bilinear form of the real diamond algebra with the metric above is (2, 2), which is one of the possible signatures of

$\mathfrak{sl}(2,\mathbb{R})\oplus\mathbb{R}$, but does not coincide with a possible signature for $\mathfrak{so}(3,\mathbb{R})\oplus\mathbb{R}$, so it could not possibly deform to that algebra. A simple computation shows that the real diamond algebra does deform to $\mathfrak{sl}(2,\mathbb{R})\oplus\mathbb{R}$.

The real diamond algebra is given as a semidirect product of the Heisenberg algebra by $\mathbb{R}$, and this fact plays a role in the applications of this algebra.

5.4 *The Real Oscillator Algebra*

The oscillator algebra is the other real form of the complex diamond algebra. The structure of this real Lie algebra can be given by the nontrivial brackets $[e_1,e_2]=e_3$, $[e_1,e_3]=-e_2$, $[e_2,e_3]=-e_4$. An invariant inner product is given by

$$\begin{bmatrix} 0\,0\,0\,1 \\ 0\,1\,0\,0 \\ 0\,0\,1\,0 \\ 1\,0\,0\,0 \end{bmatrix}.$$

The signature of this form is (3, 1), and the importance of this fact will become clear shortly.

Theorem 3 *The real oscillator algebra has metric deformation to* $\mathfrak{so}(3)\oplus\mathbb{R}$ *and* $\mathfrak{sl}(2,\mathbb{R})\oplus\mathbb{R}$.

Proof The dimension of HC^2 is 1, and the versal defomation is infinitesimal. It can be given by d_t=

$$[e_1,e_2]_t=e_3,\ [e_1,e_3]_t=-e_2,\ [e_2,e_3]_t=-e_4+te_1,\ [e_3,e_4]=te_2,\ [e_2,e_4]=-te_3.$$

It can be shown that this deformation is isomorphic to $\mathfrak{so}(3,\mathbb{R})\oplus\mathbb{R}$ when $t>0$ and to $\mathfrak{sl}(2,\mathbb{R})\oplus\mathbb{R}$ when $t<0$. It is important to note that both of these algebras have invariant inner products of signature (3, 1), and this explains why it is possible to have this deformation into two algebras, when we didn't have the same pattern for the diamond algebra.

A matrix of an isomorphism between the d_t and $\mathfrak{so}(3,\mathbb{R})\oplus\mathbb{R}$ is given by

$$G=\begin{bmatrix} 0 & 0 & 1/2 & 1/2 \\ 1/\sqrt{2t} & 0 & 0 & \\ -1/\sqrt{2t} & 0 & 0 & 0 \\ 0 & 0 & 1/(2t) & -1/(2t) \end{bmatrix}.$$

Note that t must be positive for this matrix to be real. We omit the transformation that gives the isomorphism between d_t and $\mathfrak{sl}(2,\mathbb{R})\oplus\mathbb{R}$.

6 5-Dimensional Metric Lie Algebras

6.1 The Direct Sum $\mathfrak{sl}(2,\mathbb{C})\oplus\mathbb{C}^2$ and Its Real Forms

This algebra can be given by the nontrivial brackets $[e_1, e_2] = e_3$, $[e_2, e_3] = 2e_2$, $[e_1, e_3] = -2e_1$. In this form, an invariant inner product is given by the matrix

$$B = \begin{bmatrix} 0\,1\,0\,0\,0 \\ 1\,0\,0\,0\,0 \\ 0\,0\,2\,0\,0 \\ 0\,0\,0\,1\,0 \\ 0\,0\,0\,0\,1 \end{bmatrix}.$$

In the Table below, we summararize the cohomology information. We omit any of the calculations used to obtain this information (Table 4).

Theorem 4 *$\mathfrak{sl}(2,\mathbb{C})\oplus\mathbb{C}^2$ and its real forms have no metric deformation.*

Proof As usual for the algebras with a simple part, there is a metric deformation along the algebra itself, and the nontrivial cyclic 2-cocycle is just $\mathfrak{sl}(2,\mathbb{C})\oplus\mathbb{C}^2$.

The real forms corresponding to this complex algebra are $\mathfrak{sl}(2,\mathbb{R})\oplus\mathbb{R}^2$ and $\mathfrak{so}(3,\mathbb{R})\oplus\mathbb{R}^2$. The cohomology dimensions remain the same, although the cocycles representing the cohomology and deformations change for the different algebras.

Since $\mathfrak{sl}(2,\mathbb{R})$ has an invariant metric of signature $(2, 1)$, and the signature of an invariant metric on $\mathbb{R}^2$ can be $(2, 0)$, $(1, 1)$ or $(0, 2)$, this means we can obtain invariant metrics on $\mathfrak{sl}(2,\mathbb{R})\oplus\mathbb{R}^2$ of signature $(4, 1)$, or $(3, 2)$.

Since $\mathfrak{sl}(2,\mathbb{R})$ also has an invariant metric of signature $(2, 1)$, and the signature of an invariant metric on $\mathbb{R}^2$ can be $(2, 0)$, $(1, 1)$ or $(0, 2)$, this means we can obtain invariant metrics on $\mathfrak{sl}(2,\mathbb{R})\oplus\mathbb{R}^2$ of signature $(4, 1)$, or $(3, 2)$.

6.2 The Complex Diamond Algebra Plus $\mathbb{C}$

The direct sum of the complex diamond algebra and $\mathbb{C}$ can be given by the bracket structure

Table 4 Cohomology of the 5-dimensional Lie algebra $\mathfrak{sl}(2,\mathbb{C})\oplus\mathbb{C}^2$

n	HC^n	HRC^n	H^n
0	2	2	2
1	1	1	4
2	1	0	2
3	2	2	2

Table 5 Cohomology of the 4-dimensional complex diamond Lie algebra plus $\mathbb{C}$

n	HC^n	HRC^n	H^n
0	2	1	2
1	1	1	5
2	1	1	5
3	2	2	5

$$[e_1, e_2] = e_2, [e_1, e_3] = -e_3, [e_2, e_3] = e_4.$$

It has an invariant inner product given by the matrix

$$B = \begin{bmatrix} 0\,0\,0\,1\,0 \\ 0\,0\,1\,0\,0 \\ 0\,1\,0\,0\,0 \\ 1\,0\,0\,0\,0 \\ 0\,0\,0\,0\,1 \end{bmatrix}.$$

We summarize the cohomology information in the Table below (Table 5).

Theorem 5 *The complex diamond algebra plus $\mathbb{C}$ only has metric deformation to $\mathfrak{sl}(2, \mathbb{C}) \oplus \mathbb{C}^2$.*

Proof The versal cyclic deformation of this algebra is again infinitesimal, and can be given by d_t=

$$[e_1, e_2]_t = e_2, [e_1, e_3]_t = -e_3, [e_2, e_3]_t = e_4 + te_1, [e_2, e_4] = -te_e, [e_3, e_4] = te_3.$$

This deformation is isomorphic to $\mathfrak{sl}(2, \mathbb{C}) \oplus \mathbb{C}^2$.

6.3 The Real Diamond Algebra Plus $\mathbb{R}$

The deformation we gave for the complex diamond algebra plus $\mathbb{C}$ has real coefficients, so the structure also represents the cyclic versal deformation to $\mathfrak{sl}(2, \mathbb{R}) \oplus \mathbb{R}$ with respect to the inner product given by the matrix above, which has signature $(3, 2)$. Both $\mathfrak{sl}(2, \mathbb{R}) \oplus \mathbb{R}^2$ and $\mathfrak{so}(3, \mathbb{R}) \oplus \mathbb{R}^2$ have metrics of this signature, but the deformation is only isomorphic to $\mathfrak{sl}(2, \mathbb{R}) \oplus \mathbb{R}^2$ because the matrices of the transformations which give isomorphisms with $\mathfrak{so}(3, \mathbb{R}) \oplus \mathbb{R}^2$ all have unavoidable complex coefficients.

What happens if we choose the metric given by the matrix

$$B = \begin{bmatrix} 0 & 0 & 0 & 1 & 0 \\ 0 & 0 & 1 & 0 & 0 \\ 0 & 1 & 0 & 0 & 0 \\ 1 & 0 & 0 & 0 & 0 \\ 0 & 0 & 0 & 0 & 1 \end{bmatrix}?$$

This matrix has signature $(2, 3)$ which is related to a matrix of signature $(3, 2)$ in the way we described before, because multiplying the matrix by -1 reverses the signature, but doesn't affect the deformations. It might seem that still, the cyclic versal deformation with this matrix might be different, but it turns out that the generator of HC^2 is the same for both matrices, so their versal deformations can coincide.

6.4 The Oscillator Algebra Plus $\mathbb{R}$

We can use the same structure $[e_1, e_2] = e_3, [e_1, e_3] = -e_2, [e_2, e_3] = -e_4$ for the oscillator algebra plus $\mathbb{R}$ as we used for the oscillator algebra. A matrix of an invariant inner product of signature (4, 1) is

$$\begin{bmatrix} 0 & 0 & 0 & 1 & 0 \\ 0 & 1 & 0 & 0 & 0 \\ 0 & 0 & 1 & 0 & 0 \\ 1 & 0 & 0 & 0 & 0 \\ 0 & 0 & 0 & 0 & 1 \end{bmatrix}$$

We can give the versal cyclic deformation of d by the structure d_t=

$$[e_1, e_2]_t = e_3, [e_1, e_3]_t = -e_2, [e_2, e_3]_t = -e_4 + te_1, [e_3, e_4] = te_2, e_2, e_4] = te_3.$$

We compute that $d_t \sim \mathfrak{sl}(2, \mathbb{R}) \oplus \mathbb{R}^2$ when $t < 0$ and $d_t \sim \mathfrak{so}(3, \mathbb{R}) \oplus \mathbb{R}^2$ when $t > 0$. This pattern corresponds to the pattern we observed for the deformations of the oscillator algebra.

As in the case of the oscillator algebra, if we use a metric of signature $(3, 2)$ we obtain the same versal deformation as for the one of signature $(4, 1)$, so there is no difference in the deformation pattern depending on the choice of metric.

6.5 The Nilpotent Lie Algebra W_3

The algebra W_3 can be given by the nontrivial brackets

$$[e_3, e_4] = e_2, [e_3, e_5] = e_1, [e_4, e_5] = e_3$$

Table 6 Cohomology of the 5-dimensional complex algebra W_3

n	HC^n	HRC^n
0	2	2
1	3	3
2	3	3
3	2	2

This is the first nilpotent metric Lie algebra we have encoutered.

It has an invariant inner product given by the matrix

$$B = \begin{bmatrix} 0 & 0 & 0 & -1 & 0 \\ 0 & 0 & 0 & 0 & 1 \\ 0 & 0 & 1 & 0 & 0 \\ -1 & 0 & 0 & 1 & 0 \\ 0 & 1 & 0 & 0 & 1 \end{bmatrix}$$

The cohomology of the algebra W_3 is summarized in the Table below (Table 6).

Theorem 6 *W_3 has two metric deformations, to the complex Lie algebra diamond $\oplus\mathbb{C}$, and $\mathfrak{sl}(2, \mathbb{C}) \oplus \mathbb{C}^2$.*

Proof This is the first algebra for which HC^2 is larger than 1-dimensional, and moreover, the versal deformation given by the basis above for cohomology has higher order terms. However, we can construct 1-parameter deformations corresponding to the three basis elements, and they give first order deformations, so it is easy to check what these 1-parameter deformations are equivalent to.

First consider $d_t^1 =$

$$[e_3, e_4]_t = e_2, [e_3, e_5]_t = e_1, [e_4, e_5]_t = e_3,$$

$$[e_2, e_3]_t = te_2, [e_3, e_5]_t = -te_4, [e_2, e_5]_t = -te_3, [e_3, e_5]_t = te_5.$$

This deformation is isomorphic to $\mathfrak{sl}(2, \mathbb{C}) \oplus \mathbb{C}^2$.

The second deformation, d_t^2=

$$[e_3, e_4]_t = e_2, [e_3, e_5]_t = e_1 - te_4, [e_4, e_5]_t = e_3,$$

$$[e_1, e_3]_t = te_2, [e_1, e_5]_t = -te_3,$$

the third, d_t^3=

$$[e_3, e_4]_t = e_2 + te_5, [e_3, e_5]_t = (1+t)e_1, [e_4, e_5]_t = (1+t)e_3,$$

$$[e_2, e_3]_t = -te_1, [e_2, e_4]_t = -te_3.$$

The deformations d_t^2 and d_t^3 are both isomorphic to the diamond algebra $\oplus\mathbb{C}$.

The real form of W_3 is given by the same structure, and the matrix of the metric above has signature $(3, 2)$. Using the 1-parameter deformations above, we determined that d_t^1 is equivalent to $\mathfrak{sl}(2, \mathbb{R}) \oplus \mathbb{R}^2$, but not to $\mathfrak{so}(3, \mathbb{R}) \oplus \mathbb{R}^2$. The deformation d_t^2 is equivalent to the real diamond $\oplus\mathbb{R}$ when $t < 0$ and to the oscillator algebra $\oplus\mathbb{R}$ when $t > 0$. The deformation d_t^3 is isomorphic to the diamond $\oplus\mathbb{R}$ when $-2 < t < 0$ and to the oscillator algebra $\oplus\mathbb{R}$ otherwise.

Acknowledgements The author would like to thank the Organizers of the Conference to invite her to give this lecture. She also thanks Michael Penkava for computer computations.

References

1. Astrahancev, V.V.: On the decomposability of metrizable Lie algebras. Funct. Anal. Appl. **12**(3), 210–212 (1979)
2. Benayadi, S., Elduque, A.: Classification of quadratic Lie algebras of low dimension. J. Math. Physics **55** (2014)
3. Bordemann, M.: Nondegenerate invariant bilinear forms on nonassociative algebras. Acta Math. Univ. Comenian. N.S. **66**, 151–201 (1997)
4. Casati, P., Minniti, S., Salari, V.: Indecomposable Representations of the Diamond Lie Algebra. J. Math. Phys. **51** (2010)
5. Campoamor-Stursberg, R.: Contractions and deformations of quasi-classical Lie algebras preserving a non-degenerate quadratic Casimir operator. Phys. Atom. Nucl. **71**, 830–835 (2008)
6. Fialowski, A.: An example of formal deformations of Lie algebras. NATO Conference on Deformation Theory of Algebras and Appl. Proceedings, Kluwer, pp. 375–401 (1988)
7. Fialowski, A., Fuchs, D.B.: Construction of miniversal deformations of Lie algebras. J. Funct. Anal. **161**, 76–110 (1999)
8. Favre, Santharoubane.: Symmetric, invariant, nondegenerate bilinear form on a Lie algebra. J. Algebra **105**, 451-464 (1987)
9. Kath, I.: Nilpotent metric Lie algebras of small dimension. J. Lie Theory **17**, 41–61 (2009)
10. Kath, I., Olbrich, M.: On the Structure of Pseudo-Riemannnian Symmetric Spaces. Transformation Groups (2009)
11. Kath, I., Olbrich, M.: Metric Lie algebras with maximal isotropic centre. Mathematische Zeitschrift **246**, 23–53 (2004)
12. Kath, I., Olbrich, M.: Metic Lie algebras and quadratic extensions. Transformation Groups **11**, 87–131 (2006)
13. Liu, D., Pei, Y., Xia, L.: Irreducible representations over the diamond Lie algebra. Comm. Algebra **46**, 143–148 (2017)
14. Medina, A., Revoy, P.: Algebres de Lie et product scalaire invariant, Annales scientifiques de P.N.S. serie **18**, 553–561 (1985)
15. Ovando, G.P.: Lie algebras with ad-invariant metrics. A survey guide, Rendicaonti Seminario Matematico Univ. Pol. Torino, Workshop for Sergio Console **74**, 243–268 (2016)
16. Ovando, G.P.: Naturally reductive pseudo-Riemannian spaces. J. Geom. Phys. **61**(1), 157–171 (2010)
17. Penkava, M.: L∞ algebras and their cohomology. arXiv:q-alg/9512014

Study of Three-Layer Semi-Discrete Schemes for Second Order Evolution Equations by Chebyshev Polynomials

Romeo Galdava, David Gulua, and Jemal Rogava

Abstract Three-layer semi-discrete schemes for second order evolution equations are studied in the Hilbert space using Chebyshev polynomials. A priori estimates are proved for approximate solutions, as well as for difference analogues of first and second order derivatives. Using these a priori estimates we obtain estimates of the approximate solution error and, taking into account the smoothness of the solution of the continuous problem, the rate of convergence of an approximate solution with respect to step is estimated. The paper also discusses three-layer semi-discrete schemes for a second order complete equation and for an equation with a variable operator.

Keywords Second order evolution equation · Semi-discrete scheme · A priori estimates

1 Introduction

The method that we use to study a three-layer semi-discrete scheme can be called the method of associated polynomials. The idea is as follows. To each semi-discrete scheme (for an evolution equation) there corresponds a specific class of polynomials which we call associated polynomials. In the case of a second order evolution

R. Galdava
Sokhumi State University, 61 Politkovskaya st., 0186 Tbilisi, Georgia
e-mail: romeogaldava@gmail.com

D. Gulua (✉)
Department of Computational Mathematics of Georgian Technical University, 77 M. Kostava St., 0175 Tbilisi, Georgia
e-mail: d_gulua@gtu.ge

J. Rogava
I. Vekua Institute of Applied Mathematics and Department of Mathematics of Ivane Javakhishvili Tbilisi State University, 2 University St., 0186 Tbilisi, Georgia
e-mail: jemal.rogava@tsu.ge

G. Jaiani and D. Natroshvili (eds.), *Applications of Mathematics and Informatics in Natural Sciences and Engineering*, Springer Proceedings in Mathematics & Statistics 334, https://doi.org/10.1007/978-3-030-56356-1_10

equation, the corresponding semi-discrete scheme gives Chebyshev polynomials of second kind. Using them, we construct an exact representation for the solution of a difference problem. Based on the properties of this representation and classical Chebyshev polynomials, we study the stability of the considered semi-discrete scheme and prove a priori estimates.

We should note that the following works are devoted to use of orthogonal polynomials in approximate solution schemes for differential equations: Makarov [13], Morris and Horner [16], Novikov and Demidov [17], Rastrenin [19]. In the work [13] many aspects of using orthogonal polynomials in difference problems is widely covered.

Interesting results concerning the approximate solution of the Cauchy problem for a second order evolution equation can be found in Baker [2], Baker and Bramble [3], Baker et al. [4], Bales [5], Kacur [9], Ladyzhenskaya [12], Pultar [18], Sobolevskii and Chebotareva [25].

The well known monographs Godunov and Ryaben'kii [7], Marchuk [14], Mikhlin [15], Richtmyer and Morton [20], Samarskii [24], Yanenko [27] cover many important topics related to the construction and study of algorithms for an approximate solution of evolution problems.

Our present study provides novel results in addition to those obtained in [22, 23].

2 Abstract Hyperbolic Equation with a Constant Self-Adjoint Operator

2.1 *Weighted Second-Order Scheme. A Priori Estimates of Solution to Difference Problems*

Let us consider the Cauchy problem for an abstract hyperbolic equation in the Hilbert space H:

$$\frac{d^2u(t)}{dt^2} + Au(t) = f(t), \quad t \in]0, T], \tag{1}$$

$$u(0) = \varphi_0, \quad \frac{du(0)}{dt} = \varphi_1, \tag{2}$$

where A is a self-adjoint (not depending on t), positive definite (generally unbounded) operator with the definition domain $D(A)$, which is everywhere dense in H, i.e. $\overline{D(A)} = H, \quad A = A^*$ and

$$(Au, u) \geq \alpha \|u\|^2, \quad \forall u \in D(A), \quad \alpha = const > 0,$$

where the norm and scalar product in H are respectively defined by $\|\cdot\|$ and $(\cdot, \cdot)$; φ_0 and φ_1 are given vectors from H; $u(t)$ is a sought continuous, twice continuously differentiable function with values in H, and $f(t)$ is a given continuous function with values in H.

The vector function $u(t)$ with values in H, defined on the interval $[0, T]$, is called a solution of problem (1), (2) if it satisfies the following conditions: (a) $u(t)$ is twice continuously differentiable on the interval $[0, T]$; (b) $u(t) \in D(A)$ for any t from $[0, T]$, the function $Au(t)$ is continuous; (c) $u(t)$ satisfies Eq. (1) on the interval $[0, T]$ and the initial condition (2). Here the continuity and differentiability are obtained by means of the metric H. The existence and uniqueness of the solution of problem (1), (2) are proved in [11].

Remark 1 If $f(t)$ is continuously differentiable on $[0, T]$ (or $f(t) \in D(A^{1/2})$ for any t from $[0, T]$ and the function $A^{1/2}f(t)$ is continuous), $\varphi_0 \in D(A)$ and $\varphi_1 \in D(A^{1/2})$, then there exists a unique solution $u(t)$ of problem (1), (2) that satisfies the condition: the function $u'(t)$ gets its values from $D(A^{1/2})$, and $A^{1/2}u'(t)$ is continuous on $[0, T]$ (see [11], Theorem 1.5, p. 301).

To solve problem (1), (2) we use the semidiscrete scheme

$$\frac{u_{k+1} - 2u_k + u_{k-1}}{\tau^2} + A\,\frac{u_{k+1} + \nu u_k + u_{k-1}}{2+\nu} = f_k, \quad k = 1, \dots, n-1, \tag{3}$$

where $\tau = \frac{T}{n}$ ($n > 1$ is a natural number), $f_k = f(t_k)$, $t_k = k\tau$, $\nu \neq -2$; u_k is an approximate value of the exact solution $u(t)$ at the point $t = t_k$, $u(t_k) \approx u_k$.

The following statement is true.

Theorem 1 *Let* $u_0,\ u_1 \in D(A),\ f_k \in H,\ k = 1, \dots, n-1,\ \nu \in]-2, 2[$, *then for scheme (3) the following estimates are valid:*

$$\|A^s u_{k+1}\| \le c_0\Bigg(\|A^s u_0\| + \left\| B_\tau^{1/2} A^{s-1/2} \frac{\Delta u_0}{\tau} \right\|$$

$$+\,\tau \sum_{i=1}^{k} \left\| A^{s-1/2} B_\tau^{-1/2} f_i \right\| \Bigg), \quad 0 \le s \le 1, \tag{4}$$

$$\|A^s u_{k+1}\| \le c_0\left(\|A^s u_0\| + \left\| B_\tau^{1/2} A^{s-1/2} \frac{\Delta u_0}{\tau} \right\| \right)$$

$$+\,\tau \sum_{i=1}^{k} \|A^{s-1/2} f_i\|, \quad 0 \le s \le 1, \tag{5}$$

$$\|A^s u_{k+1}\| \le c_0\left(\|A^s u_0\| + \left\| A^{s-1/2} \frac{\Delta u_0}{\tau} \right\| + \nu_0 \|A^s(\Delta u_0)\| \right)$$

$$+\,\tau \sum_{i=1}^{k} \|A^{s-1/2} f_i\|, \quad 0 \le s \le 1, \tag{6}$$

$$\|A^s u_{k+1}\| \le c_0\left(\|A^s u_0\| + \left\| A^{s-1/2}\frac{\Delta u_0}{\tau} \right\| + \nu_0 \|A^s(\Delta u_0)\| \right)$$

$$+ \widetilde{c}(1-s)\cdot \tau^{2(1-s)} \sum_{i=1}^{k} \|f_i\|, \ \ \frac{1}{2} \le s \le 1, \tag{7}$$

$$\|u_{k+1}\| \le c_0\|u_0\| + c_1\left\|\frac{\Delta u_0}{\tau}\right\| + \tau\sum_{i=1}^{k}\|A^{-1/2} f_i\|, \ \ 0 \le s \le 1, \tag{8}$$

where $\Delta u_0 = u_1 - u_0$,

$$c_0 = \frac{2}{\sqrt{2-\nu}}, \quad c_1(\tau) = 2\left(\frac{2+\nu+\alpha\tau^2}{(4-\nu^2)\alpha}\right)^{1/2},$$

$$\widetilde{c}(s) = 2^{1-2s}\left(\frac{2+\nu}{2-\nu}\right)^{1/2-s}, \quad \nu_0 = \frac{1}{\sqrt{2+\nu}}, \quad B_\tau = I + \frac{\tau^2}{2+\nu}A.$$

2.2 Estimation for Two-Variable Chebyshev Polynomials

To obtain a priori estimates for the difference Eq. (3), we need to estimate a specific class of so-called two-variable Chebyshev polynomials which are defined by the recurrent relation (see [22])

$$U_{k+1}(x, y) = xU_k(x, y) - yU_{k-1}(x, y), \qquad k = 1, 2, \ldots, \tag{9}$$

$$U_1(x, y) = x, \quad U_0(x, y) = 1.$$

We call $U_k(x, y)$ a two-variable Chebyshev polynomial because $U_k(2x, 1)$ is a Chebyshev polynomial of second kind (see, e.g., [26]).

In the sequel, we need estimates for the polynomial $U_k(x, 1)$ on the interval $]-2,\ 2[$. The formula (see, e.g., [26])

$$U_k(2x, 1) = \frac{\sin((k+1)\arccos)}{\sin(\arccos x)}, \quad |x| \le 1,$$

clearly implies

$$|U_k(x, 1)| \le \frac{4}{\sqrt{4-x^2}}, \quad x \in]-2,\ 2[\,, \tag{10}$$

which is a well-known estimate of a Chebyshev polynomial of second kind (see, e.g., [26]).

The following inequality is simply obtained from representation $U_k(2x, 1)$:

$$|U_k(x,1) - U_{k-1}(x,1)| \le \frac{2}{\sqrt{2+x}}, \qquad x \in]-2,2]. \tag{11}$$

From the recurrent relation (9), we will get by induction

$$U_k(x,y) = \sqrt{y^k}\,U_k(\xi,1), \qquad \xi = \frac{x}{\sqrt{y}},\ y > 0. \tag{12}$$

Formula (12) is an important one as it relates polynomials $U_k(x, y)$ with the classical Chebyshev polynomials, in which replace the variable x to $x/2$.

Let us introduce the following domains:

$$\Delta = \{(x,y):\ (|y| < 1)\ \&\ (|x| < y+1)\}.$$

$$\Omega^+ = \left\{(x,y):\ 4y - x^2 > 0\right\}, \quad \Omega^- = \left\{(x,y):\ 4y - x^2 < 0\right\},$$

$$\Delta^+ = \{(x,y) \in \Delta:\ x \ge 0\}, \quad \Omega_1 = \Omega^+ \cap \Delta^+, \quad \Omega_2 = \Omega^- \cap \Delta^+.$$

As is known, the roots of a classical Chebyshev polynomial lie on $]-1, 1[$ (see, e.g., [26]). Hence, according to formula (12), it follows that, for any fixed positive y the roots of the polynomial $U_k(x, y)$ will lie within $]-2\sqrt{y},\ 2\sqrt{y}[$. Moreover, if we take into consideration that $U_k(\pm 2, 1) = (-1)^k(k+1)$ and $|U_k(2\xi, 1)|$ reaches its maximum on the boundary (see, e.g., [26]), then from formula (12) we get the estimate

$$|U_k(x,y)| \le U_k(2\sqrt{y}, y) = (k+1)\sqrt{y^k},\ (x,y) \in \Omega^+. \tag{13}$$

By virtue of the above reasoning we conclude that for any positive y, $U_k(x, y)$ is an increasing function with respect to the variable x when $x \ge 2\sqrt{y}$. Moreover, the recurrent relation (9) implies that, for any fixed $y \le 0$, $U_k(x, y)$ is an increasing function with respect to the variable x when $x \ge 0$. Hence we have

$$|U_k(x,y)| \le U_k(1+y, y) = 1 + y + \ldots + y^k, \tag{14}$$

where $y \ge -1$ and $|x| \le 1 + y$.

From (14) obtain the estimate

$$|U_k(x,y)(1-y)| \le 1, \quad (x,y) \in \Delta. \tag{15}$$

We also want to estimate the polynomial $U_k(x, y) - y^m U_{k-1}(x, y)$, $m = 0, 1$, where $(x, y) \in \Delta^+$.

By formula (12) and inequality (11), the following estimate is valid:

$$\left|U_k(x,y) - \sqrt{y}U_{k-1}(x,y)\right| = \sqrt{y^k}\,|U_k(\xi,1) - U_{k-1}(\xi,1)| \le \sqrt{2y^k}\,, \tag{16}$$

where $\xi = x/\sqrt{y}\,,\ (x, y) \in \Omega_1$.

Let us estimate the difference $U_k(x, y) - yU_{k-1}(x, y)$, when $(x, y) \in \Omega_1$. By inequalities (16) and (13), we have

$$|U_k(x, y) - yU_{k-1}(x, y)| \le \sqrt{2}, \quad (x, y) \in \Omega_1. \tag{17}$$

Let us now estimate the difference $U_k(x, y) - yU_{k-1}(x, y)$ when $(x, y) \in \Omega_2$ and $y > 0$. For this estimation we use the following formulas:

$$U_k(x, y) = \sqrt{y^k} \sum_{i=0}^{k} C_{k+i+1}^{2i+1} (\xi - 2)^i, \tag{18}$$

$$U_k(x, y) - \sqrt{y} U_{k-1}(x, y) = \sqrt{y^k} \sum_{i=0}^{k} C_{k+i}^{2i} (\xi - 2)^i, \tag{19}$$

where $\xi = x/\sqrt{y}\,,\ C_k^i$ are the binomial coefficients ($C_k^0 = 1$).

By some simple transformations, from (18) we obtain formula (19).

Formula (18) can be obtained using the Taylor expansion of $U_k(\xi, 1)$ at $\xi = 2$, and also taking into account that $U_k^{(i)}(2, 1) = i! C_{k+i+1}^{2i+1}$.

Due to (18) and (19), from the equality

$$U_k(x, y) - yU_{k-1}(x, y) = \left(U_k(x, y) - \sqrt{y} U_{k-1}(x, y)\right) + \left(1 - \sqrt{y}\right) \sqrt{y} U_{k-1}(x, y),$$

it follows that, for any fixed y on the interval $]0, 1]$, $U_k(x, y) - yU_{k-1}(x, y)$ is an increasing function with respect to x when $x \ge 2\sqrt{y}$. Hence we have

$$U_k\left(2\sqrt{y}, y\right) - yU_{k-1}\left(2\sqrt{y}, y\right) \le U_k - yU_{k-1} \le U_k(1 + y, y) - yU_{k-1}(1 + y, y)\,, \tag{20}$$

where $y > 0$ and $(x, y) \in \Omega_2$.

Substituting $U_k(2\sqrt{y}, y) = (k + 1)\sqrt{y^k}$ and (14) into relation (20), we obtain the estimate

$$\sqrt{y^k}\left((k + 1)\left(1 - \sqrt{y}\right) + \sqrt{y}\right) \le U_k(x, y) - yU_{k-1}(x, y) \le 1, \tag{21}$$

where $y > 0$ and $(x, y) \in \Omega_2$.

We easily obtain the inequality

$$0 \le U_k(x, y) - yU_{k-1}(x, y) \le 1, \tag{22}$$

where $y \le 0$ and $(x, y) \in \Omega_2$.

From estimates (17), (21) and (22) we have

$$|U_k(x, y) - yU_{k-1}(x, y)| \le \sqrt{2}, \quad (x, y) \in \Delta^+. \tag{23}$$

Analogously to (23) we get

$$|U_k(x, y) - U_{k-1}(x, y)| \leq \sqrt{2}, \quad (x, y) \in \Delta^+. \tag{24}$$

It is obvious that from (14) there follows the estimate

$$|U_k(x, y)| \leq k + 1, \tag{25}$$

where $|y| \leq 1$ and $|x| \leq 1 + y$.

We also need the estimate

$$|(x - y - 1)U_k(x, y)| \leq 2, \quad (x, y) \in \Delta^+. \tag{26}$$

Let us prove inequality (26). Consider the cases with $(x, y) \in \Delta^+$ and $y > 0$. On the interval $2\sqrt{y} \leq x \leq y + 1$, $U_k(x, y)$ is an increasing function (for fixed y). This follows from formula (12). Thus we have

$$|(x - y - 1)U_k(x, y)| \leq \left|(2\sqrt{y} - y - 1)U_k(y + 1, y)\right|$$
$$= (1 - \sqrt{y})^2(1 + y + \cdots + y^k) \leq 1. \tag{27}$$

Let us show that (26) holds when $(x, y) \in \Delta^+$, $y > 0$ and $0 \leq x \leq 2\sqrt{y}$. Using (10) and (12) we obtain

$$|U_k(x, y)| \leq 2\sqrt{\frac{y^{k+1}}{4y - x^2}}, \tag{28}$$

where $4y - x^2 > 0$.

By inequalities (14) and (28) we have

$$|(x - y - 1)U_k(x, y)| \leq \left|(2\sqrt{y} - x)U_k(x, y)\right| + \left|(1 - \sqrt{y})^2 U_k(x, y)\right|$$
$$\leq 2\sqrt{\frac{y^{k+1}(2\sqrt{y} - x)}{2\sqrt{y} + x}} + \frac{1 - \sqrt{y}}{1 + \sqrt{y}} \leq 2\sqrt{y} + \frac{1 - \sqrt{y}}{1 + \sqrt{y}} \leq 2\,. \tag{29}$$

From (27) and (29) it follows that inequality (26) holds when $(x, y) \in \Delta^+$ and $y > 0$. Let us now that show (26) holds when $(x, y) \in \Delta^+$ and $y \leq 0$. By the recurrence relation (9) we have that $U_k(x, y)$ is an increasing function when $x \in]0, +\infty[$, for each fixed $y \leq 0$. Therefore if $(x, y) \in \Delta^+$ and $y \leq 0$, then

$$|U_k(x, y)| \leq |U_k(y + 1, y)| \leq 1. \tag{30}$$

It is obvious that (30) gives (26) when $(x, y) \in \Delta^+$ and $y \leq 0$. This proves estimate (26).

2.3 Proof of Theorem 1

Let us return to the proof of Theorem 1.

From (3) we obtain

$$u_{k+1} = L_\tau u_k - u_{k-1} + \tau^2 B_\tau^{-1} f_k, \quad k = 1, \dots, n-1, \tag{31}$$

where

$$B_\tau = I + \frac{\tau^2}{2+\nu} A, \quad L_\tau = (2+\nu) B_\tau^{-1} - \nu I.$$

From the recurrence relation (31) we obtain by induction

$$u_{k+1} = U_k(L_\tau, I) u_1 - U_{k-1}(L_\tau, I) u_0 + \tau^2 \sum_{i=1}^{k} U_{k-i}(L_\tau, I) B_\tau^{-1} f_i. \tag{32}$$

which, after some simple transformations, gives (for simplicity, instead of $U_k(t_\tau, I)$ we write U_k)

$$u_{k+1} = \tau U_k \frac{\Delta u_0}{\tau} + (U_k - U_{k-1}) u_0 + \tau^2 \sum_{i=1}^{k} U_{k-i} B_\tau^{-1} f_i. \tag{33}$$

If to both sides of equality (33) we apply the operator A^s $(0 \le s \le 1)$ and pass over to the norms, then we obtain

$$\|A^s u_{k+1}\| \le \tau \left\| A^s U_k \frac{\Delta u_0}{\tau} \right\| - \|U_k - U_{k-1}\| \cdot \|A^s u_0\| + \tau^2 \sum_{i=1}^{k} \left\| A^s U_{k-i} B_\tau^{-1} f_i \right\|. \tag{34}$$

For the first summand on the right-hand side of this inequality we have

$$\tau \left\| A^s U_k \frac{\Delta u_0}{\tau} \right\| \le \tau \left\| A^{1/2} B_\tau^{-1/2} U_k \right\| \cdot \left\| B_\tau^{1/2} A^{s-1/2} \frac{\Delta u_0}{\tau} \right\|. \tag{35}$$

The following representation is obvious:

$$2I - L_\tau = \tau^2 A B_\tau^{-1}.$$

By the Heinz theorem (see, e.g., [11], p.117) the equality

$$(2I - L_\tau)^s = (\tau^2 A B_\tau^{-1})^s = \tau^{2s} A^s B_\tau^{-s}, \quad 0 \le s \le 1 \tag{36}$$

is valid. Hence we obtain

$$\tau \left\| A^{1/2} B_\tau^{-1/2} U_k \right\| = \left\| (2I - L_\tau)^{1/2} U_k(L_\tau, I) \right\|.$$

As is known, when the argument is a self-adjoint bounded operator, the norm of the operator polynomial is equal to the C-norm of the corresponding scalar polynomial on the spectrum (see, e.g., [10, 21]). By virtue of this result we have

$$\tau \left\| A^{1/2} B_\tau^{-1/2} U_k \right\| = \left\| (2I - L_\tau)^{1/2} U_k(L_\tau, I) \right\| = \max_{x \in \sigma(L_\tau)} \left| (2 - x)^{1/2} U_k(x, 1) \right|. \tag{37}$$

Let us estimate the spectrum of the operator L_τ. For this, it suffices to estimate the spectrum of the operator B_τ^{-1}. Since $(Au, u) \geq \alpha \|u\|^2$, we have

$$(B_\tau u, u) \geq \left(1 + \frac{\tau^2 \alpha}{2 + \nu}\right) \|u\|^2, \quad \forall u \in D(A),$$

thence it follows that

$$0 \leq (B_\tau^{-1} u, u) \leq \left(1 + \frac{\tau^2 \alpha}{2 + \nu}\right)^{-1} (u, u).$$

This relation implies that

$$\sigma(B_\tau^{-1}) \subset \left[0, \left(1 + \frac{\tau^2 \alpha}{2 + \nu}\right)^{-1}\right].$$

From the latter relation and the representation

$$L_\tau = (2 + \nu) B_\tau^{-1} - \nu I$$

we obtain

$$\sigma(L_\tau) \subset [-\nu, \nu_\tau], \tag{38}$$

where

$$\nu_\tau = \frac{4 + 2\nu - \nu \alpha \tau^2}{2 + \nu + \alpha \tau^2}, \quad \nu \in]-2, 2[.$$

Using relation (38) and estimate (10) we obtain

$$\max_{x \in \sigma(L_\tau)} \left| (2 - x)^{1/2} U_k(x, 1) \right| \leq \max_{x \in \sigma(L_\tau)} \left(\sqrt{2 - x} \cdot \frac{2}{\sqrt{4 - x^2}} \right)$$

$$\leq \max_{x \in [-\nu, \nu_\tau]} \frac{2}{\sqrt{2 + x}} = \frac{2}{\sqrt{2 - \nu}}. \tag{39}$$

From (37) and (39) there follows

$$\tau \left\| A^{1/2} B_\tau^{-1/2} U_k(L_\tau, I) \right\| \le \frac{2}{\sqrt{2-\nu}}. \tag{40}$$

(35) and (40) imply

$$\tau \left\| A^s U_k(L_\tau, I) \frac{\Delta u_0}{\tau} \right\| \le c_0 \left\| B_\tau^{1/2} A^{s-1/2} \frac{\Delta u_0}{\tau} \right\|. \tag{41}$$

Taking into account inequality (41) we obtain

$$\begin{aligned} \left\| A^s U_{k-i} B_\tau^{-1} \varphi_i \right\| &\le \left\| A^{1/2} B_\tau^{-1/2} U_{k-i} \right\| \cdot \left\| A^{s-1/2} B_\tau^{-1/2} \varphi_i \right\| \\ &\le c_0 \left\| A^{s-1/2} B_\tau^{-1/2} \varphi \right\|, \quad \forall \varphi \in H. \end{aligned} \tag{42}$$

Let us estimate the norm of the operator polynomial $U_k - U_{k-1}$ in inequality (34). Taking into account estimate (11), we obtain

$$\begin{aligned} \left\| U_k(L_\tau, I) - U_{k-1}(L_\tau, I) \right\| &= \max_{x \in \sigma(L_\tau)} \left| U_k(x, 1) - U_{k-1}(x, 1) \right| \\ &\le \max_{x \in \sigma(L_\tau)} \frac{2}{\sqrt{2+x}} \le \max_{x \in [-\nu, \nu_\tau]} \frac{2}{\sqrt{2+x}} = c_0. \end{aligned} \tag{43}$$

If we now take into account estimates (41), (42) and (43), then from (34) we obtain inequality (6).

Let us prove inequality (7).

It is obvious that the following inequality is valid:

$$\left\| A^s U_k B_\tau^{-1} \varphi \right\| \le \left\| A^{1/2} B_\tau^{-1} U_k \right\| \cdot \| A^{s-1/2} \varphi \|, \quad \forall \varphi \in D(A^{s-1/2}). \tag{44}$$

Taking into account equality (36) and the representation

$$B_\tau^{-1} = (2+\nu)^{-1} (\nu I + L_\tau) \tag{45}$$

we obtain

$$\begin{aligned} \tau A^{1/2} B_\tau^{-1} U_k &= \tau (A^{1/2} B_\tau^{-1/2}) B_\tau^{-1/2} U_k = (\tau A B_\tau^{-1})^{1/2} B_\tau^{-1/2} U_k \\ &= (2+\nu)^{-1/2} (2I - L_\tau)^{1/2} (\nu I + L_\tau)^{1/2} U_k(L_\tau, I). \end{aligned} \tag{46}$$

Since L_τ is a self-adjoint bounded operator, (46) implies the estimate

$$\tau \left\| A^{1/2} B_\tau^{-1} U_k(L_\tau, I) \right\| =$$

$$= (2+\nu)^{-1/2} \max_{x \in \sigma(L_\tau)} \left| (2-x)^{1/2} (\nu + x)^{1/2} U_k(x, 1) \right|,$$

whence, using estimate (10) and relation (38), we obtain

$$\tau \left\| A^{1/2} B_\tau^{-1} U_k(L_\tau, I) \right\|$$

$$\leq \frac{1}{\sqrt{2+\nu}} \max_{x\in[-\nu,\nu_\tau]} \left[(2-x)^{1/2}(\nu+x)^{1/2} \cdot \frac{2}{\sqrt{4-x^2}} \right]$$

$$= \frac{2}{\sqrt{2+\nu}} \max_{x\in[-\nu,\nu_\tau]} \left(\frac{\nu+x}{2+x} \right)^{1/2} = \frac{2}{\sqrt{2+\nu}} \left(\frac{\nu+\nu_\tau}{2+\nu_\tau} \right)^{1/2}$$

$$= \frac{2}{\sqrt{2+\nu}} \frac{2+\nu}{\sqrt{2+\nu+\alpha\tau^2}} \sqrt{\frac{2+\nu+\alpha\tau^2}{4(2+\nu)+(2-\nu)\alpha\tau^2}} \leq 1. \quad (47)$$

From (44) and (47) there follows

$$\left\| A^s U_k B_\tau^{-1} \varphi \right\| \leq \frac{1}{\tau} \| A^{s-1/2} \varphi \|, \quad \forall \varphi \in D(A^{s-1/2}). \quad (48)$$

Using estimates (41), (43) and (48), from (34) we obtain (7).

Let us prove estimate (8).

By the definition of a fractional degree of the operator and Heinz' theorem (see e.g., [11], p.117) we have

$$\| B_\tau^{1/2} R_\tau^{-1} \| = \left\| B_\tau^{1/2} (R_\tau^{-2})^{1/2} \right\| = \left\| (B_\tau (R_\tau^{-2})^{1/2} \right\|, \quad (49)$$

where

$$R_\tau = I + \tau \nu_0 A^{1/2}.$$

Since $R_\tau^2 \geq B_\tau \geq 0$, the following inequality is valid:

$$\left\| B_\tau (R_\tau^2)^{-1} \right\| = \| B_\tau R_\tau^{-2} \| \leq 1.$$

This inequality and (49) imply

$$\left\| B_\tau^{1/2} (R_\tau^{-1}) \right\| \leq 1.$$

By the latter inequality we have

$$\left\| B_\tau^{1/2} A^{s-1/2} \varphi \right\| \leq \| B_\tau^{1/2} R_\tau^{-1} \| \cdot \left\| R_\tau A^{s-1/2} \varphi \right\| \leq \left\| (I + \tau \nu_0 A^{1/2}) A^{s-1/2} \varphi \right\|$$

$$\leq \| A^{s-1/2} \varphi \| + \tau \nu_0 \| A^s \varphi \|, \quad \forall \varphi \in D(A^s). \quad (50)$$

Substituting this inequality into (7), we obtain estimate (8).

Let us prove estimate (9).

We have the estimate (for its proof see Sect. 2.4)

$$\left\|(2I-L_\tau)U_k(L_\tau,I)\varphi\right\| \le \widetilde{c}(s)\tau^{2s}\|A^s\varphi\|,\ \ 0\le s\le \frac{1}{2},\quad \varphi\in D(A^s),$$

by virtue of which we obtain

$$\left\|\tau^2 A^s B_\tau^{-1} U_k(L_\tau,I)\right\| = \left\|(\tau^2 A B_\tau^{-1})A^{-(1-s)}U_k(L_\tau,I)\right\|$$

$$= \left\|(2I-L_\tau)U_k(L_\tau,I)A^{-(1-s)}\right\| \le \widetilde{c}(1-s)\tau^{2(1-s)},\ \ \frac{1}{2}\le s\le 1. \tag{51}$$

Inequalities (41) and (50) clearly imply

$$\tau\left\|A_s U_k(L_\tau,I)\varphi\right\|$$

$$\le c_0\big(\|A^{s-1/2}\varphi\| + \tau\nu_0\|A^s\varphi\|\big),\ \ 0\le s\le 1,\ \ \varphi\in D(A^s). \tag{52}$$

Taking into account inequalities (43), (51) and (52) from (34) we obtain estimate (9).

Let us prove estimate (10).

Taking into account estimate (10) we have

$$\tau\|U_k(L_\tau,I)\| = \tau\max_{x\in\sigma(L_\tau)}|U_k(x,1)| \le \tau\max_{x\in[-\nu,\nu_\tau]}\frac{2}{\sqrt{4-x^2}}$$

$$\le \frac{2\tau}{\sqrt{(2-\nu)(2-\nu_\tau)}} = 2\Big(\frac{2+\nu+\alpha\tau^2}{(4-\nu^2)\alpha}\Big)^{1/2}. \tag{53}$$

Taking into account inequality (47) we have

$$\tau\|U_k B_\tau^{-1}\varphi\| \le \tau\left\|A^{1/2}B_\tau^{-1}U_k\right\|\cdot\|A^{-1/2}\varphi\| \le \|A^{-1/2}\varphi\|,\qquad \forall\varphi\in H. \tag{54}$$

Taking into account estimates (43), (53) and (54), from (33) we obtain (10).

The inequalities obtained from (6) for $s=0$, $s=\frac{1}{2}$ and $s=1$ are proved in [23] for an abstract hyperbolic equation with a self-adjoint positive definite operator.

The method of obtaining a priori estimates for the difference problem investigated in this paper is considered in [23]. We call it the method of associated polynomials. In our opinion, this name is natural because the investigation of the stability of a many-layer scheme and the derivation of a priori estimates lead to the study of the properties of a certain class of polynomials obtained by this scheme.

2.4 Estimates for Chebyshev's Operator Polynomials

In the preceding subsection we have derived a priori estimates for the solution obtained by means of the semi-discrete scheme (3). To obtain complete information on a dynamic problem, it is necessary to know how the velocity (and, which is better, also the acceleration) changes. The next step is to obtain a priori estimates for difference analogues of first and second order derivatives. For this, we need to estimate the operator polynomials $U_k(L_\tau, I)$ (some estimates of $U_k(L_\tau, I)$ have already been obtained in the preceding subsection). It is obvious that the operator polynomials $U_k(L_\tau, I)$ satisfy the recurrence relation

$$U_{k+1}(L_\tau, I) = L_\tau U_k(L_\tau, I) - U_{k-1}(L_\tau, I), \quad k = 1, 2, \dots,$$

$$U_0(L_\tau, I) = I, \quad U_1(L_\tau, I) = L_\tau. \tag{55}$$

In our opinion, it is quite natural to call the operators $U_k(L_\tau, I)$ Chebyshev operator polynomials because the scalar polynomials $U_k(2x, 1)$ are classical Chebyshev polynomials of second kind.

Lemma 1 *Let* $\nu \in]-2, 2[$. *Then the following estimates are valid:*

$$\left\|(2I - L_\tau)U_k(L_\tau, I)\varphi\right\| \le \alpha^{1/2-s}\tau\|A^s\varphi\|, \quad \frac{1}{2} \le s \le 1, \quad \varphi \in D(A^s), \tag{56}$$

$$\left\|(2I - L_\tau)U_k(L_\tau, I)\varphi\right\| \le \widetilde{c}(s)\tau^{2s}\|A^s\varphi\|, \quad 0 \le s \le \frac{1}{2}, \quad \varphi \in D(A^s), \tag{57}$$

$$\left\|(2I - L_\tau)U_k(L_\tau, I)B_\tau^{-1}\right\| \le 1, \tag{58}$$

$$\tau^{2s}\left\|\big(U_k(L_\tau, I) - U_{k-1}(L_\tau, I)\big)A^s B_\tau^{-1}\right\| \le (2+\nu)^s, \quad 0 < s \le \frac{1}{2}, \tag{59}$$

$$\left\|\big(U_k(L_\tau, I) - U_{k-1}(L_\tau, I)\big)B_\tau^{-1}\right\| \le 1, \tag{60}$$

where

$$\widetilde{c}(s) = 2^{1-2s}\left(\frac{2+\nu}{2-\nu}\right)^{1/2-s}.$$

Proof The proof of the above estimates (like the proof of analogous estimates in the preceding subsection) rests on the properties of the scalar polynomial corresponding to an operator polynomial and also on the fact that when the argument is a self-adjoint bounded operator, the norm of the operator function is equal to the C-norm of the corresponding scalar function on the spectrum.

Let us prove estimate (56). By formulas (36) and (25) we have

$$(2I - L_\tau)U_k\varphi = \tau^2 A B_\tau^{-1} U_k\varphi = \left(\tau^2 A^{1-s} B_\tau^{-(1-s)}\right)\left(B_\tau^{-s} U_k A^s \varphi\right)$$

$$= \tau^{2s}(\tau^2 A B_\tau^{-1})^{1-s} B_\tau^{-s} U_k A^s \varphi$$

$$= \tau^{2s}(2+\nu)^{-s}(2I - L_\tau)^{1-s}(\nu I + L_\tau)^s U_k A^s \varphi, \quad \forall \varphi \in D(A^s). \tag{61}$$

Analogously to the estimates obtained for operator functions in Sect. 2.3, we obtain

$$\left\|(2I - L_\tau)^{1-s}(\nu I + L_\tau)^s U_k\right\|$$

$$= \max_{x \in \sigma(L_\tau)} \left|(2-x)^{1-s}(\nu + x)^s U_k(x, 1)\right|$$

$$\leq \max_{x \in [-\nu, \nu_\tau]} \left[(2-x)^{1-s}(\nu + x)^s \frac{2}{\sqrt{4 - x^2}}\right]$$

$$= 2 \max_{x \in [-\nu, \nu_\tau]} \psi_s(x, \nu), \tag{62}$$

where

$$\psi_s(x, \nu) = (\nu + x)^s (2-x)^{1/2-s}(2+x)^{-1/2}.$$

Let us show that the function $\psi_s(x, \nu)$ is increasing when $\frac{1}{2} \leq s \leq 1$ and $\nu \in]-2, 2[$. Indeed, we have

$$\psi_s'(x, \nu) = s(\nu + x)^{s-1}(2-x)^{1/2-s}(2+x)^{-1/2}$$

$$-\left(\frac{1}{2} - s\right)(\nu + x)^s (2-x)^{-(1/2+s)}(2+x)^{-1/2}$$

$$-\frac{1}{2}(\nu + x)^s (2-x)^{1/2-s}(2+x)^{-3/2}$$

$$= (\nu + x)^{s-1}(2-x)^{-(1/2+s)}(2+x)^{-3/2}\Big[s(2-x)(2+x)$$

$$-\left(\frac{1}{2} - s\right)(\nu + x)(2+x) - \frac{1}{2}(\nu + x)(2-x)\Big]. \tag{63}$$

Let us simplify the expression enclosed in the square brackets in the latter formula:

$$s(2-x)(2+x) - \left(\frac{1}{2} - s\right)(\nu + x)(2+x) - \frac{1}{2}(\nu + x)(2-x)$$

$$= \left(s - \frac{1}{2}\right)[(2-x)(2+x) + (\nu + x)(2+x)]$$

$$+\frac{1}{2}\big[(2-x)(2+x)-(\nu+x)(2-x)\big]$$

$$=\Big(s-\frac{1}{2}\Big)(2+x)(2+\nu)+\frac{1}{2}(2-x)(2-\nu)>0. \tag{64}$$

From (63) and (64) it follows that the function $\psi_s(x,\nu)$ is increasing. Therefore

$$\max_{x\in[-\nu,\nu_\tau]}\psi_s(x,\nu)=\psi_s(\nu_\tau,\nu)$$

$$=(\nu+\nu_\tau)^s(2-\nu_\tau)^{1/2-s}(2+\nu_\tau)^{-1/2}$$

$$=\frac{(2+\nu)^{2s}}{(2+\nu+\alpha\tau^2)^s}\cdot\Big(\frac{(2+\nu)\alpha\tau^2}{(2+\nu+\alpha\tau^2)}\Big)^{1/2-s}$$

$$\times\Big(\frac{4(2+\nu)+(2-\nu)\alpha\tau^2}{2+\nu+\alpha\tau^2}\Big)^{-1/2}$$

$$=(\alpha\tau^2)^{1/2-s}\cdot\frac{(2+\nu)^{2+1/2}}{\sqrt{4(2+\nu)+(2-\nu)\alpha\tau^2}}$$

$$\le\frac{1}{2}(\alpha\tau^2)^{1/2-s}(2+\nu)^s. \tag{65}$$

From (62) and (65) we obtain estimate (56).

Let us prove estimate (57). For this, it suffices to estimate $\psi_s(x,\nu)$ on the interval $[-\nu,2]\supset[-\nu,\nu_\tau]$ when $0<s<\frac{1}{2}$. We have to find the critical points of the function $\psi_s(x,\nu)$ on this interval. After performing some transformations of (64), from (63) we obtain

$$(2s-1)(2+x)(2+\nu)+(2-x)(2-\nu)=0,$$

from which we have a solution

$$x_0=\frac{2[(2-\nu)-(1-2s)(2+\nu)]}{(2-\nu)+(1-2s)(2+\nu)}<2.$$

Clearly, the following estimate is valid:

$$\max_{x\in[-\nu,2]}\psi_s(x,\nu)=\psi_s(x_0,\nu),\quad 0<s<\frac{1}{2}. \tag{66}$$

Let us estimate $\psi_s(x_0,\nu)$. It is obvious that

$$\psi_s(x_0,\nu)=\Big(\frac{2s(4-\nu)}{\lambda}\Big)^s\Big(\frac{4(1-2s)(2+\nu)}{\lambda}\Big)^{1/2-s}\Big(\frac{4(2-\nu)}{\lambda}\Big)^{-1/2}$$

$$< (4-\nu^2)^s(4(2+\nu))^{1/2-s}(4(2-\nu))^{-1/2} = \frac{(2+\nu)^s}{2^{2s}}\left(\frac{2+\nu}{2-\nu}\right)^{1/2-s}, \tag{67}$$

where

$$\lambda = (2-\nu)+(1-2s)(2+\nu).$$

(61), (62), (64) and (67) imply estimate (57).

Let us prove estimate (58). By analogy with (62) we have

$$\left\|(2I-L_\tau)U_k(L_\tau, I)B_\tau^{-1}\right\|$$

$$= \frac{1}{2+\nu}\left\|(2I-L_\tau)(\nu I+L_\tau)U_k(L_\tau, I)\right\|$$

$$= \frac{1}{2+\nu}\max_{x\in\sigma(L_\tau)}\left|(2-x)(\nu+x)U_k(x,1)\right|$$

$$\le \frac{1}{2+\nu}\max_{x\in[-\nu,\nu_\tau]}\left[(2-x)(\nu+x)\frac{2}{\sqrt{4-x^2}}\right]$$

$$\le \frac{2}{2+\nu}\max_{x\in[-\nu,\nu_\tau]}\frac{\nu+x}{\sqrt{2+x}} = \frac{2}{2+\nu}\cdot\frac{\nu+\nu_\tau}{\sqrt{2+\nu_\tau}}$$

$$= \frac{2}{2+\nu}\cdot\frac{(2+\nu)^2}{2+\nu+\alpha\tau^2}\cdot\sqrt{\frac{2+\nu+\alpha\tau^2}{4(2+\nu)+(2-\nu)\alpha\tau^2}} \le 1. \tag{68}$$

Let us prove estimate (59). By analogy with (61) we have

$$\tau^{2s}(U_k-U_{k-1})A^sB_\tau^{-1} = (\tau^{2s}A^sB^{-s})B^{-(1-s)}(U_k-U_{k-1}) =$$

$$= (\tau^2AB^{-1})^sB_\tau^{-(1-s)}(U_k-U_{k-1}) =$$

$$= (2+\nu)^{-(1-s)}(2I-L_\tau)^s(\nu I+L_\tau)^{1-s}(U_k-U_{k-1}).$$

Hence, in view of estimate (11), we obtain

$$\tau^{2s}\left\|(U_k-U_{k-1})A^sB_\tau^{-1}\right\|$$

$$= (2+\nu)^{-(1-s)}\left\|(2I-L_\tau)^s(\nu I+L_\tau)^{1-s}(U_k-U_{k-1})\right\|$$

$$= (2+\nu)^{-(1-s)}\max_{x\in\sigma(L_\tau)}\left|(2-x)^s(\nu+x)^{1-s}\big(U_k(x,1)-U_{k-1}(x,1)\big)\right|$$

$$\le (2+\nu)^{-(1-s)}\max_{x\in[-\nu,\nu_\tau]}\left[(2-x)^s(\nu+x)^{1-s}\frac{2}{\sqrt{2+x}}\right]$$

$$= 2(2+\nu)^{-(1-s)} \max_{x\in[-\nu,\nu_\tau]} \widetilde{\varphi}_s(x,\nu), \tag{69}$$

where

$$\widetilde{\varphi}_s(x,\nu) = (\nu+x)^{1-s}(2-x)^s(2+x)^{-1/2}, \quad 0 \le s \le \frac{1}{2}.$$

The function $\widetilde{\varphi}_s(x,\nu)$ is estimated in a standard manner. We define its derivative. Clearly, we have

$$\widetilde{\varphi}_s'(x,\nu) = (\nu+x)^{-s}(2-x)^{s-1}(2+x)^{-3/2}$$

$$\times\Big[(1-s)(2-s)(2+s) - s(\nu+x)(2+x) - \frac{1}{2}(\nu+x)(2-x)\Big].$$

The expression enclosed in the square brackets is denoted by

$$\varphi_s(x,\nu) = (1-s)(2-x)(2+x) - s(\nu+x)(2+x) - \frac{1}{2}(\nu+x)(2-x).$$

Since $0 < s \le \frac{1}{2}$ and $\nu \in]-2,2[$, we have

$$\varphi_s(2,\nu) = -4s(2+\nu) < 0,$$

$$\varphi_s(-\nu,\nu) = (1-s)(4-\nu^2) > 0.$$

Thus $\widetilde{\varphi}_s(x,\nu)$ has a unique critical point on the interval $]-\nu,2[$. Let us assume that the square function $\varphi_s(x,\nu)$ is equal to zero and find its roots. By some simple transformations we obtain ($t = 2s$)

$$x^2 + \big[(1-t)(2-\nu)+4t\big]x - 2\big[(1+t)(2-\nu)+2(1-2t)\big] = 0.$$

Then we have

$$x = -\frac{1}{2}\Big[\big((1-t)(2-\nu)+4t\big) \pm \sqrt{(1-t)^2(2+\nu)^2+16(2-\nu)}\Big].$$

The root contained in the interval $]-\nu,2[$ is

$$x_1 = -\frac{1}{2}\Big[-\big((1-t)(2-\nu)+4t\big)+\sqrt{(1-t)^2(2+\nu)^2+16(2-\nu)}\Big].$$

We easily obtain

$$2 + x_1 = \frac{1}{2}\Big[(1-t)(2+\nu) + \sqrt{(1-t)^2(2+\nu)^2+16(2-\nu)}\Big], \tag{70}$$

$$\nu + x_1 = \frac{1}{2}\Big[(1-t)(2+\nu)-2(2-\nu)+\sqrt{(1-t)^2(2+\nu)^2+16(2-\nu)}\,\Big], \tag{71}$$

$$2 - x_1 = \frac{1}{2}\Big[(1+t)(2+\nu)+2(2-\nu)-\sqrt{(1-t)^2(2+\nu)^2+16(2-\nu)}\,\Big]. \tag{72}$$

(70) and (71) imply

$$\frac{\nu + x_1}{2 + x_1} = \frac{(1-t)(2+\nu) - 2(2-\nu) + \sqrt{(1-t)^2(2+\nu)^2 + 16(2-\nu)}}{(1-t)(2+\nu) + \sqrt{(1-t)^2(2+\nu)^2 + 16(2-\nu)}}$$

$$= 1 - \frac{2(2-\nu)}{(1+t)(2+\nu) + \sqrt{(1-t)^2(2+\nu)^2 + 16(2-\nu)}}$$

$$= 1 + \frac{2(2-\nu)\big[(1-t)(2+\nu) - \sqrt{(1-t)^2(2+\nu)^2 + 16(2-\nu)}\,\big]}{16(2-\nu)}$$

$$= \frac{1}{8}\Big[4(2-\sqrt{2-\nu}) + \big((a+b) - \sqrt{a^2+b^2}\big)\Big], \tag{73}$$

where

$$a = (1-t)(2+\nu), \quad b = 4\sqrt{2-\nu}.$$

Let us estimate the expression enclosed in the square brackets. By some simple transformations we have

$$4(2-\sqrt{2-\nu}) + \big(a + b - \sqrt{a^2+b^2}\big)$$

$$= \frac{4(2+\nu)}{2+\sqrt{2-\nu}} + \frac{8(1-t)(2+\nu)\sqrt{2-\nu}}{a+b+\sqrt{a^2+b^2}}$$

$$= 2(2+\nu)\Big[\frac{2}{2+\sqrt{2-\nu}} + \frac{4(1-t)\sqrt{2-\nu}}{a+b+\sqrt{a^2+b^2}}\Big]. \tag{74}$$

Since the second summand enclosed in the square brackets is decreasing for $t \in [0, 1]$, we have

$$\frac{4(1-t)\sqrt{2-\nu}}{a+b+\sqrt{a^2+b^2}}$$

$$\leq \frac{4\sqrt{2-\nu}}{(2+\nu) + 4\sqrt{2-\nu} + \sqrt{(2+\nu)^2 + 16(2-\nu)}}$$

$$= \frac{4\sqrt{2-\nu}}{(2+\nu) + 4\sqrt{2-\nu} + (6-\nu)} = \frac{\sqrt{2-\nu}}{2+\sqrt{2+\nu}}. \tag{75}$$

(73), (74) and (75) imply

$$\frac{\nu + x_1}{2 + x_1} \leq \frac{1}{4}\,(2 + \nu). \tag{76}$$

from which it follows that

$$\nu + x_1 \leq \frac{1}{4}\,(2 + \nu)(2 + x_1) \leq \frac{1}{4}\,(2 + \nu) \cdot 4 = 2 + \nu. \tag{77}$$

From (72), clearly,

$$2 - x_1 \leq \frac{1}{2}\left[2(2 + \nu) + 2(2 - \nu) - 4\sqrt{2 - \nu}\right]$$

$$= 2\big(2 - \sqrt{2 - \nu}\big) = \frac{2(2 + \nu)}{2 + \sqrt{2 - \nu}} \leq 2 + \nu. \tag{78}$$

Taking into account estimates (76)–(78), we obtain

$$\max_{x \in [-\nu, 2]} \widetilde{\varphi}_s(x, \nu) = \widetilde{\varphi}_s(x_1, \nu) = (\nu + x_1)^{1-s}(2 - x_1)^s (2 + x_1)^{1/2}$$

$$= (\nu + x_1)^{1/2-s}(2 - x_1)^s \left(\frac{\nu + x_1}{2 + x_1}\right)^{1/2}$$

$$\leq (2 + \nu)^{1/2-s}(2 + \nu)^s \cdot \frac{1}{2}\,(2 + \nu)^{1/2} = \frac{1}{2}\,(2 + \nu). \tag{79}$$

From (69) and (78) we have

$$\tau^{2s}\left\|(U_k - U_{k-1})A^s B_\tau^{-1}\right\| \leq (2 + \nu)^s, \quad 0 < s \leq \frac{1}{2}.$$

Let us estimate (60). Clearly, the following representation is true:

$$(U_k - U_{k-1})B_\tau^{-1} = (2 + \nu)^{-1}(\nu I + L)(U_k - U_{k-1}).$$

Analogously to (68), the latter equality implies with estimate (11), taken into account that

$$\left\|(U_k - U_{k-1})B_\tau^{-1}\right\| = (2 + \nu)^{-1}\left\|(\nu I + L)(U_k - U_{k-1})\right\| \leq$$

$$\leq (2 + \nu)^{-1} \max_{x \in [-\nu, \nu_\tau]} \left|(\nu + x) \cdot \frac{2}{\sqrt{2 + x}}\right| = \frac{2}{2 + \nu} \cdot \frac{\nu + \nu_\tau}{\sqrt{2 + \nu_\tau}} \leq 1.$$

2.5 A Priori Estimates for Difference Analogues of First and Second Order Derivatives

Theorem 2 *Let the conditions of Theorem 1 be fulfilled. Then for scheme (3) the following estimates are true:*

$$\left\|\frac{\Delta u_k}{\tau}\right\| \le \widetilde{c}(s)\tau^{2s-1}\|A^s u_0\| + c_0\left\|\frac{\Delta u_0}{\tau}\right\| + \tau\sum_{i=1}^{k}\|f_i\|, \tag{80}$$

$$\left\|A^s\frac{\Delta u_k}{\tau}\right\| \le \left\|A^{s+1/2}u_0\right\| + c_0\left\|A^s\frac{\Delta u_0}{\tau}\right\| + \widetilde{c}_0(s)\tau^{1-2s}\sum_{i=1}^{k}\|f_i\|, \tag{81}$$

$$\left\|A^s\frac{\Delta u_k}{\tau}\right\| \le \left\|A^{s+1/2}u_0\right\| + c_0\left\|A^s\frac{\Delta u_0}{\tau}\right\| + \tau\sum_{i=1}^{k}\|A^s f_i\|, \tag{82}$$

$$\left\|A^s\frac{\Delta u_k}{\tau}\right\| \le \left\|A^{s+1/2}u_0\right\| + c_0\left\|A^s\frac{\Delta u_0}{\tau}\right\| + \tau\sum_{i=1}^{k}\left\|A^{s-1/2}\frac{\Delta f_{i-1}}{\tau}\right\|, \tag{83}$$

where $k = 1, \ldots, n-1$, $\Delta u_k = u_{k+1} - u_k$, $0 \le s \le \frac{1}{2}$,

$$\widetilde{c}_0(s) = \frac{1}{(2+\nu)^s}, \quad \widetilde{c}(s) = 2^{1-2s}\left(\frac{2+\nu}{2-\nu}\right)^{1/2-s}, \quad f_0 = 0.$$

Proof Let us rewrite formula (32) as

$$u_{k+1} = (U_k - U_{k-1})u_0 + U_k(u_1 - u_0) + \tau^2\sum_{i=1}^{k}U_{k-i}B_\tau^{-1}f_i. \tag{84}$$

By virtue of the latter formula it is obvious that

$$u_k = (U_{k-1} - U_{k-2})u_0 + U_{k-1}(u_1 - u_0) + \tau^2\sum_{i=1}^{k}U_{k-i-1}B_\tau^{-1}f_i. \tag{85}$$

Subtracting (85) from equality (84) and assuming that $U_{-1} = 0$, we obtain

$$\Delta u_k = \big(U_k + U_{k-2} - 2U_{k-1}\big)u_0 + (U_k - U_{k-1})(u_1 - u_0)$$
$$+ \tau^2\sum_{i=1}^{k}(U_{k-i} - U_{k-i-1})B_\tau^{-1}f_i. \tag{86}$$

The recurrence relation (55) implies

$$U_k(L_\tau, I) + U_{k-2}(L_\tau, I) - 2U_{k-1}(L_\tau, I)$$

$$= L_\tau U_{k-1} - 2U_{k-1} = (L_\tau - 2I)U_{k-1}. \tag{87}$$

Using (87) in (86) and dividing both sides of the equality by τ we obtain

$$\frac{\Delta u_k}{\tau} = \tau^{-1}(L_\tau - 2I)U_{k-1}(L_\tau, I)u_0 + (U_k - U_{k-1})\frac{\Delta u_0}{\tau}$$

$$+ \tau \sum_{i=1}^{k} (U_{k-i} - U_{k-i-1}) B_\tau^{-1} f_i, \tag{88}$$

from which it obviously follows that

$$\left\| \frac{\Delta u_k}{\tau} \right\| \le \tau^{-1} \left\| (2I - L_\tau)U_{k-1}u_0 \right\| + \|U_k - U_{k-1}\| \cdot \left\| \frac{\Delta u_0}{\tau} \right\| +$$

$$\tau \sum_{i=1}^{k} \left\| (U_{k-i} - U_{k-i-1}) B_\tau^{-1} f_i \right\|.$$

Taking into account estimates (57), (60) and (43), the above formula implies inequality (80).

Let us prove inequality (81). If to both sides of equality (88) we apply the operator A^s $(0 \le s \le \frac{1}{2})$ and pass over to the norms, then we have

$$\left\| A^s \frac{\Delta u_k}{\tau} \right\| \le \tau^{-1} \left\| (2I - L_\tau)U_{k-1}(L_\tau, I)(A^s u_0) \right\|$$

$$+ \|U_k - U_{k-1}\| \cdot \left\| A^s \frac{\Delta u_0}{\tau} \right\| + \tau \sum_{i=1}^{k} \left\| (U_{k-i} - U_{k-i-1}) A^s B_\tau^{-1} \right\| \cdot \|f_i\| \tag{89}$$

which, in view of estimates (43), (56) and (59), implies inequality (81).

Let us prove inequality (82). If the expression (88) under the summation sign is replaced by $\|(U_{k-i} - U_{k-i-1})B_\tau^{-1}\| \cdot \|A^s f_i\|$, then the resulting inequality will obviously be valid. The latter inequality, in view of estimates (43), (56) and (60), implies inequality (82).

Let us prove inequality (83). Subtracting (85) from equality (84) and taking into account (87), we obtain

$$\Delta u_k = (L_\tau - 2I)U_{k-1}u_0 + (U_k - U_{k-1})\Delta u_0 + \tau^2 \sum_{i=1}^{k} U_{k-1} B_\tau^{-1} \Delta f_{i-1}, \tag{90}$$

where $f_0 = 0$.

If we divide both sides of equality (90) by τ, apply the operator A^s $(0 \le s \le \frac{1}{2})$ and pass over to the norms, then we have

$$\left\| A^s \frac{\Delta u_k}{\tau} \right\| \le \tau^{-1} \left\| (2I - L_\tau) U_{k-1}(L_\tau, I) A^s u_0 \right\|$$

$$+ \|U_k - U_{k-1}\| \cdot \left\| A^s \frac{\Delta u_0}{\tau} \right\| + \tau^2 \sum_{i=1}^{k} \left\| U_{k-i} A^{1/2} B_\tau^{-1} \right\| \cdot \left\| A^{s-1/2} \frac{\Delta f_{i-1}}{\tau} \right\|$$

which, in view of estimates (43), (47) and (56), implies inequality (83).

Now, we will derive estimates which take place for the corresponding difference analogue of a second order derivative.

Theorem 3 *Let the conditions of Theorem 1 be fulfilled. Then for scheme (3) the following estimates are true:*

$$\left\| \frac{\Delta^2 u_k}{\tau^2} \right\| \le \tau^{2s-1} \left[\left\| A^{1/2+s} u_0 \right\| + \widetilde{c}(s) \left\| A^s \frac{\Delta u_0}{\tau} \right\| \right] + \tau \sum_{i=2}^{k+1} \left\| \frac{\Delta f_{i-1}}{\tau} \right\| + \|f_1\|, \quad (91)$$

$$\left\| \frac{\Delta^2 u_k}{\tau^2} \right\| \le \tau^{2s-1} \left[\left\| A^{1/2+s} u_0 \right\| + \widetilde{c}(s) \left\| A^s \frac{\Delta u_0}{\tau} \right\| \right]$$

$$+ \widetilde{c}(s) \tau^{2s} \sum_{i=1}^{k} \left\| A^s B_\tau^{-1} f_i \right\| + \|B_\tau^{-1} f_{k+1}\|, \quad (92)$$

$$\left\| \frac{\Delta^2 u_k}{\tau^2} \right\| \le \|A u_0\| + \left\| A^{1/2} \frac{\Delta u_0}{\tau} \right\| + \sum_{i=1}^{k+1} \|f_i\|, \quad (93)$$

where $k = 1, \ldots, n-2$, $0 \le s \le \frac{1}{2}$.

Proof Taking into account (87), from (86) we have

$$\Delta u_k = (L_\tau - 2I) U_{k-1}(L_\tau, I) u_0 + (U_k - U_{k-1}) \Delta u_0$$

$$+ \tau^2 \sum_{i=1}^{k} (U_{k-i} - U_{k-i-1}) B_\tau^{-1} f_i. \quad (94)$$

If k in (94) is replaced by $k+1$, then we obtain

$$\Delta u_{k+1} = (L_\tau - 2I) U_k u_0 + (U_{k+1} - U_k) \Delta u_0$$

$$+\tau^2 \sum_{i=1}^{k+1} (U_{k-i+1} - U_{k-i}) B_\tau^{-1} f_i. \tag{95}$$

With (87) taken into account, (94) and (95) imply the formulas:

$$\Delta^2 u_k = (L_\tau - 2I)\big[(U_k - U_{k-1})u_0 + U_k \Delta u_0\big]$$

$$+\tau^2 \sum_{i=1}^{k+1} (U_{k-i+1} - U_{k-i}) B_\tau^{-1} (f_i - f_{i-1}), \tag{96}$$

$$\Delta^2 u_k = (L_\tau - 2I)\Big[(U_k - U_{k-1})u_0 + U_k \Delta u_0 + \tau^2 \sum_{i=1}^{k} U_{k-i} B_\tau^{-1} f_i\Big] + \tau^2 B_\tau^{-1} f_{k+1}, \tag{97}$$

where $k = 1, \ldots, n-2$, $f_0 = 0$, $U_{-1} = 0$, $\Delta u_k = u_{k+1} - u_k$, $\Delta^2 u_k = \Delta(\Delta u_k)$.

It is obvious that the following representation is true:

$$(L_\tau - 2I)(U_k - U_{k-1})u_0 = -\tau^2 A B_\tau^{-1} (U_k - U_{k-1})u_0 =$$

$$= -\tau^{1+2s}\big[\tau^{1-2s}(U_k - U_{k-1}) A^{1/2-s} B_\tau^{-1}\big](A^{1/2+s} u_0).$$

By (59) the latter formula yields the estimate

$$\big\|(L_\tau - 2I)(U_k - U_{k-1})u_0\big\|$$

$$\le \tau^{1+2s} \big\|A^{1/2+s} u_0\big\|, \quad u_0 \in D(A^{1/2+s}), \quad 0 \le s \le \frac{1}{2}. \tag{98}$$

Dividing both sides of equality (96) by τ^2 and passing over to the norms, we obtain

$$\left\|\frac{\Delta^2 u_k}{\tau^2}\right\| \le\le \big\|(L_\tau - 2I)(U_k - U_{k-1})u_0\big\| + \tau^{-1} \left\|(L_\tau - 2I) U_k \frac{\Delta u_0}{\tau}\right\|$$

$$+\tau \sum_{i=1}^{k+1} \big\|(U_{k-i+1} - U_{k-i}) B_\tau^{-1}\big\| \cdot \left\|\frac{\Delta f_{i-1}}{\tau}\right\|.$$

Hence, taking into account estimates (57), (60), and (98), we obtain estimate (91).

Analogously, taking into account estimates (98) and (57), from (97) we obtain estimate (91).

Estimate (93) follows from (97) with equalities (98), (56), (58) and $\|B_\tau^{-1}\| \le 1$ taken into account.

2.6 Theorems on the Convergence of a Semi-Discrete Scheme

Let us introduce the following spaces. If we define the Hermite norm $\|u\|_1 = \|A^{1/2}u\|$ in $D(A^{1/2})$, then we obtain the Hilbert space which is denoted by W^1. Analogously, by defining the Hermite norm $\|u\|_2 = \|Au\|$ in $D(A)$ we obtain the Hilbert space which is denoted by W^2. We denote by $C([0,T];H)$ the set of vector functions $u(t)$ continuous on $[0,T]$ and taking their values from H. We denote by $C^m([0,T];H)$ $(m \geq 1)$ the set of differentiable vector functions, continuous on $[0,T]$ up to order m from $C([0,T];H)$. $C([0,T];W^i)$ and $C^m([0,T];W^i)$, $i = 1, 2$, are defined analogously.

Depending on the smoothness of a solution of a continuous problem, we will next establish the order of convergence in τ for an approximate solution obtained by the semidiscrete scheme (3).

Rewrite Eq. (1) at the point $t = t_k$ as follows:

$$\frac{\Delta^2 u(t_{k-1})}{\tau^2} + A\frac{u(t_{k+1}) + \nu u(t_k) + u(t_{k-1})}{2+\nu}$$

$$= f(t_k) + \Big(\frac{\Delta^2 u(t_{k-1})}{\tau^2} - u''(t_k)\Big) + (2+\nu)^{-1} A\big(\Delta^2 u(t_{k-1})\big). \tag{99}$$

Equation (1) clearly implies

$$A\big(\Delta^2 u(t_{k-1})\big) = \Delta^2 f(t_{k-1}) - \Delta^2 u''(t_{k-1}). \tag{100}$$

If we subtract equality (99) from (3) and take into account (100), then for the error $z_k = u(t_k) - u_k$ we obtain the equation

$$\frac{z_{k+1} - 2z_k + z_{k+1}}{\tau^2} + A\,\frac{z_{k+1} + \nu z_k + z_{k-1}}{2+\nu} = r_\tau(t_k), \tag{101}$$

where $k = 1, \ldots, n-1$,

$$r_\tau(t_k) = r_{0,\tau}(t_k) + (2+\nu)^{-1}\big(r_{2,\tau}(t_k) - r_{1,\tau}(t_k)\big),$$

$$r_{0,\tau}(t) = \frac{\Delta^2 u(t-\tau)}{\tau^2} - u''(t),$$

$$r_{1,\tau}(t) = \Delta^2 u''(t-\tau), \quad r_{2,\tau}(t) = \Delta^2 f(t-\tau),$$

$$\Delta^2 u(t-\tau) = \Delta\big(\Delta u(t-\tau)\big), \quad t,\ t-\tau,\ t+\tau \in [0,T].$$

We have the following statement.

Theorem 4 *Let problem (1), (2) have a solution* $u(t) \in C^2([0,T];H) \cap C([0,T];W^2)$, $f(t) \in C([0,T];H)$, $u_0, u_1 \in W^2$ *and* $\nu \in]-2,2[$. *Then for the errors* $z_k = u(t_k) - u_k$ *the following estimates are valid:*

$$\|z_{k+1}\| \le c_0\|z_0\| + c_1(\tau)\left\|\frac{\Delta z_0}{\tau}\right\| + \tau_0 \sum_{i=1}^{k} \left\|A^{-1/2} r_\tau(t_i)\right\|, \tag{102}$$

$$\|A^{1/2} z_{k+1}\| \le c_0\left(\|A^{1/2} z_0\| + \left\|\frac{\Delta z_0}{\tau}\right\| + \nu_0 \left\|A^{1/2}(\Delta z_0)\right\|\right) + \tau \sum_{i=1}^{k} \|r_\tau(t_i)\|, \tag{103}$$

$$\|A z_{k+1}\| \le c_0\left(\|A z_0\| + \left\|A^{1/2}\frac{\Delta z_0}{\tau}\right\| + \nu_0 \left\|A(\Delta z_0)\right\|\right) + c_2 \sum_{i=1}^{k} \|r_\tau(t_i)\|, \tag{104}$$

$$\left\|\frac{\Delta z_k}{\tau}\right\| \le c_2 \tau^{-1}\|z_0\| + c_0\left\|\frac{\Delta z_0}{\tau}\right\| + \tau \sum_{i=1}^{k} \|r_\tau(t_i)\|, \tag{105}$$

$$\left\|\frac{\Delta z_k}{\tau}\right\| \le \left\|A^{1/2} z_0\right\| + c_0\left\|\frac{\Delta z_0}{\tau}\right\| + \tau \sum_{i=1}^{k} \|r_\tau(t_i)\|, \tag{106}$$

$$\left\|A^{1/2}\frac{\Delta z_k}{\tau}\right\| \le \|A z_0\| + c_0\left\|A^{1/2}\frac{\Delta z_0}{\tau}\right\| + \nu_0 \sum_{i=1}^{k} \|r_\tau(t_i)\|, \tag{107}$$

$$\begin{aligned} \left\|A^{1/2}\frac{\Delta z_k}{\tau}\right\| &\le \|A z_0\| + c_0\left\|A^{1/2}\frac{\Delta z_0}{\tau}\right\| \\ &+ \left(\|r_\tau(t_1)\| + \sum_{i=2}^{k} \left\|r_\tau(t_i) - r_\tau(t_{i-1})\right\|\right), \end{aligned} \tag{108}$$

$$\left\|\frac{\Delta^2 z_k}{\tau^2}\right\| \le \|A z_0\| + \left\|A^{1/2}\frac{\Delta z_0}{\tau}\right\| + \sum_{i=1}^{k+1} \|r_\tau(t_i)\|, \tag{109}$$

$$\begin{aligned} \left\|\frac{\Delta^2 z_k}{\tau^2}\right\| &\le \|A z_0\| + \left\|A^{1/2}\frac{\Delta z_0}{\tau}\right\| \\ &+ \sum_{i=2}^{k+1} \left\|r_\tau(t_i) - r_\tau(t_{i-1})\right\| + \|r_\tau(t_1)\|, \end{aligned} \tag{110}$$

where $k = 1, \dots, n-1$ *(in (109) and (110)* $k = 1, \dots, n-2$*);* c_0*,* c_1 *and* ν_0 *are the same constants as in the preceding section,*

$$c_2 = 2\Big(\frac{2+\nu}{2-\nu}\Big)^{1/2}.$$

It is obvious that the a priori estimates obtained in Sects. 2.1 and 2.5 hold automatically and for system (101) as well if u_k is replaced by z_k, and f_i by $r_\tau(t_i)$).

Now, if we insert $s = 1/2$ into estimates (10), (8), (9), (80)–(83), (93) and (91), we will respectively obtain estimates (102)–(104), (106)–(110). The substitution of $s = 0$ into the estimate corresponding to (91) gives (105).

Theorem 4 is a basis for proving theorems on the convergence of an approximate solution obtained by means of the semi-discrete scheme (3).

The following theorem is true (in the sequel c denotes a positive constant).

Theorem 5 *Let* $u_0 = \varphi_0$, $u_1 = \varphi_0 + \tau\varphi_1$, $\varphi_0, \varphi_1 \in W^2$ *and* $\nu \in]-2, 2[$. *Then*

(a) *if* $u(t) \in C^2([0,T]; H) \cap C([0,T]; W^2)$ *and* $f(t) \in C([0,T]; H)$*, then*

$$\max_{1\le k\le n-1}\Big(\|z_{k+1}\| + \Big\|\frac{\Delta z_k}{\tau}\Big\| + \|A^{1/2} z_{k+1}\|\Big) \to 0,\ \tau \to 0;$$

(b) *if the conditions of the item* (a) *are fulfilled and the functions* $f(t)$ *and* $u''(t)$ *satisfy the Hölder condition with index* λ $(0 < \lambda \le 1)$*, then*

$$\Big(\|z_{k+1}\| + \Big\|\frac{\Delta z_k}{\tau}\Big\| + \|A^{1/2} z_{k+1}\|\Big) \le c\tau^\lambda,\quad k = 1, \dots, n-1;$$

(c) *if* $u(t) \in C^3([0,T]; H) \cap C([0,T]; W^2)$ *and* $f(t) \in C^1([0,T]; H)$*, then*

$$\max_{1\le k\le n-1}\Big(\Big\|A^{1/2}\frac{\Delta z_k}{\tau}\Big\| + \Big\|\frac{\Delta^2 z_k}{\tau^2}\Big\| + \|A z_{k+1}\|\Big) \to 0,\ \tau \to 0;$$

(d) *if the conditions of the item* (c) *are fulfilled and the functions* $f'(t)$ *and* $u'''(t)$ *satisfy the Hölder condition with index* λ $(0 < \lambda \le 1)$*, then*

$$\Big(\Big\|A^{1/2}\frac{\Delta z_k}{\tau}\Big\| + \Big\|\frac{\Delta^2 z_k}{\tau^2}\Big\| + \|A z_{k+1}\|\Big) \le c\tau^\lambda,\quad k = 1, \dots, n-1.$$

Proof The validity of the following formulas depends on how smooth the functions $u(t)$ and $f(t)$ are:

$$\frac{\Delta^2 u(t_{k-1})}{\tau^2} - u''(t_k) = \frac{1}{\tau^2}\int_{t_k}^{t_{k+1}}\int_{t_k}^{t}\big(u''(s) - u''(t_k)\big)\,ds\,dt$$

$$+\frac{1}{\tau^2}\int\limits_{t_{k-1}}^{t_k}\int\limits_{t_{k-1}}^{t}\big(u''(s)-u''(t_k)\big)\,ds\,dt, \tag{111}$$

$$\frac{\Delta^2 u(t_{k-1})}{\tau^2}-u''(t_k)=\frac{1}{\tau^2}\int\limits_{t_k}^{t_{k+1}}\int\limits_{t_k}^{t}\int\limits_{t_k}^{s}\big(u'''(\xi)-u'''(t_k)\big)\,d\xi\,ds\,dt$$

$$+\frac{1}{\tau^2}\int\limits_{t_{k-1}}^{t_k}\int\limits_{t_{k-1}}^{t}\int\limits_{t_{k-1}}^{s}\big(u'''(t_k)-u'''(\xi)\big)\,d\xi\,ds\,dt, \tag{112}$$

$$\Delta^2 f(t_{k-1})=\int\limits_{t_k}^{t_{k+1}}\big(f'(t)-f'(t_k)\big)\,dt+\int\limits_{t_{k-1}}^{t_k}\big(f'(t_k)-f'(t)\big)\,dt, \tag{113}$$

$$\frac{\Delta u(0)}{\tau}=u'(0)+\frac{1}{\tau}\int\limits_0^{\tau}\big(u'(t)-u'(0)\big)\,dt=u'(0)+\frac{1}{\tau}\int\limits_0^{\tau}\int\limits_0^{t}u''(s)\,ds\,dt, \tag{114}$$

$$\frac{\Delta u(0)}{\tau}=u'(0)+\frac{\tau}{2}u''(0)+\frac{1}{\tau}\int\limits_0^{\tau}\int\limits_0^{t}\big(u''(s)-u''(0)\big)\,ds\,dt. \tag{115}$$

From (111) we obtain

$$\max_{1\le k\le n-1}\left\|\frac{\Delta^2 u(t_{k-1})}{\tau^2}-u''(t_k)\right\|\to 0,\ \tau\to 0 \tag{116}$$

if $u(t)\in C^2([0,T];H)$;

$$\left\|\frac{\Delta^2 u(t_{k-1})}{\tau^2}-u''(t_k)\right\|\le c\tau^{\lambda},\quad k=1,\dots,n-1, \tag{117}$$

if $u(t)\in C^2([0,T];H)$ and $u''(t)$ satisfies the Hölder condition with index λ $(0<\lambda\le 1)$.

From (112) we obtain

$$\frac{1}{\tau}\max_{1\le k\le n-1}\left\|\frac{\Delta^2 u(t_{k-1})}{\tau^2}-u''(t_k)\right\|\to 0,\ \tau\to 0 \tag{118}$$

if $u(t)\in C^3([0,T];H)$;

$$\left\| \frac{\Delta^2 u(t_{k-1})}{\tau^2} - u''(t_k) \right\| \le c\tau^{1+\lambda}, \quad k = 1, \dots, n-1, \tag{119}$$

if $u(t) \in C^3([0, T]; H)$ and $u'''(t)$ satisfies the Hölder condition with index λ $(0 < \lambda \le 1)$.

From (113) we obtain

$$\frac{1}{\tau} \max_{1 \le k \le n-1} \left\| \Delta^2 f(t_{k-1}) \right\| \to 0, \tau \to 0 \tag{120}$$

if $f(t) \in C^1([0, T]; H)$;

$$\|\Delta^2 u(t_{k-1})\| \le c\tau^{1+\lambda}, \quad k = 1, \dots, n-1, \tag{121}$$

if $f(t) \in C^1([0, T]; H)$ and $f'(t)$ satisfies the Hölder condition with index λ $(0 < \lambda \le 1)$.

From (114) we obtain

$$\left\| \frac{\Delta z_0}{\tau} \right\| \le c\tau \tag{122}$$

if $u(t) \in C^2([0, T]; H)$;

$$\left\| A^{1/2} \frac{\Delta z_0}{\tau} \right\| \to 0, \tau \to 0 \tag{123}$$

if $u(t) \in C^1([0, T]; W^1)$;

$$\left\| A^{1/2} \frac{\Delta z_0}{\tau} \right\| \le c\tau^\lambda \tag{124}$$

if $u(t) \in C^1([0, T]; W^1)$ and $A^{1/2}u'(t)$ satisfies the Hölder condition with index λ $(0 < \lambda \le 1)$.

In connection with relations (123) and (124) note that if the conditions of the item (c) are fulfilled, then Eq. (1) implies that $Au(t) \in C^1([0, T]; H)$. Hence, since A is a self-adjoint positive definite operator, we obtain $u'(t) \in C([0, T]; W^2)$ and $(Au(t))' = Au'(t)$.

It is obvious that the following estimates are valid:

$$\|A(\Delta z_0)\| = \left\| A\big(u(\tau) - u(0)\big) - \tau A\varphi_1 \right\| \to 0, \tau \to 0 \tag{125}$$

if $u(t) \in C([0, T]; W^2)$;

$$\|A(\Delta z_0)\| \le c\tau^\lambda \tag{126}$$

if $u(t) \in C([0, T]; W^2)$ and $Au(t)$ satisfies the Hölder condition with index λ $(0 < \lambda \le 1)$

$$\max_{1 \le k \le n-1} |\Delta^2 f(t_{k-1})\| \to 0, \tau \to 0 \tag{127}$$

if $f(t) \in C([0, T]; H)$;

$$\|\Delta^2 f(t_{k-1})\| \le c\tau^{\lambda}, \quad k = 1, \ldots, n-1, \tag{128}$$

if on $[0, T]$ the function $f(t)$ satisfies the Hölder condition with index λ $(0 < \lambda \le 1)$.

By virtue of the above estimates, inequalities (102)–(110) imply the estimates for the error $z_k = u(t_k) - u_k$.

The conclusion of the item (a) follows from inequalities (102), (103) and (106) with estimates (116), (122), (125) and (127) taken into account.

The conclusion of the item (b) follows from inequalities (102), (103) and (106) with estimates (117), (122), (126) and (128) taken into account.

The conclusion of the item (c) follows from inequalities (104), (107) and (109) with estimates (118), (120), (123) and (125) taken into account.

The conclusion of the item (d) follows from inequalities (104), (107) and (109) with estimates (119), (121), (124) and (126) taken into account.

The following theorem is valid in a smoother class of solutions.

Theorem 6 *Let* $u_0 = \varphi_0$, $\varphi_0 \in W^2$, $u_1 = \varphi_0 + \tau\varphi_1 + \frac{\tau^2}{2}\varphi_2$, $\varphi_2 = f(0) - A\varphi_0$, φ_1, $A\varphi_0$, $f(0) \in W^2$ *and* $\nu \in]-2, 2[$. *Then*

(a) *if* $u(t) \in C^3([0, T]; H) \cap C([0, T]; W^2)$, $f(t) \in C^1([0, T]; H)$, *and the functions* $u'''(t)$ *and* $f'(t)$ *satisfy the Hölder conditions with index* λ $(0 < \lambda \le 1)$, *then*

$$\|z_{k+1}\| + \left\|\frac{\Delta z_k}{\tau}\right\| + \|A^{1/2} z_{k+1}\| \le c\tau^{1+\lambda}, \quad k = 1, \ldots, n-1;$$

(b) *if* $u(t) \in C^4([0, T]; H) \cap C([0, T]; W^2)$, $f(t) \in C^2([0, T]; H)$, *and the functions* $u^{IV}(t)$ *and* $f''(t)$ *satisfy the Hölder condition with index* λ $(0 < \lambda \le 1)$, *then*

$$\left\|A^{1/2}\frac{\Delta z_k}{\tau}\right\| + \left\|\frac{\Delta^2 z_k}{\tau^2}\right\| \le c\tau^{1+\lambda}, k = 1, \ldots, n-2.$$

Proof From (115) we have

$$\left\|\frac{\Delta z_0}{\tau}\right\| \le c\tau^2 \tag{129}$$

if $u(t) \in C^3([0, T]; H)$;

$$\left\|A^{1/2}\frac{\Delta z_0}{\tau}\right\| \le c\tau^{1+\lambda} \tag{130}$$

if $u(t) \in C^2([0, T]; W^1)$ and $A^{1/2}u''(t)$ satisfies the Hölder condition with index λ $(0 < \lambda \le 1)$.

From (114) it follows that

$$\|A^{1/2}(\Delta z_0)\| \le \int_0^{\tau} \|A^{1/2}(u'(t) - u'(0))\| dt + \frac{\tau^2}{2}\|A^{1/2}\varphi_2\| \le c\tau^{1+\lambda}, \tag{131}$$

if $u(t) \in C^1([0,T]; W^1)$ and $A^{1/2}u'(t)$ satisfies the Hölder condition with index λ $(0<\lambda \le 1)$.

Let the conditions of the item (a) of the theorem be fulfilled. Then Eq. (1) implies that $Au(t) \in C^1([0,T]; H)$ and $(Au(t))'$ satisfies the Hölder condition. Since A is a self-adjoint and positive definite operator, the condition $Au(t) \in C^1([0,T]; H)$ implies that $u'(t) \in C([0,T]; W^2)$ and $(Au(t))' = Au'(t)$. Therefore if the conditions of the item (a) are fulfilled, then inequality (131) is valid. If the conditions of the item (b) are fulfilled, then analogously we establish that $Au(t) \in C^2([0,T]; H)$ and $(Au(t))''$ satisfies the Hölder condition. From this fact there follows inequality (130).

If the functions $u(t_{k+1}) = u(t_k+\tau)$ and $u(t_{k-1}) = u(t_k-\tau)$ are expanded using the Taylor formula and the remainder term is written in the integral form, then we have

$$r_{0,\tau}(t_k) = \frac{\Delta^2 u(t_{k-1})}{\tau^2} - u''(t_k)$$

$$= \frac{1}{\tau^2}\left(\frac{1}{4!}\int_{t_k}^{t_{k+1}} (t_{k+1}-t)^3 u^{IV}(t)\,dt + \frac{1}{4!}\int_{t_{k-1}}^{t_k} (t-t_{k-1})^3 u^{IV}(t)dt\right)$$

which clearly implies

$$r_{0,\tau}(t_k) - r_{0,\tau}(t_{k-1}) = \frac{1}{\tau^2}\left(\frac{1}{4!}\int_{t_k}^{t_{k+1}} (t_{k+1}-t)^3\big(u^{IV}(t) - u^{IV}(t_k)\big)dt\right.$$

$$+\frac{1}{4!}\int_{t_{k-1}}^{t_k} (t_k-t)^3\big(u^{IV}(t_k) - u^{IV}(t)\big)dt$$

$$+\frac{1}{4!}\int_{t_{k-1}}^{t_k} (t-t_{k-1})^3\big(u^{IV}(t) - u^{IV}(t_{k-1})\big)dt$$

$$\left.+\frac{1}{4!}\int_{t_{k-2}}^{t_{k-1}} (t-t_{k-2})^3\big(u^{IV}(t_{k-1}) - u^{IV}(t)\big)dt\right). \tag{132}$$

If in equality (132) we pass over to the norms and take into account that $u^{IV}(t)$ satisfies the Höder condition with index λ $(0 < \lambda \le 1)$, then we obtain

$$\left\| r_{0,\tau}(t_k) - r_{0,\tau}(t_{k-1}) \right\| \le c\tau^{2+\lambda}, \quad k = 2, \ldots, n-1. \tag{133}$$

It is obvious that the following formula is true:

$$\Delta^2 f(t_{k-1}) = \int_{t_k}^{t_{k+1}} \int_{t_k}^{t} f''(s)\, ds\, dt + \int_{t_{k-1}}^{t_k} \int_{t}^{t_k} f''(s)\, ds\, dt.$$

Analogously to (133), the latter formula yields

$$\left\| \Delta^2 f(t_k) - \Delta^2 f(t_{k-1}) \right\| \leq c\tau^{2+\lambda}, \quad k = 1, \ldots, n-1. \tag{134}$$

We have already obtained all those estimates the use of which in inequalities (102), (103), (106), (108) and (110) gives the estimates obtained in the items (a) and (b).

The estimate of the item (a) follows from inequalities (102), (103) and (106) with estimates (119), (121), (129) and (131) taken into account.

The estimate of the item (b) follows from inequalities (108) and (110) with estimates (130), (133) and (134) taken into account.

2.7 Approximation with Splines

In this subsection we study the convergence of the abstract cubic spline $\widetilde{S}_\tau(t)$, satisfying the conditions

$$\widetilde{S}(t_i) = u_i, \qquad \widetilde{S}'(t_i) = \delta u_i = \frac{u_{i+1} - u_i}{\tau},$$

where $i = 0, 1, \ldots, n-1$, $\tau = T/n$, $t_i = i\tau$, $\delta u_n = 0$, to solution $u(t)$ of problem (1), (2). At the same time depending on the smoothness of the solution $u(t)$, the order of convergence will be set with respect to τ.

Below we formulate some auxiliary lemmas, which are the extension of the results from [1].

Lemma 2 *Let* $\|u_i\| \leq a$ *and* $\|\delta u_i\| \leq b$, $i = 0, 1, \ldots, n$. *Then the following estimates are true:*

$$\|\widetilde{S}_\tau(t)\| \leq a + \frac{\tau}{4} b, \quad \|\widetilde{S}'_\tau(t)\| \leq 2b, \qquad t \in [0, T].$$

The estimate for $\widetilde{S}_\tau(t)$ is implied by the representation

$$\widetilde{S}_\tau(t) = \delta u_{i-1}(1 - \sigma_i)^2 \sigma_i - \delta u_i \tau \sigma_i^2 (1 - \sigma_i) +$$

$$u_{i-1}(1 - \sigma_i)^2 (2\sigma_i + 1) + u_i \sigma_i^2 (2(1 - \sigma_i) + 1),$$

and for the derivative, by the representation

$$\widetilde{S}'_\tau(t) = \delta u_{i-1}(1-\sigma_i)(1+3\sigma_i) - \delta u_i(2-3\sigma_i)\sigma_i,$$

where

$$\sigma_i = \sigma_i(t) = \frac{t}{\tau} - i + 1, \quad t \in [t_{i-1}, t_i], \quad i = 1, \dots, n.$$

Lemma 3 *Let $u(t) \in C^2(H)$ and $\hat{S}_\tau(t)$ be the cubic spline satisfying the conditions*

$$\hat{S}_\tau(t_i) = u(t_i), \quad \hat{S}'_\tau(t_i) = u'(t_i), \quad t_i = i\tau, \quad i = 0, 1, \dots, n.$$

Then the following estimates are true:

$$\|u(t) - \hat{S}_\tau(t)\| \leq \sup_t \|u''(t)\|\tau^2, \tag{135}$$

$$\|u'(t) - \hat{S}'_\tau(t)\| \leq 2\sup_t \|u''(t)\|\tau, \quad t \in [0, T]. \tag{136}$$

Proof First, we will prove estimate (136). It is obvious that

$$u'(t) - \hat{S}'_\tau(t) = \int_{t_{i-1}}^{t} [u''(t) - \hat{S}''_\tau(t)]dt, \tag{137}$$

where

$$t \in [t_{i-1},\ t_{i-1} + \frac{\tau}{2}].$$

Since

$$\hat{S}''_\tau(t) = \delta u'(t_{i-1})(6\sigma_i(t) - 2),$$

$$\delta u'(t_{i-1}) = \frac{u'(t_i) - u'(t_{i-1})}{\tau},$$

we have

$$u''(t) - \hat{S}''_\tau(t) = [u''(t) - \delta u'(t_{i-1})] + 3\delta u'(t_{i-1})(1 - 2\delta_i).$$

After substitution into (137) we get

$$\|u'(t) - \hat{S}'_\tau(t)\| \leq \int_{t_{i-1}}^{t} (\|u''(t)\| + \|\delta u'(t_{i-1})\|)dt + 3\|\delta u'(t_{i-1})\| \int_{t_{i-1}}^{t} (1 - 2\sigma_i(t))dt.$$

From here, taking into account the estimate

$$\int_{t_{i-1}}^{t} (1 - 2\sigma_i(t))dt \leq \frac{\tau}{4},$$

$$\|\delta u'(t_{i-1})\| = \|\tau^{-1} \int_{t_{i-1}}^{t_i} u''(t)dt\| \le \sup_{0\le t\le T} \|u''(t)\|,$$

follows (136) when $t \in [t_{i-1},\ t_{i-1} + \frac{\tau}{2}]$. Estimate (136) is proved similarly, when $t \in [t_{i-1} + \frac{\tau}{2}, t_i]$.

Estimate (135) follows from the identity

$$u(t) - \hat{S}_\tau(t) = \int_{t_{i-1}}^{t} (u'(t) - \hat{S}'_\tau(t))dt.$$

Lemma 4 *Let $u(t) \in C^2(H)$ and $S_\tau(t)$ be the cubic spline satisfying the conditions:*

$$S_\tau(t_i) = u(t_i), \quad S'_\tau(t_i) = \delta u(t_i) = \frac{u(t_{i+1}) - u(t_i)}{\tau}.$$

where $i = 0, 1, \dots, n-1,\ \ t_i = i\tau,\ \ \delta u(t_n) = 0$. Then the following estimates are true:

$$\|\hat{S}_\tau(t) - S_\tau(t)\| \le c\tau^2, \tag{138}$$

$$\|\hat{S}'_\tau(t) - S'_\tau(t)\| \le 2c\tau, \tag{139}$$

$$c = \sup_t \|u''(t)\|, \quad t \in [0, T].$$

Proof First, we prove the estimate (139). It is obvious that

$$\hat{S}'_\tau(t) - S'_\tau(t) = (u'(t_{i-1}) - \delta u(t_{i-1}))(1 - \sigma_i)(1 + 3\sigma_i) -$$

$$(u'(t_i) - \delta u(t_i))\sigma_i(2(1 - \sigma_i) - \sigma_i), \quad 0 \le \sigma_i(t) \le 1, \quad t \in [t_{i-1}, t_i].$$

Hence

$$\|\hat{S}'_\tau(t) - S'_\tau(t)\| \le \|u'(t_{i-1}) - \delta u(t_{i-1})\|(1 - \sigma_i)(1 + 3\sigma_i)$$

$$+ \|u'(t_i) - \delta u(t_i)\|\sigma_i(2 - \sigma_i). \tag{140}$$

Next, by the inclusion $u(t) \in C^2(H)$, it follows that

$$\|u'(t_i) - u'(t)\| \le c\tau, \quad c = \sup_t \|u''(t)\|, \quad t \in [t_{i-1}, t_i].$$

Then the following estimate is true

$$\|u'(t_i) - \delta u(t_i)\| = \|u'(t_i) - \frac{\|u(t_{i+1}) - u(t_i)}{\tau}\|$$

$$\leq \frac{1}{\tau}\int_{t_i}^{t_{i+1}} \|u_i'(t) - u'(t)\|dt \leq c\tau.$$

If we now substitute this estimate into (140) and take into account that

$$(1-\sigma_i)(1+3\sigma_i) + \sigma_i(2-\sigma_i) = 1 + 4\sigma_i - 4\sigma_i^2 \leq 2,$$

then we get estimate(139).

Estimate (138) follows from the identity

$$\hat{S}_\tau(t) - S_\tau(t) = \int_{i-1}^{t} [\hat{S}_\tau'(t) - S_\tau'(t)]dt, \quad t \in [t_{i-1}, \ t_{i-1} + \frac{\tau}{2}].$$

The following convergence theorem holds.

Theorem 7 *Let the conditions of Theorem 6 be fulfilled and* $u(t) \in C^4(H) \cap C(W^2)$. *Then the following estimates are true:*

$$\|u(t) - \widetilde{S}_\tau(t)\| \leq c\tau^2, \tag{141}$$

$$\|u'(t) - \widetilde{S}_\tau'(t)\| \leq c\tau. \tag{142}$$

Proof By virtue of the conclusion (a) of Theorem 6 the following estimates are true

$$\|z_{k+1}\| + \|\frac{\Delta z_k}{\tau}\| \leq c\tau^2, \quad k = 1, \ldots, n-1. \tag{143}$$

By Lemma 2 from (143) it follows that

$$\|S_\tau(t) - \widetilde{S}_\tau(t)\| \leq (1 + \frac{\tau}{4})c\tau^2, \tag{144}$$

$$\|S_\tau'(t) - \widetilde{S}_\tau'(t)\| \leq 2c\tau^2. \tag{145}$$

Now, taking into account (135), (138), and (144), the identity

$$u(t) - \widetilde{S}_\tau(t) = [u(t) - \hat{S}_\tau(t)] + [\hat{S}_\tau(t) - S_\tau(t)] + [S_\tau(t) - \widetilde{S}_\tau(t)],$$

implies estimate (141).

The estimate (142) it follows from identity

$$u'(t) - \widetilde{S}_\tau'(t) = [u'(t) - \hat{S}_\tau'(t)] + [\hat{S}_\tau'(t) - S_\tau'(t)] + [S_\tau'(t) - \widetilde{S}_\tau'(t)],$$

taking into account (136), (139) and (145).

The next theorem is proved similarly.

Theorem 8 *Let the conditions of Theorem 5 be fulfilled Then:*
(a) if $f(t) \in C^1(H)$ *(or* $f(t) \in C(W^2)$*), then*

$$\|u(t) - \widetilde{S}_\tau(t)\| + \|u'(t) - \widetilde{S}'_\tau(t)\| \to 0;$$

(b) if $f(t) \in C^1(H)$ *and* $u(t) \in C^3(H) \cap C(W^2)$*, then*

$$\|u(t) - \widetilde{S}_\tau(t)\| + \|u'(t) - \widetilde{S}'_\tau(t)\| \leq c\tau.$$

3 Second Order Complete Equation

Let us consider the Cauchy problem for an abstract hyperbolic equation in the Hilbert space H:

$$\frac{d^2u(t)}{dt^2} + B\frac{du}{dt} + Au\,(t) = f\,(t)\,, \quad t \in]0,\ T], \tag{146}$$

$$u\,(0) = \varphi_0, \quad u'\,(t)\,|_{t=0} = \varphi_1. \tag{147}$$

where A and B are self-adjoint, positively defined (generally unbounded) operators with the definition domains $D\,(A)$ and $D\,(B)$ which are everywhere dense in H; moreover, $D\,(A) \subset D\,(B)$; φ_0 and φ_1 are given vectors from H; $u\,(t)$ is a sought continuous, twice continuously differentiable function with values in H and $f\,(t)$ is a given continuous function with values in H.

We seek a searching solution of the problem (146), (147) by the following semi-discrete scheme:

$$\frac{u_{k+1} - 2u_k + u_{k-1}}{\tau^2} + B\frac{u_{k+1} - u_{k-1}}{2\tau} + A\frac{u_{k+1} + u_{k-1}}{2} = f_k, \tag{148}$$

where $f_k = f\,(t_k)$, $k = 1, ..., n-1$, $t_k = k\tau$, $\tau = T/n$ $(n > 1)$.

The following theorem holds.

Theorem 9 *Let* A *and* B *be self-adjoint, positive-definite (generally unbounded) operators with the definition domains* $D\,(A)$ *and* $D\,(B)$ *which are everywhere dense in* H*; besides,* $D\,(A) \subset D\,(B)$ *and* $BA^{-1}\varphi = A^{-1}B\varphi$, $\forall \varphi \in D(A)$*. Then for scheme (148) the following a priori estimates are valid:*

$$\|u_{k+1}\| \leq t_{k+1}\left\|\frac{\Delta u_0}{\tau}\right\| + \sqrt{2}\,\|u_0\| + t_k\tau\sum_{i=1}^{k}\|f_i\|\,, \tag{149}$$

$$\left\|\frac{\Delta u_k}{\tau}\right\| \leq \sqrt{2}\left\|\frac{\Delta u_0}{\tau}\right\| + \frac{2}{\tau}\,\|u_0\| + \sqrt{2}\tau\sum_{i=1}^{k}\|f_i\|\,, \tag{150}$$

where $k = 1, \ldots, n-1,\ \ \Delta u_k = u_{k+1} - u_k$.

Proof From (148) we have

$$u_{k+1} = Lu_k - Su_{k-1} + \frac{\tau^2}{2} Lf_k, \tag{151}$$

where

$$L = 2\left(I + \frac{\tau}{2}B + \frac{\tau^2}{2}A\right)^{-1},$$

$$S = \left(I - \frac{\tau}{2}B + \frac{\tau^2}{2}A\right)\left(I + \frac{\tau}{2}B + \frac{\tau^2}{2}A\right)^{-1}.$$

From the recurrence relation (151), we obtain by induction

$$u_{k+1} = U_k(L,S)u_1 - SU_{k-1}(L,S)u_0 + \frac{\tau^2}{2}\sum_{i=1}^{k} U_{k-i}(L,S)Lf_i, \tag{152}$$

in which the operator polynomials $U_k(L,S)$ satisfy the following recurrence relation:

$$U_k(L,S) = LU_{k-1}(L,S) - SU_{k-2}(L,S), \quad k = 1, 2, ...,$$

$$U_0(L,S) = I, \quad U_{-1}(L,S) = 0. \tag{153}$$

Let us rewrite (152) as

$$u_{k+1} = \tau U_k(L,S)\frac{\Delta u_0}{\tau} + (U_k(L,S) - SU_{k-1}(L,S))\,u_0$$

$$+ \frac{\tau^2}{2}\sum_{i=1}^{k} U_{k-i}(L,S)Lf_i. \tag{154}$$

If we consider the norms in equality (154), then we obtain

$$\|u_{k+1}\| \le \tau \left\| U_k(L,S)\frac{\Delta u_0}{\tau} \right\| + \|U_k(L,S) - SU_{k-1}(L,S)\|\,\|u_0\|$$

$$+ \frac{\tau^2}{2}\sum_{i=1}^{k} \|U_{k-i}(L,S)Lf_i\|. \tag{155}$$

We can write the operator S as

$$S = BA^{-1}\left(BA^{-1}+\tau I\right)^{-1}L - \left(BA^{-1}-\tau I\right)\left(BA^{-1}+\tau I\right)^{-1}$$

$$= \left(I - \tau\left(BA^{-1}+\tau I\right)^{-1}\right)L - \left(I - 2\tau\left(BA^{-1}+\tau I\right)^{-1}\right)$$

$$= (I-M)\,L - (I-2M)\ , \qquad (156)$$

where $M = \tau\left(BA^{-1}+\tau I\right)^{-1}$.

Indeed,

$$S = BA^{-1}\left(BA^{-1}+\tau I\right)^{-1}L - \left(BA^{-1}-\tau I\right)\left(BA^{-1}+\tau I\right)^{-1}$$

$$= 2BA^{-1}\left(BA^{-1}+\tau I\right)^{-1}\left(I+\frac{\tau}{2}B+\frac{\tau^2}{2}A\right)^{-1} - \left(BA^{-1}-\tau I\right)\left(BA^{-1}+\tau I\right)^{-1}$$

$$= \left(2BA^{-1}\left(BA^{-1}+\tau I\right)^{-1} - \left(BA^{-1}-\tau I\right)\left(BA^{-1}+\tau I\right)^{-1}\left(I+\frac{\tau}{2}B+\frac{\tau^2}{2}A\right)\right) \times\left(I+\frac{\tau}{2}B+\frac{\tau^2}{2}A\right)^{-1}$$

$$= \left(2BA^{-1}\left(BA^{-1}+\tau I\right)^{-1} - \left(BA^{-1}-\tau I\right)\left(BA^{-1}+\tau I\right)^{-1}\left(I+\frac{\tau}{2}\left(BA^{-1}+\tau I\right)A\right)\right) \times\left(I+\frac{\tau}{2}B+\frac{\tau^2}{2}A\right)^{-1}$$

$$= \left(2BA^{-1}\left(BA^{-1}+\tau I\right)^{-1} - \left(\left(BA^{-1}-\tau I\right)\left(BA^{-1}+\tau I\right)^{-1}+\frac{\tau}{2}\left(BA^{-1}-\tau I\right)A\right)\right) \times\left(I+\frac{\tau}{2}B+\frac{\tau^2}{2}A\right)^{-1}$$

$$= \left(\left(BA^{-1}+\tau I\right)\left(BA^{-1}+\tau I\right)^{-1} - \frac{\tau}{2}\left(BA^{-1}-\tau I\right)A\right) \times\left(I+\frac{\tau}{2}B+\frac{\tau^2}{2}A\right)$$

$$= \left(I-\frac{\tau}{2}B+\frac{\tau^2}{2}A\right)\left(I+\frac{\tau}{2}B+\frac{\tau^2}{2}A\right)^{-1}.$$

When L and S are commutative, self-adjoint, bounded operators (in our case these conditions are satisfied, since B and A^{-1} are commutative on $D(A)$), the operator-polynomial norm $U_k(L, S)$ is less than or equal to the corresponding scalar polynomial's $U_k(x, y)$, the C-norm on $\sigma(L) \times \sigma(S)$, where $\sigma(L)$ is the spectrum of the operator L, while $\sigma(S)$ is the spectrum of S (this result is a concrete case of a more general theorem (see Garnir [6])). This fact and representation (156) give

$$\|U_k(L, S)\| \leq \max_{(x,y)} |U_k(x, y)|, \quad (x, y) \in G, \tag{157}$$

where

$$G = \{(x, y) : x \in \sigma(L),\ y = (1-\lambda)x - (1-2\lambda),\ \lambda \in \sigma(M)\}.$$

Let us estimate the spectrum of the operator L. Since, according to the hypothesis, the operators A and B are self-adjoint and positive-definite, we have:

$$\sigma(L) \subset [0, 2]. \tag{158}$$

Because B and A^{-1} are commutative on $D(A)$, we have $M = M^* \geq 0$ (this can be proved easily). This yields

$$\sigma(M) \subset [0, 1]. \tag{159}$$

If we consider relations (158) and (159) then we get

$$G \subset \Delta^+. \tag{160}$$

Together with (160), inequalities (157) and (25) give

$$\|U_k(L, S)\| \leq k + 1. \tag{161}$$

In a similar manner, (23) yields

$$\|U_k(L, S) - SU_{k-1}(L, S)\| \leq \sqrt{2}. \tag{162}$$

If we substitute estimates: (161), (162) and $\|L\| \leq 2$ into inequality (155), we get (149).

Let us prove a priori estimate (150). From (154) we have:

$$u_k = \tau U_{k-1}(L, S)\frac{\Delta u_0}{\tau} + (U_{k-1}(L, S) - SU_{k-2}(L, S))\, u_0$$

$$+ \frac{\tau^2}{2}\sum_{i=1}^{k} U_{k-i-1}(L, S)Lf_i. \tag{163}$$

If we subtract (163), where $U_{-1} = 0$, from inequality (154), we will get

$$\Delta u_k = (U_k(L,S) + SU_{k-2}(L,S) - (S+I)U_{k-1}(L,S))\,u_0$$
$$+\tau\,(U_k(L,S) - U_{k-1}(L,S))\,\frac{\Delta u_0}{\tau}$$
$$+\frac{\tau^2}{2}\sum_{i=1}^{k}(U_{k-i}(L,S) - U_{k-i-1}(L,S))\,Lf_i. \tag{164}$$

Using the recurrence relation (153) we have

$$U_k(L,S) + SU_{k-2}(L,S) - (S+I)U_{k-1}(L,S) = (L-S-I)U_{k-1}(L,S). \tag{165}$$

If in (164) we consider (165) and then divide both sides of the resulting equality by τ, we obtain

$$\frac{\Delta u_k}{\tau} = \tau^{-1}(L-S-I)U_{k-1}(L,S)u_0 + (U_k(L,S) - U_{k-1}(L,S))\,\frac{\Delta u_0}{\tau}$$
$$+\frac{\tau}{2}\sum_{i=1}^{k}(U_{k-i}(L,S) - U_{k-i-1}(L,S))\,Lf_i.$$

Continuing our discussion in a manner similar to (149), estimates (24) and (26) give (150).

4 Remark Concerning Equations with a Variable Operator

Let us consider the Cauchy problem for an abstract hyperbolic equation in the Hilbert space H:

$$\frac{d^2u(t)}{dt^2} + A(t)u\,(t) = f\,(t)\,,\quad t\in]0,\ T], \tag{166}$$

$$u\,(0) = \varphi_0,\quad u'\,(t)\,|_{t=0} = \varphi_1. \tag{167}$$

where $A(t)$ is a self-adjoint, positive-definite (generally unbounded) operator with the definition domain $D\,(A)$ ($D(A)$ does not depend on t), which is everywhere dense in H ($\overline{D\,(A)} = H$); φ_0 and φ_1 are given vectors from H; $u\,(t)$ is a sought continuous, twice continuously differentiable function with values in H and $f\,(t)$ is given continuous function with values in H.

We seek a solution of problem (166), (167) by the following semi-discrete scheme

$$\frac{u_{k+1} - 2u_k + u_{k-1}}{\tau^2} + A_k\frac{u_{k+1} + u_{k-1}}{2} = f_k, \tag{168}$$

where $k = 1, ..., n-1$, $\tau = T/n$ $(n > 1)$, $A_k = A(t_k)$, $f_k = f(t_k)$, $t_k = k\tau$, $u_0 = \varphi_0$.

As an approximate solution $u(t)$ of problem (166), (167) at the point $t_k = k\tau$ we assume u_k, $u(t_k) \approx u_k$.

The following theorem is true (everywhere below c denotes a positive constant).

Theorem 10 *Let A be a self-adjoint, positive-definite (generally unbounded) operator with the definition domains $D(A)$, which is everywhere dense in H, and, besides, the following conditions be fulfilled*

$$\left\| (A(t) - A(s)) A^{-1}(s) \right\| \leq c\,|t-s|\,, \quad \forall s, t \in [0,\ T]\,.$$

Then for scheme (168) the following a priori estimates are valid

$$\left\| \frac{\Delta u_k}{\tau} \right\| + \left\| A_k^{1/2} u_{k+1} \right\|$$

$$\leq c \left(\left\| \frac{\Delta u_0}{\tau} \right\| + \tau \left\| A_0^{1/2} \frac{\Delta u_0}{\tau} \right\| + \left\| A_1^{1/2} u_0 \right\| + \tau \sum_{i=1}^{k} \|f_i\| \right), \tag{169}$$

where $k = 1, \ldots, n-1$, $\Delta u_k = u_{k+1} - u_k$.

Lemma 5 *Let the operator $A(t)$ satisfy the conditions of Theorem 10 , then the following inequality is valid:*

$$|(A(t)u, u) - (A(s)u, u)| \leq c\,|t-s|\,(A(s)u, u), \quad u \in D(A). \tag{170}$$

Proof Let us introduce the operators:

$$A_n(t) = A(t) \left(I + \frac{1}{n} A(t) \right)^{-1},$$

$$B_n = (A_n(t) - A_n(s)) \left(I + \frac{1}{n} A(s) \right) A^{-1}(s),$$

where n is a natural number.

Clearly, we have

$$B_n = \left[A(t) \left(I + \frac{1}{n} A(t) \right)^{-1} - A(s) \left(I + \frac{1}{n} A(s) \right)^{-1} \right] \left(I + \frac{1}{n} A(s) \right) A^{-1}(s)$$

$$= A(t) \left[\left(I + \frac{1}{n} A(t) \right)^{-1} - \left(I + \frac{1}{n} A(s) \right)^{-1} \right] \left(I + \frac{1}{n} A(s) \right) A^{-1}(s)$$

$$+\left(A(t)-A(s)\right)A^{-1}(s)=A(t)\left(I+\frac{1}{n}A(t)\right)^{-1}$$

$$\times\left[\left(I+\frac{1}{n}A(s)\right)-\left(I+\frac{1}{n}A(t)\right)\right]A^{-1}(s)+\left(A(t)-A(s)\right)A^{-1}(s)$$

$$=\frac{1}{n}A(t)\left(I+\frac{1}{n}A(t)\right)^{-1}\left(A(t)-A(s)\right)A^{-1}(s)+\left(A(t)-A(s)\right)A^{-1}(s). \quad (171)$$

The following representation

$$A_n(t)=A(t)\left(I+\frac{1}{n}A(t)\right)^{-1}=n\left[I-\left(I+\frac{1}{n}A(t)\right)^{-1}\right]$$

implies that $A_n(t)$ is a self-adjoint, nonnegative, bounded operator, which in turn implies that $B_nA_n(s)=A_n(t)-A_n(s)$ is a self-adjoint, bounded operator, i.e. $(B_nA_n(s))^*=B_nA_n(s)$. On the other hand, we have $(B_nA_n(s))^*=A_n(s)B_n^*$. Thus, we get $B_nA_n(s)=A_n(s)B_n^*$. Therefore, according to the famous Reid inequality, we have (see [8], Problem 82):

$$|(B_nA_n(s)u,u)|=\left|\left(A_n(s)B_n^*u,u\right)\right|\le\left\|B_n^*\right\|(A_n(s)u,u)=\|B_n\|\,(A_n(s)u,u)$$

or , equivalently

$$|((A_n(t)-A_n(s))\,u,u)|\le\|B_n\|\,(A_n(s)u,u). \quad (172)$$

In accordance with the hypothesis of Theorem 10, (171) yields the estimate

$$\|B_n\|\le\left[\left\|\frac{1}{n}A(t)\left(I+\frac{1}{n}A(t)\right)^{-1}\right\|+1\right]\left\|\left(A(t)-A(s)\right)A^{-1}(s)\right\|\le c\,|t-s|\,.$$

By this inequality, (172) implies

$$|((A_n(t)-A_n(s))\,u,u)|\le c\,|t-s|\,(A_n(s)u,u). \quad (173)$$

It is an easy exercise to prove that

$$\lim_{n\to\infty}I_n(t)u=\lim_{n\to\infty}\left(I+\frac{1}{n}A(t)\right)^{-1}u=u,\qquad u\in D(A).$$

Applying this relation, we arrive at

$$\lim_{n\to\infty}(A_n(s)u,u)=\lim_{n\to\infty}(A(s)I_n(s)u,u)=\lim_{n\to\infty}(I_n(s)u,A(s)u)$$

$$= (u, A(s)u) = (A(s)u, u), \quad u \in D(A). \tag{174}$$

It is obvious that with (174) taken into account, (170) can be derived from (173).

Let us return to the proof of Theorem 10.

Proof If both sides of equality (168) multiply scalarly on the vector $u_{k+1} - u_{k-1} = (u_{k+1} - u_k) + (u_k - u_{k-1})$, we obtain:

$$\left\| \frac{u_{k+1} - u_k}{\tau} \right\|^2 + \frac{1}{2} \left\| A_k^{1/2} u_{k+1} \right\|^2 = \left\| \frac{u_k - u_{k-1}}{\tau} \right\|^2 + \frac{1}{2} \left\| A_k^{1/2} u_{k-1} \right\|^2$$

$$+ (f_k, (u_{k+1} - u_k)) + (f_k, (u_k - u_{k-1})) . \tag{175}$$

Let us introduce the notations

$$\alpha_k = \left\| \frac{u_k - u_{k-1}}{\tau} \right\|^2, \quad \gamma_k^+ = \left\| A_k^{1/2} u_{k+1} \right\|^2, \quad \gamma_k^- = \left\| A_k^{1/2} u_{k-1} \right\|^2.$$

Then, according to the Schwartz inequality, from (175) it follows that

$$\alpha_{k+1} + \frac{1}{2}(\gamma_k^+ + \gamma_{k+1}^-) \le \alpha_k + \frac{1}{2}(\gamma_{k-1}^+ + \gamma_k^-)$$

$$+ \left[\frac{1}{2} \left(\gamma_{k+1}^- - \gamma_{k-1}^+ \right) + \tau \left(\sqrt{\alpha_{k+1}} + \sqrt{\alpha_k} \right) \| f_k \| \right]. \tag{176}$$

The following inequality is valid:

$$\left| \gamma_{k+1}^- - \gamma_{k-1}^+ \right| = \left| \left\| A_{k+1}^{1/2} u_k \right\|^2 - \left\| A_{k-1}^{1/2} u_k \right\|^2 \right|$$

$$= \left| (A_{k+1} u_k, u_k) - (A_{k-1} u_k, u_k) \right| \le c\tau \left\| A_{k-1}^{1/2} u_k \right\|^2 = c\tau \gamma_{k-1}^+ .$$

Taking into account this inequality, from (176) it follows that

$$\lambda_{k+1} \le \lambda_k + \varepsilon_k, \tag{177}$$

where

$$\lambda_k = \alpha_k + \frac{1}{2}(\gamma_{k-1}^+ + \gamma_k^-), \quad \varepsilon_k = \tau \left[c\gamma_{k-1}^+ + \left(\sqrt{\alpha_{k+1}} + \sqrt{\alpha_k} \right) \| f_k \| \right].$$

From (177) we obtain

$$\lambda_{k+1} \le \lambda_1 + (\varepsilon_1 + \varepsilon_2 + ... + \varepsilon_k) = \lambda_1 + c\tau \sum_{i=1}^{k} \gamma_{i-1}^{+} + \tau \sum_{i=1}^{k} \left(\sqrt{\alpha_i} + \sqrt{\alpha_{i+1}}\right) \|f_i\| .$$

Hence, we obvuously obtain

$$\delta_{k+1}^2 \le \delta_1^2 + c\tau \sum_{i=1}^{k} \delta_i^2 + \tau \sum_{i=1}^{k} (\delta_i + \delta_{i+1}) \|f_i\| , \quad \delta_k = \sqrt{\lambda_k} .$$

From here follows following inequality

$$\delta_{k+1} \le \delta_1 + c\tau \sum_{i=1}^{k} \delta_i + 2\tau \sum_{i=1}^{k} \|f_i\| .$$

Using this, the discrete analogue of Gronwall's lemma yields

$$\delta_{k+1} \le e^{ct_{k-1}} \left((1 + c\tau)\, \delta_1 + 2\tau \sum_{i=1}^{k} \|f_i\| \right) . \tag{178}$$

Simple transformations of (178) result in (169).

Acknowledgements This work was supported by the Shota Rustaveli National Science Foundation of Georgia (SRNFG), grant no. FR17-252.

References

1. Ahlberg, J., Nilson, E., Walsh, J.: The Theory of Splines and its Application. Elsevier Science, Amsterdam (1967)
2. Baker, G.A.: Error estimates for finite element methods for second order hyperbolic equations. SIAM J. Numer. Anal. **13**(4), 564–576 (1976)
3. Baker, G.A., Bramble, J.H.: Semidiscrete and single step fully discrete approximations for second order hyperbolic equations. RAIRO Anal. Numer. **13**(2), 75–100 (1979)
4. Baker, G.A., Dougalis, V.A., Serbin, S.M.: An approximation theorem for second-order evolution equations. Numer. Math. **35**(2), 127–142 (1980)
5. Bales, L.A.: Semidiscrete and single step fully discrete finite element approximations for second order hyperbolic equations with nonsmooth solutions. RAIRO Model. Math. Anal. Numer. **27**(1), 55–63 (1993)
6. Garnir, H.G.: Demonstration elementaire d'une inegalite relative aux operateurs normaux commutatifs. (French) Collection of articles dedicated to Mauro Picone on the occasion of his ninetieth birthday, II. Rend. Mat. **8**(2–6), 473–480 (1975)
7. Godunov, S.K., Ryaben'kii, V.S.: Finite Difference Schemes. Nauka, Moscow, [in Russian] (1973)
8. Halmos, P.R.: A Hilbert Space Problem Book. Mir (1970)
9. Kacur, J.: Application of Rothe's method to perturbed linear hyperbolic equations and variational inequalities. Czechoslovak Math. J. **34(109)**(1), 92–106 (1984)

10. Kantorovich, L.V., Akilov, G.P.: Functional Analysis. Nauka, Moscow (1977), Pergamon, Oxford (1982)
11. Krein, S.G.: Linear differential equations in Banach space. M.: Nauka (1967)
12. Ladyzhenskaya, O.A.: On the solution of non-stationary operator equations. (Russian) Mat. Sb. N.S. **39**(81), 491–524 (1956)
13. Makarov, V.L.: Orthogonal polynomials and difference schemes with exact and explicit spectrum Doctoral dissertation, Physics and Mathematics Sciences, Kiev, p. 286 (1974)
14. Marchuk, G.I.: Methods of Numerical Mathematics, 2nd edn. (Nauka, Moscow (1977) Springer, New York (1982)
15. Mikhlin, S.G.: Numerical realization of variational methods, M.: Nauka (1966)
16. Morris, A.G., Horner, T.S.: Chebyshev polynomials in the numerical solution of differential equations. Math. Comput. **31**(140), 881–891 (1977)
17. Novikov, V.A., Demidov, G.V.: A remark on a certain method of constructing schemes of high accuracy. (Russian) Chisl. Metody Meh. Sploshnoi Sredy **3**(4), 89–91 (1972)
18. Pultar, M.: Solutions of abstract hyperbolic equations by Rothe method. Apl. Mat. **29**(1), 23–39 (1984)
19. Rastrenin, V.A.: The application of a certain difference method to abstract hyperbolic equations. (Russian) Differencial'nye Uravnenija, **IX**(12), 2222–2226 (1973)
20. Richtmyer, R.D., Morton, K.W.: Difference Methods for Initial-Value Problems (Wiley, New York (1967). Mir, Moscow (1972)
21. Read, M., Simon, B.: Methods of Modern Mathematical Physics. I. Functional Analysis, 2nd edn. Academic Press Inc., New York (1980)
22. Rogava, J.: The study of the stability of semi-discrete scheme by means of Chebyshev orthogonal polynomials. GSSR Mecn. Akad. Moambe **83**(3), 545–548 (1976)
23. Rogava, J.: Semidiscrete schemes for operator-differential equations. Izdatelstvo Tekhnicheskogo Universiteta, p. 288 (1995)
24. Samarskii, A.A.: Theory of Finite Difference Schemes. Nauka, Moscow (1977); Marcel Dekker, New York (2001)
25. Sobolevskii, P.E., Chebotareva, L.M.: Approximate solution of the Cauchy problem for an abstract hyperbolic equation by the method of lines. (Russian) Izv. Vysh. Uchebn. Zaved. Matematika, **5**(180), 103–116 (1977)
26. Szego, S.: Orthogonal polynomials. Amer. Math. Soc. Colloq. Publ. **23**, p. 421 (1959)
27. Yanenko, N.N.: The Method of Fractional Steps. Nauka, Novosibirsk (1967); Springer, Berlin (1971)

Notes on Sub-Gaussian Random Elements

George Giorgobiani, Vakhtang Kvaratskhelia, and Vaja Tarieladze

Abstract We give a short survey concerning sub-Gaussian random elements in a Banach space and prove a statement about the induced operator of a bounded random element in a Hilbert space.

1 Sub-Gaussian and Related Random Variables

The sub-Gaussian random variables were explicitly defined by Kahane in [1] (see also [2]). They were further studied by Buldygin and Kozachenko in [3, 4] (see also [5, Chap. 3] and [6]).

A real valued random variable ξ given on a probability space $(\Omega, \mathcal{A}, \mathbb{P})$ is called *sub-Gaussian* if there exists $a \geq 0$ such that

$$\mathbb{E}\, e^{t\xi} \leq e^{\frac{1}{2}t^2a^2}, \quad \text{for every} \quad t \in \mathbb{R}\,.$$

To a random variable ξ let us associate a quantity $\tau(\xi) \in [0, +\infty]$ defined by the equality:

$$\tau(\xi) = \inf\{a \geq 0 : \mathbb{E}\, e^{t\xi} \leq e^{\frac{1}{2}t^2a^2} \;\; \textit{for every} \;\; t \in \mathbb{R}\},$$

and call it the Gaussian standard of ξ [3] (it is called *the Gaussian deviation* ("écart de Gauss") of ξ in [1]).

Dedicated to 90th birthday anniversary of Nicholas Vakhania (1930–2014).

G. Giorgobiani (✉) · V. Kvaratskhelia · V. Tarieladze
Muskhelishvili Institute of Computational Mathematics, 0159 Tbilisi, Georgia
e-mail: giorgobiani.g@gtu.ge

V. Kvaratskhelia
e-mail: v.kvaratskhelia@gtu.ge

V. Tarieladze
e-mail: v.tarieladze@gtu.ge

G. Jaiani and D. Natroshvili (eds.), *Applications of Mathematics and Informatics in Natural Sciences and Engineering*, Springer Proceedings in Mathematics & Statistics 334, https://doi.org/10.1007/978-3-030-56356-1_11

Lemma 1 ([1, 4]; see also, [6, Proposition 2.1 and Corollary 2.1]) *For a real valued random variable ξ the following statements are equivalent:*

(i) ξ is sub-Gaussian.

(ii) $\tau(\xi) < +\infty$ and $\mathbb{E}\,\xi = 0$.

(iii) There is $\lambda > 0$ such that $\mathbb{E}\exp(\lambda\xi^2) < +\infty$ and $\mathbb{E}\,\xi = 0$.

Moreover, if (i) holds, then

$$\mathbb{E}\,e^{\lambda\xi^2} \leq \frac{1}{\sqrt{1-2\lambda\tau^2(\xi)}} < \infty \quad \textit{for every} \quad \lambda \in \left[0, \frac{1}{2\tau^2(\xi)}\right[,$$

and

$$(\mathbb{E}\,|\xi|^p)^{\frac{1}{p}} \leq \beta_p \tau(\xi) \quad \textit{for every} \quad p \in]0, \infty[\,,$$

where $\beta_p = 1$ if $p \in]0, 2]$ and $\beta_p = 2^{\frac{1}{p}}(\frac{p}{e})^{\frac{1}{2}}$ if $p \in]2, \infty[$.

In particular we have

$$\mathbb{E}\,\xi = 0 \quad \textit{and} \quad \mathbb{E}\,\xi^2 \leq \tau^2(\xi)\,.$$

Remark 1 An interesting application of implication $(i) \Longrightarrow (iii)$ of Lemma 1 is the following observation: *if ξ is sub-Gaussian random variable with infinitely divisible distribution, then ξ is (possibly degenerate) Gaussian.* This can be derived e.g. from [7, Theorem 2], or from [8, Theorem 1(a)] or (more directly) from [9, Theorem 2] which asserts in particular that if for a random variable ξ *with infinitely divisible distribution* we have

$$\mathbb{E}\,\exp(\alpha|\xi|\ln(|\xi|+1)) < \infty \quad \text{for every} \quad \alpha > 0\,,$$

then it is Gaussian.

A sub-Gaussian random variable ξ with $\tau(\xi) \leq 1$ is called in [2, p. 67] *subnormal*. For a centered Gaussian random variable ξ clearly $\tau^2(\xi) = \mathbb{E}\,\xi^2$.

A random variable ξ is called *strictly sub-Gaussian* if it is sub-Gaussian and $\tau^2(\xi) = \mathbb{E}\,\xi^2$.

Let $SG(\Omega)$ be the set of all sub-Gaussian random variables $\xi : \Omega \to \mathbb{R}$. It is known that $SG(\Omega)$ is a vector space with respect to the natural point-wise operations, the functional $\tau(\cdot)$ is a norm on $SG(\Omega)$ (provided the random variables which coincide a.s. are identified) and, moreover, $(SG(\Omega), \tau(\cdot))$ is a Banach space [3, 4]. It follows, that if ξ_1 and ξ_2 *are centered Gaussian random variables* (not necessarily jointly Gaussian) then the random variable $\xi_1 + \xi_2$ is sub-Gaussian, but in general $\xi_1 + \xi_2$ *may not be strictly sub-Gaussian* (even if $\mathbb{E}\,\xi_1\xi_2 = 0$) [6, Example 3.7 (d)].

From Lemma 1 we can conclude that for every $p \in]0, +\infty[$ we have

$$SG(\Omega) \subset L_p(\Omega)$$

and the norm of the inclusion mapping $\leq \beta_p$.

2 Sub-Gaussian Random Elements

Below X will be a real normed space with the dual space X^*.

We recall that a mapping $\eta : \Omega \to X$ is a *random element* (in X) if

$$\langle x^*, \eta\rangle := x^* \circ \eta$$

is a random variable for every $x^* \in X^*$.

A random element $\eta : \Omega \to X$ is called *Gaussian* if for every $x^* \in X^*$ the random variable $\langle x^*, \eta\rangle$ is Gaussian.

Such a definition of a Gaussian random element goes back to Kolmogorov [10] and Fréchet [11]. For a Gaussian random element we have the following important integrability result (Vakhania [12] for $X = lp$, $1 \le p < +\infty$; Fernique [13], Landau-Shepp [14], Skorokhod [15] in general; see [16, Corollary 2 of Proposition V.5.5, p. 329–330] for a proof):

Theorem 1 *Let η be a separably valued Gaussian random element in a normed space X. Then there is $\lambda > 0$ such that $\mathbb{E}\exp(\lambda\|\eta\|^2) < +\infty$.*

A random element $\eta : \Omega \to X$ is called *weakly sub-Gaussian* if for every $x^* \in X^*$ the random variable $\langle x^*, \eta\rangle$ is sub-Gaussian (cf. [6, 17]).

In [17] it was shown that an analogue of Theorem 1 may fail for weakly sub-Gaussian random elements (see also [6, Theorem 4.2 and Remark 4.1]).

Let us call a random element $\eta : \Omega \to X$ *strictly sub-Gaussian* if for every $x^* \in X^*$ the random variable $\langle x^*, \eta\rangle$ is strictly sub-Gaussian.

Definition 1 ([18]) A random element $\eta : \Omega \to X$ is called *sub-Gaussian*, if there is a finite constant $C_\eta \ge 0$ such that

$$\tau(\langle x^*, \eta\rangle) \le C_\eta \left(\mathbb{E}\,|\langle x^*, \eta\rangle|^2\right)^{\frac{1}{2}} < +\infty \quad \text{for every} \quad x^* \in X^* .$$

We call a random element $\eta : \Omega \to X$ satisfying conditions of Definition 1 *sub-Gaussian in Fukuda's sense*, or *F-sub-Gaussian.*

An analogue of Theorem 1 remains true for F-sub-Gaussian random elements with values in $X = L_p$ with $1 \le p < +\infty$ [18, Theorem 4.3]; however, it may fail for $X = c_0$ (S. Kwapien, personal communication).

In [18] (motivating by [19, Theorem 15 (p. 120)], where a similar concept is implicitly used) a random element $\eta : \Omega \to X$ is called *γ-sub-Gaussian* if there exists a centered Gaussian random element ζ in X such that

$$\mathbb{E}\, e^{\langle x^*, \eta\rangle} \le \mathbb{E}\, e^{\langle x^*, \zeta\rangle} \quad \text{for every} \quad x^* \in X^* .$$

We call a γ-sub-Gaussian random element *sub-Gaussian in Talagrand's sense* or *T-sub-Gaussian.* In [20, Remark 4] the definition of a γ-sub-Gaussian random element in a Hilbert space is attributed to [19].

An analogue of Theorem 1 remains true for γ-sub-Gaussian random elements in a Banach space [18, Theorem 3.4].

If $X = \mathbb{R}$ then the notion of a T-sub-Gaussian, as well as the notion of a F-sub-Gaussian random element coincides with the notion of a sub-Gaussian random variable and the notion of a F-sub-Gaussian random variable ξ with the constant $C_\xi = 1$ coincides with the notion of a strictly sub-Gaussian random variable.

If X is a finite-dimensional Banach space then weakly sub-Gaussian random elements are γ-sub-Gaussian (see [6, Proposition 4.4]). In every infinite-dimensional Banach space there exists a weakly sub-Gaussian random element, which is not γ-sub-Gaussian (see [6, Theorem 4.4]).

In what follows H will denote an infinite-dimensional separable Hilbert space with the inner product $\langle \cdot, \cdot \rangle$.

Definition 2 ([20, Definition 2.1]) Let $\mathbf{e} := \{e_n, n \in \mathbb{N}\}$ be an orthonormal basis of H. A random element η with values in H is *subgaussian with respect to* $\mathbf{e}$ if the following conditions are satisfied:

(1) For every $x \in H$ the real valued random variable $\langle x, \eta \rangle$ is sub-Gaussian (i.e. η is weakly sub-Gaussian),

(2) $\sum_{n=1}^{\infty} \tau^2(\langle e_n, \eta \rangle) < \infty$.

Using the terminology of the definition we have obtained (see [21, Theorem 1.6]) the following characterization of weakly sub-Gaussian random elements in a separable Hilbert space which are γ-sub-Gaussian.

Theorem 2 *For a random element η with values in H the following statements are equivalent:*

(i) η is γ-sub-Gaussian.

(ii) For every orthonormal basis $\mathbf{e} := \{e_n, n \in \mathbb{N}\}$ of H the random element η is subgaussian with respect to $\mathbf{e}$.

For a weakly sub-Gaussian random element η in a Banach space X let

$$T_\eta : X^* \to SG(\Omega)$$

be *the induced operator*, which sends each $x^* \in X^*$ to the element $\langle x^*, \eta \rangle \in SG(\Omega)$ (the continuity and other related properties of induced operators can be seen in [6, Proposition 4.2]).

Theorem 2 in [21] is derived from the following general result (the definitions of a 2-summing operator and a type 2 space can be seen e.g.. in [16]):

Theorem 3 *For a weakly sub-Gaussian random element η with values in a Banach space X consider the assertions:*

(i) η is γ-sub-Gaussian;

(ii) $T_\eta : X^* \to SG(\Omega)$ is a 2-summing operator.

Then $(i) \Rightarrow (ii)$. The implication $(ii) \Rightarrow (i)$ is true when X is a reflexive type 2 space.

The following statement, which is a refinement of a similar assertion contained in [5, Chap. 3], shows in particular that the implication $(i) \Longrightarrow (ii)$ of Theorem 2 may fail for a bounded symmetrically distributed elementary random element η.

Proposition 1 *Let* $\mathbf{e} := \{e_n, n \in \mathbb{N}\}$ *be an orthonormal basis of* H. *Then there exists a symmetric bounded random element* $\eta : \Omega \to H$ *with a countable range, such that*

(a) $\sum_{i=1}^{\infty} \|\langle \eta, e_i \rangle\|_{L_p}^2 < \infty$ *for every* $p \in]0, \infty[$;

(b) $\sum_{i=1}^{\infty} (\tau(\langle \eta, e_i \rangle))^2 = \infty$ *and hence* η *is not subgaussian with respect to* $\mathbf{e}$.

Proof (a). Denote

$$I_n = \{2^n - 1, \ldots, 2^{n+1} - 2\}, \;\; n = 1, 2, \ldots$$

and

$$b_n = 2^{-n} \sum_{k \in I_n} e_k, \;\; n = 1, 2, \ldots.$$

Observe that

$$\sum_{k=1}^{\infty} \|b_k\|^2 = \sum_{n=1}^{\infty} \sum_{k \in I_n} \|b_k\|^2 = \sum_{n=1}^{\infty} 2^{-2n} \cdot 2^n = 1 \,.$$

Thus we can define a probability measure $\mathbb{P}$ on $\Omega := \mathbb{N}$ and a random element $\eta : \Omega \to H$ by setting:

$$\mathbb{P}(\{2n-1\}) = \mathbb{P}(\{2n\}) = \frac{1}{2}\|b_n\|^2, \;\; n = 1, 2, \ldots$$

and

$$\eta(2n-1) = -\frac{b_n}{\|b_n\|}, \;\; \eta(2n) = \frac{b_n}{\|b_n\|}, \;\; n = 1, 2, \ldots.$$

Fix now $p \in]0, \infty[$ and $i \in \mathbb{N}$. Clearly,

$$\mathbb{E}|\langle \eta, e_i \rangle|^p = \sum_{n=1}^{\infty} \left(\sum_{k \in I_n} \langle e_k, e_i \rangle \right) \frac{1}{2^{n(1+p/2)}} \,.$$

Hence

$$\mathbb{E}|\langle \eta, e_i \rangle|^p = \frac{1}{2^{n(1+p/2)}} \quad \text{for every} \quad i \in I_n, \;\; n = 1, 2, \ldots$$

and so

$$\sum_{i=1}^{\infty} \|T_\eta e_i\|_{L_p}^2 = \left(\mathbb{E}|\langle \eta, e_i \rangle|^p\right)^{2/p} = \sum_{n=1}^{\infty} \sum_{k \in I_n} \frac{1}{2^{n(1+2/p)}} =$$

$$\sum_{n=1}^{\infty} \frac{2^n}{2^{n(1+2/p)}} = \sum_{n=1}^{\infty} \frac{1}{2^{2n/p}} < \infty .$$

(*b*). To a (real-valued) random variable ξ let us associate a quantity $\vartheta_2(\xi) \in [0, +\infty]$ defined by the equality:

$$\vartheta_2(\xi) = \sup_{m \in \mathbb{N}} \frac{\left(\mathbb{E}\,|\xi|^{2m}\right)^{1/2m}}{\sqrt{m}} .$$

According to [6, Proposition 2.9(b)] we have:

$$\vartheta_2(\xi) \le \frac{2}{\sqrt{e}} \tau(\xi) \qquad \text{for every} \qquad \xi \in SG(\Omega).$$

So, it is sufficient to show that

$$\sum_{i=1}^{\infty} \left(\vartheta_2(T_\eta e_i)\right)^2 = \infty . \tag{2.1}$$

We have for every $n \in \mathbb{N}$ and $i \in I_n$:

$$\vartheta_2(\langle \eta, e_i \rangle) = \sup_m \frac{\left(\mathbb{E}\,|\langle \eta, e_i \rangle|^{2m}\right)^{1/2m}}{\sqrt{m}} = \sup_m \frac{1}{2^{n(1/2+1/2m)}\sqrt{m}} \ge \frac{1}{2^{n(1/2+1/2n)}\sqrt{n}} .$$

Hence

$$\sum_{i=1}^{\infty} \vartheta_2^2(\langle \eta, e_i \rangle) = \sum_{n=1}^{\infty} \sum_{i \in I_n} \vartheta_2^2(\langle \eta, e_i \rangle) \ge \sum_{n=1}^{\infty} 2^n \left(\frac{1}{2^{n(1/2+1/2n)}\sqrt{n}} \right)^2 = \frac{1}{2} \sum_{n=1}^{\infty} \frac{1}{n} = \infty$$

and (2.1) is proved.

The authors do not know whether the following conjecture related with Proposition 1 is true.

Conjecture 1 *There exists a symmetric bounded random element $\eta : \Omega \to H$ such that*

(*a*) $\sum_{i=1}^{\infty} \|\langle \eta, e_i \rangle\|_{L_p}^2 < \infty$ for every $p \in]0, \infty[$ *and for every orthonormal basis* $\mathbf{e} := \{e_n, n \in \mathbb{N}\}$ of H;

(*b*) $\sum_{i=1}^{\infty} (\tau(\langle \eta, e_i \rangle))^2 = \infty$ *for some orthonormal basis* $\mathbf{e} := \{e_n, n \in \mathbb{N}\}$ of H.

Acknowledgements 1. The authors are grateful to Professor Sergei Chobanyan for useful remarks. 2. The third author was partially supported by Shota Rustaveli National Science Foundation of Georgia (SRNSFG) grant no. DI-18-1429: "Application of probabilistic methods in discrete optimization and scheduling problems".

References

1. Kahane, J.P.: Propriétés locales des fonctions à séries de Fourier aleatoires. Stud. Math. **19**, 1–25 (1960)
2. Kahane, J.P.: Some Random Series of Functions, 2nd edn. Cambridge University Press, Cambridge (1985)
3. Buldygin, V.V., Kozachenko, YuV.: Subgaussian random variables. Ukrainian Math. J. **32**, 723–730 (1980)
4. Buldygin, V.V., Kozachenko, Yu.V.: Metric Characterization of Random Variables and Random Processes. Translations of Mathematical Monographs, vol. 188. AMS, Providence R.I. (2000)
5. Kvaratskhelia, V.V.: Unconditional Convergence of Functional Series in Problems of Infinite-Dimensional Probability Theory. Doctoral Dissertation, Tbilisi (2002)
6. Vakhania, N.N., Kvaratskhelia, V.V., Tarieladze, V.I.: Weakly Sub-Gaussian random elements in Banach spaces. Ukrainian Math. J. **57**, 1187–1208 (2005)
7. Ruegg, A.: A characterization of certain infinitely divisible laws. Ann. Math. Stat. **41**, 1354–1356 (1974)
8. Horn, R.A.: On necessary and sufficient conditions for an infinitely divisible distribution to be normal or degenerate. Z. Wahrscheinlichkeitstheorie verw. Geb. **21**, 179–187 (1972)
9. Kruglov, V.M.: Characterization of a class of infinitely divisible distributions in a Hilbert space. Mat. Zametki **16**, 777–782 (1974)
10. Kolmogorov, A.N.: La transformation de Laplace dans les espaces linéaires. C.R. Acad. Sc. Paris **200**, 1717–1718 (1935)
11. Fréchet, M.: Généralisation de la loi de probabilité de Laplace. Ann. Inst. H. Poincare **12**(1), 1–29 (1951)
12. Vakhania, N.N.: Sur les répartitions de probabilités dans les espaces de suites numériques. C.R. Acad. Sc. Paris **260**, 1560–1562 (1965)
13. Fernique, X.: Intégrabilité des vecteurs gaussiens. C.R. Acad. Sc. Paris, Série A **270**, 1698–1699 (1970)
14. Landau, H.J., Shepp, L.A.: On the supremum of Gaussian process. Sankhya, Série A **32**(4), 369–378 (1970)
15. Skorokhod, A.V.: A note on Gaussian measures in a Banach space. Theory Probab. Appl. **15**(3), 508–508 (1970)
16. Vakhania, N.N., Tarieladze, V.I., Chobanyan, S.A.: Probability distributions on Banach spaces. D. Reidel Publishing Company, Dordrecht (1987)
17. Vakhania, N.: Subgaussian random vectors in normed spaces. Bull. Georg. Acad. Sci. **163**, 8–11 (2001)
18. Fukuda, R.: Exponential integrability of sub-Gaussian vectors. Probab. Theory Relat. Fields **85**, 505–521 (1990)
19. Talagrand, M.: Regularity of gaussian processes. Acta Math. **159**, 99–149 (1987)
20. Antonini, R.G.: Subgaussian random variables in Hilbert spaces. Rend. Sem. Mat. Univ. Padova **98**, 89–99 (1997)
21. Kvaratskhelia, V., Tarieladze, V., Vakhania, N.: Characterization of γ-subgaussian random elements in a Banach space. J. Math. Sci. **216**, 564–568 (2016)

Localized Boundary-Domain Integro-Differential Equations Approach for Stationary Heat Transfer Equation

Sveta Gorgisheli, Maia Mrevlishvili, and David Natroshvili

Abstract Localized boundary-domain integro-differential equations (LBDIDE) systems associated with the Dirichlet and Robin boundary value problems (BVP) for the stationary heat transfer partial differential equation (PDE) with a variable coefficient are obtained and analysed. Localization is performed by a non-smooth parametrix represented as the product of a global parametrix and the characteristic function of a ball centered at a reference point. The equivalence of the LBDIDE systems to the original variable-coefficient BVPs and unique solvability of the LBDIDE systems in appropriate Sobolev spaces are the main results of the present paper.

Keywords Elliptic problems with variable coefficients · Localized parametrix · Localized Boundary-Domain Integral Equations

1 Introduction

Partial Differential Equations (PDEs) with variable coefficients arise naturally in mathematical modelling of inhomogeneous media in solid mechanics, electromagnetics, thermo-conductivity, fluid flows through porous media, and other areas of physics and engineering. The Boundary Integral Equation Method/Boundary Element Method (BIEM/BEM) is a well established tool for Boundary Value Problems (BVPs) with constant coefficients. The main ingredient for reducing a BVP to a BIE is a fundamental solution to the original PDE. But for PDEs with variable coefficients a fundamental solution is generally not available in an analytical form. However, in

S. Gorgisheli · M. Mrevlishvili · D. Natroshvili (✉)
Georgian Technical University, 77 M.Kostava st., 0160 Tbilisi, Georgia
e-mail: natrosh@hotmail.com

S. Gorgisheli
e-mail: 18barabare@gmail.com

M. Mrevlishvili
e-mail: m_mrevlishvili@yahoo.com

G. Jaiani and D. Natroshvili (eds.), *Applications of Mathematics and Informatics in Natural Sciences and Engineering*, Springer Proceedings in Mathematics & Statistics 334, https://doi.org/10.1007/978-3-030-56356-1_12

this case, one can use a global or a localized parametrix (Levi function) as a substitute for the fundamental solution. This approach reduces the BVP not to a boundary integral equation but to a system of Boundary-Domain Integral Equations (BDIE) or to a system of Localized Boundary-Domain Integral Equations (LBDIE), see e.g. [1–3, 13, 14]. A discretization of the BDIE constructed by a global parametrix leads then to a system of algebraic equations of the similar size as in the Finite Element Method (FEM), however the matrix of the system is not sparse as in the FEM but dense and thus less efficient for numerical solution.

If instead of a global parametrix, we will use specially constructed localized parametrix then BVPs with variable coefficients are reduced to Localized Boundary-Domain Integral or Integro-Differential Equations. After a locally-supported mesh-based or mesh-less discretization this approach leads to systems of algebraic equations with sparse matrices. For smooth localizing cut-off functions this method is theoretically studied and substantiated in [2, 4], where the BVPs are reduced to systems of *Localized boundary-domain integral equations*.

In the present paper, we consider the Dirichlet and Robin BVPs for a divergence type elliptic differential equation with one variable coefficient (arising in the theory of heat transfer in isotropic inhomogeneous medium). We employ a parametrix which is localized by the characteristic function of a ball with an arbitrary radius centered at a reference point. Such a localized parametrix has a simple structure, but it is discontinuous in the whole space, which leads to essential difficulties in our analysis. With the help of the localized parametrix, the BVPs are reduced to systems of *Localized boundary-domain integro-differential equations* (LBDIDEs). We prove equivalence of the LBDIDEs to the original BVPs and establish unique solvability of the systems of LBDIDEs in appropriate Sobolev spaces.

2 Localised Green's Formula and Boundary-Domain Integro-Differential Relations

Let Ω be a bounded region of $\mathbb{R}^3$ surrounded by a simply connected Lipschitz surface $S = \partial\Omega$. Let $B(y, \varepsilon) := \{x \in \mathbb{R}^3 \,:\, |x - y| \leqslant \varepsilon\}$ be a ball centered at y and radius ε, where ε is a fixed positive number, and $\Sigma(y, \varepsilon) := \partial B(y, \varepsilon)$. Further, let

$$\begin{aligned} &\Omega(y,\varepsilon) := \Omega \cap B(y,\varepsilon), \qquad &&S(y,\varepsilon) := S \cap B(y,\varepsilon), \\ &\Sigma_1(y,\varepsilon) := \Sigma(y,\varepsilon) \cap \Omega, \qquad &&\ell(y,\varepsilon) := \partial\Sigma_1(y,\varepsilon) = \partial S(y,\varepsilon). \end{aligned} \tag{1}$$

It is evident that if the distance from the point y to the boundary $S = \partial\Omega$ is grater than ε, $\mathrm{dist}(y; S) > \varepsilon$, then $S(y, \varepsilon) = \varnothing$ and $\Sigma_1(y, \varepsilon) = \Sigma(y, \varepsilon)$. Note also that for $y \in \overline{\Omega}$ the part of the spherical surface $\Sigma_1(y, \varepsilon)$ always possesses a positive measure.

We assume that for a given domain Ω there is $\varepsilon_0 > 0$, such that for arbitrary $y \in \overline{\Omega}$ and $0 < \varepsilon < \varepsilon_0$ the corresponding domain $\Omega(y, \varepsilon)$ is a piecewise smooth Lipschitz domain. Notice that this condition is satisfied for a convex domain and for a domain with a smooth Lyapunov boundary $S = \partial\Omega \in C^{2,\alpha}$, $\alpha > 0$. We need this condition to write the corresponding Green identities in the domain $\Omega(y,\varepsilon)$, $y \in \overline{\Omega}$, and also to establish mapping properties of potential type integral operators involved in our analysis.

By $H^s(\Omega) = H_2^s(\Omega)$ and $H^s(S) = H_2^s(S)$, $s \in \mathbb{R}$, we denote the L_2-based Bessel potential spaces of functions on an open domain $\Omega \subset \mathbb{R}^3$ and on a closed manifold $S = \partial\Omega$. Recall that $H^0(\Omega) = L_2(\Omega)$ is a space of square integrable functions on Ω and $H^r(\Omega) = W_2^r(\Omega)$ for $r \geqslant 0$, where $W_2^r(\Omega)$ is the Sobolev space.

We consider the following stationary heat transfer elliptic differential equation with variable coefficient

$$A(x,\partial_x)u(x) := \sum_{k=1}^{3} \frac{\partial}{\partial x_k}\Big(a(x)\frac{\partial u(x)}{\partial x_k}\Big) = f(x), \quad x \in \Omega, \tag{2}$$

where $f \in H^0(\Omega)$ and

$$a \in C^2(\overline{\Omega}), \quad 0 < a_0 \leqslant a(x) \leqslant a_1, \quad \forall x \in \overline{\Omega}, \tag{3}$$

with some constants a_0 and a_1. We employ the notation $\partial_x = (\partial_{x_1}, \partial_{x_2}, \partial_{x_3})$ with $\partial_{x_k} = \partial/\partial x_k$, $k = 1, 2, 3$.

A solution function u is sought in the space

$$H^{1,0}(\Omega, A) = \{v \in H^1(\Omega) \,:\, Av \in H^0(\Omega)\}.$$

Due to the interior regularity property of solutions to elliptic equations, any solution to Eq. (2) with $f \in H^0(\Omega)$ belongs to $H^2(\Omega^*)$ for any open region Ω^* with $\overline{\Omega^*} \subset \Omega$.

For an arbitrary domain $\Omega_1 \subset \mathbb{R}^3$ with Lipschitz boundary and a function $u \in H^s(\Omega_1)$ with $s > \frac{3}{2}$ the conormal derivative can be calculated in the *conventional Sobolev trace sense*,

$$T^+(x,\partial_x)u(x) = a(x)\,\gamma^+\left(\frac{\partial u(x)}{\partial n(x)}\right) = \sum_{k=1}^{3} a(x)\,n_k(x)\,\gamma^+\Big(\frac{\partial u(x)}{\partial x_k}\Big), \quad x \in \partial\Omega_1, \tag{4}$$

where $\gamma^+ = \gamma^+_{\partial\Omega_1}$ is the trace operator on $\partial\Omega_1$, $n(x)$ is the unit normal vector at the point $x \in \partial\Omega_1$ directed outward Ω_1, and $\frac{\partial}{\partial n(x)}$ denotes the usual directional normal derivative. Due to the Lipschitz character of the boundary $\partial\Omega_1$, the components of the normal vector are essentially bounded measurable functions, $n_k \in L_\infty(\partial\Omega_1)$. Moreover, for $u \in H^s(\Omega_1)$, $s > \frac{3}{2}$, we have $\gamma^+_{\partial\Omega_1}(\partial_k u) \in H^\lambda(\partial\Omega_1)$ with $\lambda = \min\{1,\ s - \frac{3}{2}\}$ for $s \neq \frac{5}{2}$, while for $s = \frac{5}{2}$ in the role of the parameter λ can be taken any number less than 1. Consequently, in this case, $T^+u \in L_2(\partial\Omega_1)$.

With the help of Green's first identity for an arbitrary Lipschitz domain $\Omega_1 \subseteq \Omega$ and an arbitrary function $u \in H^{1,0}(\Omega_1, A)$ we can define on $\partial\Omega_1$ the *canonical conormal derivative* $T^+u \equiv T^+_{\partial\Omega_1}u \in H^{-\frac{1}{2}}(\partial\Omega_1)$ by the relation

$$\langle T^+u\,,\,g\rangle_{\partial\Omega_1} \equiv \langle T^+_{\partial\Omega_1}u\,,\,g\rangle_{\partial\Omega_1} := \int\limits_{\Omega_1}\sum_{k=1}^{3} a\,\frac{\partial u}{\partial x_k}\,\frac{\partial v}{\partial x_k}\,dx + \int\limits_{\Omega_1} v\,Au\,dx \quad \forall\, g \in H^{\frac{1}{2}}(\partial\Omega_1), \tag{5}$$

where $v \in H^1(\Omega_1)$ with $\gamma^+_{\partial\Omega_1}v = g$ (for details see [5], [11, Chap. 4], [2, 12]). Below we drop the subscript $\partial\Omega_1$ in the notation of trace operator and conormal derivative operator when it does not lead to misunderstanding.

For arbitrary functions $u, v \in H^{1,0}(\Omega_1, A)$ we have the following Green first and second identities (cf. [11, Chap. 4])

$$\int\limits_{\Omega_1} v\,Au\,dx + \int\limits_{\Omega_1}\sum_{k=1}^{3} a\,\frac{\partial u}{\partial x_k}\,\frac{\partial v}{\partial x_k}\,dx = \langle T^+u\,,\,\gamma^+v\rangle_{\partial\Omega_1}, \tag{6}$$

$$\int\limits_{\Omega_1}\big(v\,Au - u\,Av\big)\,dx = \langle T^+u\,,\,\gamma^+v\rangle_{\partial\Omega_1} - \langle T^+v\,,\,\gamma^+u\rangle_{\partial\Omega_1}. \tag{7}$$

Remark 1 Here and in what follows the angled brackets should be understood as duality pairing of $H^{-\frac{1}{2}}(\partial\Omega_1)$ with $H^{\frac{1}{2}}(\partial\Omega_1)$. In the case of a proper submanifold $S_1 \subset \partial\Omega_1$ with Lipschitz boundary curve $\partial S_1 \neq \varnothing$ (e.g., $S_1 = S(y, \varepsilon)$ or $S_1 = \Sigma_1(y, \varepsilon)$ for $\text{dist}(y, S) < \varepsilon$) the angled brackets denote duality pairing of either $H^{-\frac{1}{2}}(S_1)$ with $\widetilde{H}^{\frac{1}{2}}(S_1)$ or $\widetilde{H}^{-\frac{1}{2}}(S_1)$ with $H^{\frac{1}{2}}(S_1)$, where

$$\widetilde{H}^s(S_1) := \big\{g \in H^s(\partial\Omega_1)\ :\ \text{supp}g \subseteq \overline{S}_1\big\},$$
$$H^s(S_1) := \big\{r_{S_1}g\ :\ g \in H^s(\partial\Omega_1)\big\}.$$

Here r_{S_1} is the restriction operator onto S_1. Note that $\widetilde{H}^s(S_1)$ and $H^{-s}(S_1)$ are mutually adjoint spaces (see, e.g., [11]): $\widetilde{H}^s(S_1) := [H^{-s}(S_1)]^*$ and $H^s(S_1) := [\widetilde{H}^{-s}(S_1)]^*$.

Throughout the paper, all surface integrals, which do not exist in the usual classical sense, should be understood in the duality pairing sense.

Introduce a harmonic localized parametrix

$$P_\chi(x) := -\frac{\chi(x)}{4\pi\,|x|}, \tag{8}$$

where χ is the characteristic function of the ball $B(O, \varepsilon)$,

$$\chi(x) := \begin{cases} 1 & \text{for } |x| < \varepsilon, \\ 0 & \text{for } |x| > \varepsilon. \end{cases} \tag{9}$$

Using Green's second formula (7) for the domain $\Omega_1 = \Omega(y, \varepsilon) \setminus B(y, \delta)$ with $\delta \in (0, \varepsilon)$ and for functions $u \in H^{1,0}(\Omega, A)$ and $P_\chi(y - \cdot) \in H^{1,0}(\Omega \setminus B(y, \delta), A)$, and passing to the limit as $\delta \to 0$, by standard arguments one can derive Green's third formula (cf, e.g., [2])

$$a(y)\, u(y) + \mathscr{R}_\varepsilon\, u(y) - V_\varepsilon(T^+u)(y) + W_\varepsilon(\gamma^+u)(y) = \mathscr{P}_\varepsilon(Au)(y) \quad \forall\, y \in \Omega, \tag{10}$$

where γ^+u and T^+u are respectively the trace of u and the canonical conormal derivative of u on the boundary $\partial\Omega(y, \varepsilon) = S(y, \varepsilon) \cup \Sigma_1(y, \varepsilon) \cup \ell(y, \varepsilon)$,

$$\gamma^+u \in H^{\frac{1}{2}}(\partial\Omega(y, \varepsilon)), \qquad T^+u \in H^{-\frac{1}{2}}(\partial\Omega(y, \varepsilon)), \tag{11}$$

$\mathscr{R}_\varepsilon$ is a localized weakly singular integral operator

$$\mathscr{R}_\varepsilon\, u(y) := \lim_{\delta \to 0} \int\limits_{\Omega(y,\varepsilon)\setminus B(y,\delta)} [A(x, \partial_x)P_\chi(x - y)]u(x)\, dx = \int\limits_{\Omega(y,\varepsilon)} R(x, y)\, u(x)\, dx, \tag{12}$$

$$R(x, y) := -\frac{1}{4\pi} \sum_{k=1}^{3} \frac{\partial a(x)}{\partial x_k} \frac{\partial}{\partial x_k} \frac{1}{|x - y|} = \mathscr{O}(|x - y|^{-2}) \quad \text{for} \quad x \in \Omega(y, \varepsilon); \tag{13}$$

V_ε, W_ε, and $\mathscr{P}_\varepsilon$ are the localized single layer, double layer, and Newtonian volume type potentials respectively,

$$\begin{aligned} V_\varepsilon\,(T^+u)(y) &:= -\int\limits_{\partial\Omega(y,\varepsilon)} P_\chi(x - y)\; T^+u(x)\, dS_x \\ &= \frac{1}{4\pi} \int\limits_{S(y,\varepsilon)\cup\Sigma_1(y,\varepsilon)} \frac{1}{|x - y|}\, T^+u(x)\, dS_x, \end{aligned} \tag{14}$$

$$\begin{aligned} W_\varepsilon\,(\gamma^+u)(y) &:= -\int\limits_{\partial\Omega(y,\varepsilon)} \big[T(x, \partial_x)\, P_\chi(x - y)\big]\, \gamma^+u(x)\, dS_x \\ &= \frac{1}{4\pi} \int\limits_{S(y,\varepsilon)\cup\Sigma_1(y,\varepsilon)} \left[T(x, \partial_x) \frac{1}{|x - y|}\right] \gamma^+u(x)\, dS_x, \end{aligned} \tag{15}$$

$$\mathscr{P}_\varepsilon\,(Au)(y) := \int\limits_{\Omega(y,\varepsilon)} P_\chi(x - y)\, Au(x)\, dx = -\frac{1}{4\pi} \int\limits_{\Omega(y,\varepsilon)} \frac{1}{|x - y|}\, Au(x)\, dx. \tag{16}$$

Note that the above layer type potentials are not standard classical potentials, since the surfaces of integration depend on the variable y and contain a spherical subsurface $\Sigma_1(y,\varepsilon)$ located in the interior part of the domain Ω. Therefore the corresponding densities in the layer potentials (14) and (15) must be well-defined for all possible integration surfaces $\partial\Omega(y,\varepsilon) = S(y,\varepsilon) \cup \Sigma_1(y,\varepsilon) \cup \ell(y,\varepsilon)$ with $y \in \overline{\Omega}$. Evidently, the potentials (14)–(16) are well defined due to the inclusion $u \in H^{1,0}(\Omega, A)$ implying (11).

Keeping in mind the agreement about the direction of the normal vector, it is easy to see that formulas (14)–(16) can be rewritten in the following form

$$V_\varepsilon(T^+u)(y) = \frac{1}{4\pi}\int\limits_S \frac{1}{|x-y|}\, T^+u(x)\, dS_x - \frac{1}{4\pi}\int\limits_{\partial[\Omega\setminus B(y,\varepsilon)]} \frac{1}{|x-y|}\, T^+u(x)\, dS_x \tag{17}$$

$$\begin{aligned} W_\varepsilon(\gamma^+u)(y) = {} & \frac{1}{4\pi}\int\limits_S \Big[T(x,\partial_x)\frac{1}{|x-y|}\Big]\gamma^+u(x)\, dS_x \\ & - \frac{1}{4\pi}\int\limits_{\partial[\Omega\setminus B(y,\varepsilon)]} \Big[T(x,\partial_x)\frac{1}{|x-y|}\Big]\gamma^+u(x)\, dS_x, \end{aligned} \tag{18}$$

$$\mathscr{P}_\varepsilon h(y) = -\frac{1}{4\pi}\int\limits_\Omega \frac{1}{|x-y|}\, Au(x)\, dx + \frac{1}{4\pi}\int\limits_{\Omega\setminus\Omega(y,\varepsilon)} \frac{1}{|x-y|}\, Au(x)\, dx. \tag{19}$$

In these relations the normal vector to the surface $\partial[\Omega \setminus B(y,\varepsilon)]$ is directed outward the domain $\Omega \setminus B(y,\varepsilon)$.

The first summands in (17)–(18) are the classical harmonic single and double layer potentials,

$$V_\Delta\, \psi(y) := \frac{1}{4\pi}\int\limits_S \frac{1}{|x-y|}\, \psi(x)\, dS_x, \tag{20}$$

$$W_\Delta\, \varphi(y) := \frac{1}{4\pi}\int\limits_S \Big[T(x,\partial_x)\frac{1}{|x-y|}\Big]\varphi(x)\, dS_x, \tag{21}$$

possessing the mapping properties (see, e.g., [5])

$$V_\Delta : H^{-\frac{1}{2}}(S) \to H^1(\mathbb{R}^3), \qquad W_\Delta : H^{\frac{1}{2}}(S) \to H^1(\Omega). \tag{22}$$

The first term in (19) is the classical Newtonian volume potential associated with the Laplace operator

$$\mathscr{P}_\Delta\, h(y) := -\frac{1}{4\pi}\int\limits_\Omega \frac{1}{|x-y|}\, h(x)\, dx \tag{23}$$

with the mapping property (see e.g. [11, Chaps. 4, 6])

$$\mathscr{P}_{\Delta} : H^0(\Omega) \to H^2_{loc}(\mathbb{R}^3). \tag{24}$$

The kernels in the last summands in (17)–(19) are continuous functions with respect to x for arbitrary y. Therefore the corresponding potential type operators with integration regions depending on y are continuous with respect to $y \in \overline{\Omega}$ and the traces of functions (17)–(19) on S exist in the classical sense.

Note that relation (16) can be also rewritten as

$$\mathscr{P}_{\varepsilon} h(y) = \int\limits_{\Omega(y,\varepsilon)} P_{\chi}(x-y)\, h(x)\, dx = \int\limits_{\mathbb{R}^3} P_{\chi}(x-y)\, \widetilde{h}(x)\, dx, \quad y \in \mathbb{R}^3, \tag{25}$$

where $\widetilde{h} \in \widetilde{H}^0(\Omega)$ is the extension by zero of the function $h \in H^0(\Omega)$ from Ω to $\mathbb{R}^3 \setminus \overline{\Omega}$. Evidently $\operatorname{supp} \widetilde{h} \cap \operatorname{supp} \chi(\cdot - y) \subset \overline{\Omega(y,\varepsilon)}$.

Thus $\mathscr{P}_{\varepsilon} h$ is a convolution type pseudodifferential operator and applying the same arguments as in [2] we find the corresponding symbol

$$\begin{aligned}
\widehat{P}_{\chi}(\xi) &:= \mathscr{F}_{x\to\xi}\Big[-\frac{1}{4\pi}\,\frac{\chi(x)}{|x|}\Big] = -\frac{1}{4\pi} \int\limits_{\mathbb{R}^3} \frac{\chi(x)}{|x|}\, e^{i\,x\cdot\xi}\, dx \\
&= -\frac{1}{4\pi} \int\limits_0^{\varepsilon}\int\limits_0^{\pi}\int\limits_0^{2\pi} e^{i\,\rho\,|\xi|\cos\theta}\, \rho\, \sin\theta\, d\varphi\, d\theta\, d\rho \\
&= -\frac{1}{|\xi|} \int\limits_0^{\varepsilon} \sin(\rho\,|\xi|)\, d\rho = -\frac{1-\cos(\varepsilon|\xi|)}{|\xi|^2} \\
&= -\frac{\varepsilon^2}{2} \left(\frac{\sin\frac{\varepsilon|\xi|}{2}}{\frac{\varepsilon|\xi|}{2}} \right)^2,
\end{aligned} \tag{26}$$

where $\mathscr{F}_{x\to\xi}$ denotes the distributional Fourier transform operator which for a Lebesgue integrable function f reads as

$$\mathscr{F}_{x\to\xi}[f] = \int\limits_{\mathbb{R}^3} e^{i\,x\cdot\xi}\, f(x)\, dx.$$

In view of (26), it is evident that $|\widehat{P}_{\chi}(\xi)| \leqslant \frac{\varepsilon^2}{2}$ and $|\xi|^2 |\widehat{P}_{\chi}(\xi)| \leqslant 2$ for arbitrary $\xi \in \mathbb{R}^3$. Consequently we have the estimate

$$|\widehat{P}_{\chi}(\xi)| \le \frac{C}{1+|\xi|^2}, \quad C = 2 + \frac{\varepsilon^2}{2}, \quad \forall \xi \in \mathbb{R}^3, \tag{27}$$

implying (see, e.g., [8, Lemma 4.4])

$$\mathscr{P}_\varepsilon : H^t(\mathbb{R}^3) \to H^{t+2}(\mathbb{R}^3) \qquad \forall\, t \in \mathbb{R}. \tag{28}$$

In particular, $\mathscr{P}_\varepsilon \widetilde{h} \in H^2(\mathbb{R}^3)$ holds for arbitrary $\widetilde{h} \in H^0(\Omega)$ and therefore for an arbitrary Lipschitz domain $\Omega_1 \subseteq \Omega$ we have the inclusions (see, e.g., [5, 7, 9])

$$\gamma^{\pm}_{\partial\Omega_1} \mathscr{P}_\varepsilon \widetilde{h} \in H^1(\partial\Omega_1), \qquad \gamma^{\pm}_{\partial\Omega_1}\Big[\frac{\partial \mathscr{P}_\varepsilon \widetilde{h}}{\partial y_k}\Big] \in H^{\frac{1}{2}}(\partial\Omega_1), \ \ k = 1, 2, 3, \tag{29}$$

where $\gamma^{\pm}_{\partial\Omega_1}$ denotes one-sided trace operators on $\partial\Omega_1$.

Further, from (14)–(15), using the representations of type (17)–(18) and the jump properties of layer potentials on smooth and piecewise smooth Lipschitz manifolds, we deduce (see, e.g., [5], [11, Chap. 6], [2])

$$\gamma_S^{\pm} V_\varepsilon\,(T^+u)(y) = \mathscr{V}_\varepsilon\,(T^+u)(y), \quad y \in S, \tag{30}$$

$$\gamma_S^{\pm} W_\varepsilon\,(\gamma^+u)(y) = \mp\frac{1}{2}\,a(y)\gamma_S^+u(y) + \mathscr{W}_\varepsilon\,(\gamma^+u)(y), \quad y \in S, \tag{31}$$

where

$$\begin{aligned}\mathscr{V}_\varepsilon\,(T^+u)(y) &:= -\int\limits_{\partial\Omega(y,\varepsilon)} P_\chi(x-y)\, T^+u(x)\, dS_x \\ &= \frac{1}{4\pi}\int\limits_{S(y,\varepsilon)\cup\Sigma_1(y,\varepsilon)} \frac{1}{|x-y|}\, T^+u(x)\, dS_x, \quad y \in S,\end{aligned} \tag{32}$$

$$\begin{aligned}\mathscr{W}_\varepsilon\,(\gamma^+u)(y) &:= -\int\limits_{\partial\Omega(y,\varepsilon)} \big[T(x,\partial_x)\, P_\chi(x-y)\big]\, \gamma^+u(x)\, dS_x \\ &= \frac{1}{4\pi}\int\limits_{S(y,\varepsilon)\cup\Sigma_1(y,\varepsilon)} \Big[T(x,\partial_x)\frac{1}{|x-y|}\Big]\, \gamma^+u(x)\, dS_x, \quad y \in S,\end{aligned} \tag{33}$$

From the above arguments, it follows that the trace on S of Green's third formula (10) exists and reads as

$$\gamma_S^+ \mathscr{R}_\varepsilon\, u(y) - \mathscr{V}_\varepsilon(T^+u)(y) + \frac{1}{2}\,a(y)\,\gamma_S^+u(y) + \mathscr{W}_\varepsilon(\gamma^+u)(y) = \gamma_S^+ \mathscr{P}_\varepsilon(Au)(y) \ \text{ on } S. \tag{34}$$

Now let us prove the following auxiliary lemma which plays a crucial role in our further analysis.

Lemma 1 *Let* ε *be a fixed positive number and* $\Omega(y,\varepsilon)$ *be the domain defined in* (1). *Let* $g \in \widetilde{H}^0(\Omega)$ *and*

$$\int\limits_{\Omega(y,\varepsilon)} \frac{1}{|x-y|}\, g(x)\, dx = 0 \qquad \forall y \in \Omega. \tag{35}$$

Then $g = 0$ in Ω.

Proof Let us put

$$\Phi(y) := \int\limits_{\mathbb{R}^3} \frac{\chi(x-y)}{|x-y|}\, g(x)\, dx \qquad \forall y \in \mathbb{R}^3, \tag{36}$$

where the cut-off function χ is given by (9). Since supp $\chi(\cdot - y) \cap \operatorname{supp} g \subset \Omega(y,\varepsilon)$, in view of (35) we have

$$\Phi(y) = \int\limits_{\Omega(y,\varepsilon)} \frac{1}{|x-y|}\, g(x)\, dx = 0 \quad \text{for} \quad y \in \Omega. \tag{37}$$

Evidently, if $y \in \mathbb{R}^3 \setminus \overline{\Omega}$ and $\operatorname{dist}(y, \partial\Omega) > \varepsilon$, then supp $\chi(\cdot - y) \cap \operatorname{supp} g = \varnothing$ and $\Phi(y) = 0$. Therefore supp Φ is located in the closure of the one-side exterior ε-neighbourhood of the boundary $\partial\Omega$,

$$\operatorname{supp}\Phi \subset \Omega_\varepsilon^- := \{x \in \mathbb{R}^3 \setminus \Omega \, : \operatorname{dist}(x, \partial\Omega) \leqslant \varepsilon\}$$

and evidently $\Phi \in \widetilde{H}^0(\Omega_\varepsilon^-)$.

Since both functions Φ and g belong to the space $L_2(\mathbb{R}^3)$ and the intersection supp $g \cap$ supp Φ does not contain interior points in $\mathbb{R}^3$, by the Plancherel theorem we deduce

$$\int\limits_{\mathbb{R}^3} \Phi(x)\, g(x)\, dx = \int\limits_{\mathbb{R}^3} \widehat{\Phi}(\xi)\, \overline{\widehat{g}(\xi)}\, d\xi = 0 \tag{38}$$

where $\widehat{\Phi}(\xi) = \mathscr{F}_{x\to\xi}[\Phi]$ and $\widehat{g}(\xi) = \mathscr{F}_{x\to\xi}[g]$ stand for the Fourier transform of the square integrable functions Φ and g, respectively, and the over bar denotes complex conjugation.

Since Φ, defined in (36), is a convolution, we have

$$\widehat{\Phi}(\xi) = \mathscr{F}_{x\to\xi}\Big[\frac{\chi(x)}{|x|}\Big]\, \mathscr{F}_{x\to\xi}[g]. \tag{39}$$

In view of (26) we find

$$\widehat{\Phi}(\xi) = 2\pi\varepsilon^2 \left(\frac{\sin \frac{\varepsilon|\xi|}{2}}{\frac{\varepsilon|\xi|}{2}} \right)^2 \widehat{g}(\xi)\, . \tag{40}$$

Therefore from (38) and (40) we get

$$2\pi\varepsilon^2 \int\limits_{\mathbb{R}^3} \left(\frac{\sin \frac{\varepsilon|\xi|}{2}}{\frac{\varepsilon|\xi|}{2}} \right)^2 |\widehat{g}(\xi)|^2 \, d\xi = 0. \tag{41}$$

Due to non-negativity of the integrand we conclude

$$\left(\frac{\sin \frac{\varepsilon|\xi|}{2}}{\frac{\varepsilon|\xi|}{2}} \right)^2 |\widehat{g}(\xi)|^2 = 0 \ \text{ almost everywhere in } \mathbb{R}^3. \tag{42}$$

The first multiplier in the last equation is a real analytic function in $|\xi|$ which vanishes on the two-dimensional spheres of radius $|\xi| = \frac{2k\pi}{\varepsilon}$, $k = 0, 1, 2, \cdots$ Consequently, since $\widehat{g} \in L_2(\mathbb{R}^3)$, we have $\widehat{g}(\xi) = 0$ almost everywhere in $\mathbb{R}^3$, implying $g(x) = 0$ almost everywhere in $\mathbb{R}^3$, which completes the proof.

3 Reduction to Systems of Boundary Domain Integro-Differential Equations

Here we reformulate the basic boundary value problems for Eq. (2) in the form of equivalent boundary domain integro-differential equations, in particular, we consider in detail the Dirichlet and Robin boundary value problems.

3.1 The Dirichlet Problem

Let us consider the Dirichlet problem for the operator $A(x, \partial)$ defined in (2):

Find a function $u \in H^{1,0}(\Omega, A)$ such that

$$A(x, \partial_x)u = f \ \text{ in } \Omega, \quad f \in H^0(\Omega), \tag{43}$$

$$\gamma^+ u = \varphi_0 \ \text{ on } S, \qquad \varphi_0 \in H^{\frac{1}{2}}(S). \tag{44}$$

It is a well known classical result that the Dirichlet problem (43)–(44) is uniquely solvable (see, e.g., [6, 10, 11]).

Now we reformulate the problem with the help of localizing approach based on the localized piecewise smooth parametrix P_χ introduced by (8).

Substituting the data of the problem under consideration into Green's third formula (10) and into its trace formula (34) on S, we obtain the following system of localized boundary-domain integro-differential equations with respect to unknown function u,

$$a\,u + \mathscr{R}_\varepsilon\,u - V_\varepsilon(T^+u) + W_\varepsilon(\gamma^+u) = \mathscr{P}_\varepsilon f \quad \text{in} \quad \Omega, \tag{45}$$

$$\gamma_S^+\mathscr{R}_\varepsilon\,u - \mathscr{V}_\varepsilon(T^+u) + \frac{1}{2}\,a(y)\,\varphi_0 + \mathscr{W}_\varepsilon(\gamma^+u) = \gamma_S^+\mathscr{P}_\varepsilon(f) \quad \text{on} \quad S, \tag{46}$$

where the densities of the layer potentials are the corresponding one-sided traces on the integration surface $\partial\Omega(y,\varepsilon) = S(y,\varepsilon) \cup \Sigma_1(y,\varepsilon) \cup \ell(y,\varepsilon)$. In more extended and detailed form these relations can be rewritten as follows,

$$\begin{aligned}
a(y)\,u(y) &+ \int\limits_{\Omega(y,\varepsilon)} R(x,y)\,u(x)\,dx - \frac{1}{4\pi}\int\limits_{\partial\Omega(y,\varepsilon)} \frac{1}{|x-y|}\,T^+_{\partial\Omega(y,\varepsilon)}(x,\partial x)u(x)\,dS_x \\
&+ \frac{1}{4\pi}\int\limits_{\Sigma_1(y,\varepsilon)} \Big[T(x,\partial_x)\frac{1}{|x-y|}\Big]\,\gamma^+_{\Sigma_1(y,\varepsilon)}u(x)\,dS_x = -\frac{1}{4\pi}\int\limits_{\Omega(y,\varepsilon)} \frac{1}{|x-y|}\,f(x)\,dx \\
&- \frac{1}{4\pi}\int\limits_{S(y,\varepsilon)} \Big[T(x,\partial_x)\frac{1}{|x-y|}\Big]\varphi_0(x)\,dS_x, \qquad y \in \Omega,
\end{aligned} \tag{47}$$

$$\begin{aligned}
\int\limits_{\Omega(y,\varepsilon)} & R(x,y)\,u(x)\,dx - \frac{1}{4\pi}\int\limits_{\partial\Omega(y,\varepsilon)} \frac{1}{|x-y|}\,T^+_{\partial\Omega(y,\varepsilon)}(x,\partial x)u(x)\,dS_x \\
&+ \frac{1}{4\pi}\int\limits_{\Sigma_1(y,\varepsilon)} \Big[T(x,\partial_x)\frac{1}{|x-y|}\Big]\,\gamma^+_{\Sigma_1(y,\varepsilon)}u(x)\,dS_x = -\frac{1}{4\pi}\int\limits_{\Omega(y,\varepsilon)} \frac{1}{|x-y|}\,f(x)\,dx \\
&- \frac{1}{2}\,a(y)\,\varphi_0(y) - \frac{1}{4\pi}\int\limits_{S(y,\varepsilon)} \Big[T(x,\partial_x)\frac{1}{|x-y|}\Big]\varphi_0(x)\,dS_x, \qquad y \in S.
\end{aligned} \tag{48}$$

Note that the trace of the conormal derivative function $T^\pm_{\Sigma_1^*}u$ on an arbitrary compact part Σ_1^* of the surface $\Sigma_1(y,\varepsilon)$ exists in the usual Sobolev trace sense in view of the interior regularity property of solutions of elliptic equations. Indeed, for an arbitrary compact sub-domain Ω^* of the domain Ω the solution u of Eq. (43) with $f \in H^0(\Omega)$ belongs to the space $H^2(\Omega^*)$. In particular, this implies that $T^\pm_{\Sigma_1^*}u \in H^{\frac{1}{2}}(\Sigma_1^*)$.

Now we prove the following equivalence theorem.

Theorem 1 *Let $f \in H^0(\Omega)$ and $\varphi_0 \in H^{\frac{1}{2}}(\partial\Omega)$. The Dirichlet problem* (43)–(44) *and the system of localized boundary-domain integro-differential equations* (47)–(48) *are equivalent in the following sense:*

(i) If $u \in H^{1,0}(\Omega, A)$ solves the Dirichlet problem (43)–(44)*, then u is a solution to the system of localized boundary-domain integro-differential equations* (47)–(48)*, and vice versa,*

(ii) If $u \in H^{1,0}(\Omega, A)$ solves the system of localized boundary domain integro-differential equations (47)–(48)*, then u is a solution to the Dirichlet problem* (43)–(44).

Proof The proof of item (i) directly follows from the derivation of system (47)–(48). To prove the reverse item we proceed as follows. Let $u \in H^{1,0}(\Omega, A)$ be a solution of the system of localized boundary-domain integro-differential equations (47)–(48).

As it has been shown above, the trace on the boundary S of the summands involved in Eq. (47) exist. Using jump relations (30)–(31) and comparing the trace on S of Eq. (47) with Eq. (48), we find

$$a\,\gamma_S^+ u = a\,\varphi_0 \quad \text{on} \quad S,$$

implying the Dirichlet condition on S,

$$\gamma_S^+ u = \varphi_0 \quad \text{on} \quad S. \tag{49}$$

Now, let us write Green's third formula (10) for the function u taking into account condition (49)

$$\begin{aligned} a(y)\,u(y) &+ \int\limits_{\Omega(y,\varepsilon)} R(x,y)\,u(x)\,dx - \frac{1}{4\pi}\int\limits_{\partial\Omega(y,\varepsilon)} \frac{1}{|x-y|}\,T^+_{\partial\Omega(y,\varepsilon)}(x,\partial_x)u(x)\,dS_x \\ &+ \frac{1}{4\pi}\int\limits_{\Sigma_1(y,\varepsilon)} \Big[T(x,\partial_x)\frac{1}{|x-y|}\Big]\,\gamma^+_{\Sigma_1(y,\varepsilon)}u(x)\,dS_x = -\frac{1}{4\pi}\int\limits_{\Omega(y,\varepsilon)} \frac{1}{|x-y|}\,A(x,\partial_x)u(x)\,dx \\ &- \frac{1}{4\pi}\int\limits_{S(y,\varepsilon)} \Big[T(x,\partial_x)\frac{1}{|x-y|}\Big]\varphi_0(x)\,dS_x \qquad y \in \Omega. \end{aligned} \tag{50}$$

Subtracting Eq. (47) from Eq. (50) we arrive at the relation

$$-\frac{1}{4\pi}\int\limits_{\Omega(y,\varepsilon)} \frac{1}{|x-y|}\,[A(x,\partial_x)u(x) - f(x)]\,dx = 0 \quad \forall y \in \Omega. \tag{51}$$

Since $Au - f \in H^0(\Omega)$, from (51) by Lemma 1 we conclude $Au - f = 0$ almost everywhere in Ω which completes the proof.

Corollary 1 *Let* $f \in H^0(\Omega)$ *and* $\varphi_0 \in H^{\frac{1}{2}}(\partial\Omega)$. *Then the system of localized boundary-domain integro-differential equations* (47)–(48) *is uniquely solvable in the space* $H^{1,0}(\Omega, A)$.

Proof First we show that if the right hand side functions of system (47)–(48) vanish, then $f = 0$ in Ω and $\varphi_0 = 0$ on S. Indeed, let

$$-\frac{1}{4\pi}\int\limits_{\Omega(y,\varepsilon)} \frac{1}{|x-y|}\,f(x)\,dx - \frac{1}{4\pi}\int\limits_{S(y,\varepsilon)} \Big[T(x,\partial_x)\frac{1}{|x-y|}\Big]\varphi_0(x)\,dS_x = 0, \qquad y \in \Omega, \tag{52}$$

$$-\frac{1}{4\pi}\int\limits_{\Omega(y,\varepsilon)} \frac{1}{|x-y|}\,f(x)\,dx - \frac{1}{2}\,a\,\varphi_0 - \frac{1}{4\pi}\int\limits_{S(y,\varepsilon)} \Big[T(x,\partial_x)\frac{1}{|x-y|}\Big]\varphi_0(x)\,dS_x = 0, \quad y \in S. \tag{53}$$

Keeping in mind formula (31) and comparing (53) with the trace on S of Eq. (52), we get $a\varphi_0 = 0$ on S implying $\varphi_0 = 0$ on S since $a > 0$ on $\overline{\Omega}$. Relation (52) takes then the form

$$-\frac{1}{4\pi}\int\limits_{\Omega(y,\varepsilon)} \frac{1}{|x-y|}\, f(x)\, dx = 0, \qquad y \in \Omega, \tag{54}$$

and by Lemma 1 we conclude $f = 0$ in Ω.

Now the proof of the theorem follows form equivalence Theorem 1 and unique solvability of the Dirichlet problem (43)–(44) in the space $H^{1,0}(\Omega, A)$.

We can rewrite equivalently system (47)–(48) in more convenient form. Indeed, if $u \in H^{1,0}(\Omega, A)$ solves system (47)–(48), by the equivalence Theorem 1, then u solves the Dirichlet problem and using the integration by parts formula we have

$$\int\limits_{\Omega(y,\varepsilon)} A(x,\partial_x)u(x)\, dx = \int\limits_{\partial\Omega(y,\varepsilon)} T^+_{\partial\Omega(y,\varepsilon)}(x,\partial_x)u(x)\, dS_x, \tag{55}$$

implying the identity

$$\begin{aligned}
&\int\limits_{\partial\Omega(y,\varepsilon)} \frac{1}{|x-y|}\, T^+_{\partial\Omega(y,\varepsilon)}(x,\partial_x)u(x)\, dS_x + \int\limits_{\partial\Omega(y,\varepsilon)} \frac{1}{\varepsilon}\, T^+_{\partial\Omega(y,\varepsilon)}(x,\partial_x)u(x)\, dS_x \\
&\quad = \int\limits_{\partial\Omega(y,\varepsilon)} \frac{1}{|x-y|}\, T^+_{\partial\Omega(y,\varepsilon)}(x,\partial_x)u(x)\, dS_x + \int\limits_{\Omega(y,\varepsilon)} \frac{1}{\varepsilon}\, A(x,\partial_x)u(x)\, dx
\end{aligned} \tag{56}$$

Since $Au = f$ in Ω and $|x-y| = \varepsilon$ for $x \in \Sigma_1(y,\varepsilon)$, we finally get

$$\begin{aligned}
&\int\limits_{\partial\Omega(y,\varepsilon)} \frac{1}{|x-y|}\, T^+_{\partial\Omega(y,\varepsilon)}(x,\partial_x)u(x)\, dS_x \\
&\quad = \int\limits_{\partial\Omega(y,\varepsilon)} \left[\frac{1}{|x-y|} - \frac{1}{\varepsilon}\right] T^+_{\partial\Omega(y,\varepsilon)}(x,\partial_x)u(x)\, dS_x + \int\limits_{\Omega(y,\varepsilon)} \frac{1}{\varepsilon}\, f(x)\, dx \\
&\quad = \int\limits_{S(y,\varepsilon)} \left[\frac{1}{|x-y|} - \frac{1}{\varepsilon}\right] T^+_{S(y,\varepsilon)}(x,\partial_x)u(x)\, dS_x + \int\limits_{\Omega(y,\varepsilon)} \frac{1}{\varepsilon}\, f(x)\, dx.
\end{aligned} \tag{57}$$

Further, if we replace the third summand in Eq. (47) and the second summand in Eqs. (48) by relation (57) and take into consideration that

$$T(x,\partial_x)\frac{1}{|x-y|} = -\frac{a(x)}{\varepsilon^2} \text{ for } x \in \Sigma_1(y,\varepsilon), \tag{58}$$

we arrive at the following system

$$
\begin{aligned}
&a(y)\,u(y) + \int\limits_{\Omega(y,\varepsilon)} R(x,y)\,u(x)\,dx - \frac{1}{4\pi}\int\limits_{S(y,\varepsilon)} \Big[\frac{1}{|x-y|} - \frac{1}{\varepsilon}\Big]\, T^{+}_{S(y,\varepsilon)}(x,\partial_x)u(x)\,dS_x \\
&\quad - \frac{1}{4\pi\varepsilon^2}\int\limits_{\Sigma_1(y,\varepsilon)} a(x)\,\gamma^{+}_{\Sigma_1(y,\varepsilon)}u(x)\,dS_x = -\frac{1}{4\pi}\int\limits_{\Omega(y,\varepsilon)} \Big[\frac{1}{|x-y|} - \frac{1}{\varepsilon}\Big]\, f(x)\,dx \\
&- \frac{1}{4\pi}\int\limits_{S(y,\varepsilon)} \Big[T(x,\partial_x)\frac{1}{|x-y|}\Big]\varphi_0(x)\,dS_x \qquad y\in\Omega,
\end{aligned}
\tag{59}
$$

$$
\begin{aligned}
&\int\limits_{\Omega(y,\varepsilon)} R(x,y)\,u(x)\,dx - \frac{1}{4\pi}\int\limits_{S(y,\varepsilon)} \Big[\frac{1}{|x-y|} - \frac{1}{\varepsilon}\Big]\, T^{+}_{S(y,\varepsilon)}(x,\partial_x)u(x)\,dS_x \\
&\quad - \frac{1}{4\pi\varepsilon^2}\int\limits_{\Sigma_1(y,\varepsilon)} a(x)\,\gamma^{+}_{\Sigma_1(y,\varepsilon)}u(x)\,dS_x = -\frac{1}{4\pi}\int\limits_{\Omega(y,\varepsilon)} \Big[\frac{1}{|x-y|} - \frac{1}{\varepsilon}\Big]\, f(x)\,dx \\
&\quad - \frac{1}{2}a(y)\,\varphi_0(y) - \frac{1}{4\pi}\int\limits_{S(y,\varepsilon)} \Big[T(x,\partial_x)\frac{1}{|x-y|}\Big]\varphi_0(x)\,dS_x, \qquad y\in S.
\end{aligned}
\tag{60}
$$

Let us show that this system is also equivalent to the Dirichlet problem.

Let $u \in H^{1,0}(\Omega, A)$ solve system (59)–(60). Taking the difference of the trace on S of Eqs. (59) and (60) we find $\gamma_S^{+}u = \varphi_0$ on S. Consequently, Green's formula (50) holds and with the help of (55) and (56) it can be rewritten as

$$
\begin{aligned}
&a(y)\,u(y) + \int\limits_{\Omega(y,\varepsilon)} R(x,y)\,u(x)\,dx - \frac{1}{4\pi}\int\limits_{S(y,\varepsilon)} \Big[\frac{1}{|x-y|} - \frac{1}{\varepsilon}\Big]\, T^{+}_{S(y,\varepsilon)}(x,\partial_x)u(x)\,dS_x \\
&- \frac{1}{4\pi\varepsilon^2}\int\limits_{\Sigma_1(y,\varepsilon)} a(x)\,\gamma^{+}_{\Sigma_1(y,\varepsilon)}u(x)\,dS_x = -\frac{1}{4\pi}\int\limits_{\Omega(y,\varepsilon)} \Big[\frac{1}{|x-y|} - \frac{1}{\varepsilon}\Big]\, A(x,\partial_x)u\,dx \\
&- \frac{1}{4\pi}\int\limits_{S(y,\varepsilon)} \Big[T(x,\partial_x)\frac{1}{|x-y|}\Big]\varphi_0(x)\,dS_x \qquad y\in\Omega,
\end{aligned}
\tag{61}
$$

Comparing relations (59) and (61) leads to the equality

$$
-\frac{1}{4\pi}\int\limits_{\Omega(y,\varepsilon)} \Big[\frac{1}{|x-y|} - \frac{1}{\varepsilon}\Big]\,[A(x,\partial_x)u(x) - f(x)]\,dx = 0, \quad y\in\Omega, \tag{62}
$$

which can be rewritten as

$$
\widetilde{\mathscr{P}}_{\widetilde{\chi}}\big(A(x,\partial_x)u - f\big)(y) := \int\limits_{\Omega} \widetilde{P}_{\widetilde{\chi}}(x-y)\,[Au(x) - f(x)]\,dS_x = 0, \qquad y\in\Omega, \tag{63}
$$

where

$$\widetilde{P}_{\widetilde{\chi}}(x) := -\frac{1}{4\pi}\frac{\widetilde{\chi}(x)}{|x|} \quad \text{with} \quad \widetilde{\chi}(x) := \begin{cases} \left(1 - \frac{|x|}{\varepsilon}\right) & \text{for } |x| < \varepsilon, \\ 0 & \text{for } |x| \geq \varepsilon. \end{cases} \tag{64}$$

It is easy to see that $\widetilde{\chi}$ belongs to the class of cut-off functions X^1_+ introduced in [2]. By Lemma 6.3 in [2] we then conclude that the density function $Au - f = 0$ in Ω.

Corollary 2 *Let* $f \in H^0(\Omega)$ *and* $\varphi_0 \in H^{\frac{1}{2}}(\partial\Omega)$*. Then the system of localized boundary-domain integro-differential equations* (59)–(60) *is equivalent to system* (47)–(48) *and to the Dirichlet problem* (43)–(44) *and is uniquely solvable in the space* $H^{1,0}(\Omega, A)$.

3.2 *The Robin Problem*

Let us now consider the Robin problem for the operator $A(x, \partial)$:

Find a function $u \in H^{1,0}(\Omega, A)$ such that

$$A(x, \partial_x)u = f \quad \text{in} \quad \Omega, \qquad f \in H^0(\Omega), \tag{65}$$

$$T_S^+ u + \varkappa\gamma^+ u = \psi_0 \quad \text{on} \quad S, \quad \psi_0 \in H^{-\frac{1}{2}}(S), \quad \varkappa \in C^{0,1}(S). \tag{66}$$

It is a classical result that the Robin problem (65)–(66) is uniquely solvable if $\varkappa \leq 0$ and $\varkappa \not\equiv 0$ (see, e.g., [6, 10, 11]). If $\varkappa \equiv 0$, then we have the Neumann problem which is solvable if and only if

$$\int_{\partial\Omega} \psi_0 \, dS = \int_{\Omega} f(x)\, dx \tag{67}$$

and a solution is defined modulo a constant summand.

Rewrite Green's third formula (10) and its trace (34) on S in the extended form,

$$\begin{aligned} a(y)\,u(y) &+ \int_{\Omega(y,\varepsilon)} R(x, y)\, u(x)\, dx - \frac{1}{4\pi}\int_{\partial\Omega(y,\varepsilon)} \frac{1}{|x-y|}\, T^+_{\partial\Omega(y,\varepsilon)}(x, \partial_x)u(x)\, dS_x \\ &+ \frac{1}{4\pi}\int_{\partial\Omega(y,\varepsilon)} \left[T(x, \partial_x)\frac{1}{|x-y|}\right] \gamma^+_{\partial\Omega(y,\varepsilon)} u(x)\, dS_x \\ &= -\frac{1}{4\pi}\int_{\Omega(y,\varepsilon)} \frac{1}{|x-y|}\, A(x, \partial_x)u(x)\, dx, \qquad y \in \Omega, \end{aligned} \tag{68}$$

$$\int_{\Omega(y,\varepsilon)} R(x,y)\,u(x)\,dx - \frac{1}{4\pi}\int_{\partial\Omega(y,\varepsilon)} \frac{1}{|x-y|}\,T^{+}_{\partial\Omega(y,\varepsilon)}(x,\partial_x)u(x)\,dS_x$$
$$+\frac{1}{2}\,a(y)\,\gamma^{+}_{S}u(x) + \frac{1}{4\pi}\int_{\partial\Omega(y,\varepsilon)} \Big[T(x,\partial_x)\frac{1}{|x-y|}\Big]\gamma^{+}_{\partial\Omega(y,\varepsilon)}u(x)\,dS_x$$
$$= -\frac{1}{4\pi}\int_{\Omega(y,\varepsilon)} \frac{1}{|x-y|}\,A(x,\partial_x)u(x)\,dx, \qquad y\in S. \tag{69}$$

We need to represent the third term in the left hand side of (68), which is to be treated as the duality relation between the spaces $H^{-\frac{1}{2}}(\partial\Omega(y,\varepsilon))$ and $H^{\frac{1}{2}}\big(\partial\Omega(y,\varepsilon)\big)$, as the sum of two well defined dualities over the submanifolds $S(y,\varepsilon)$ and $\Sigma_1(y,\varepsilon)$.

To this end, we use formula (55) which leads to the following identity for arbitrary $u\in H^{1,0}(\Omega,A)$ and $y\in\Omega$ (cf. (56))

$$-\frac{1}{4\pi}\int_{\partial\Omega(y,\varepsilon)} \frac{1}{|x-y|}\,T^{+}_{\partial\Omega(y,\varepsilon)}(x,\partial_x)u(x)dS_x = -\frac{1}{4\pi\,\varepsilon}\int_{\Omega(y,\varepsilon)} A(x,\partial_x)u(x)\,dx$$
$$-\frac{1}{4\pi}\int_{\partial\Omega(y,\varepsilon)} \Big(\frac{1}{|x-y|}-\frac{1}{\varepsilon}\Big)\,T^{+}_{\partial\Omega(y,\varepsilon)}(x,\partial_x)u(x)dS_x. \tag{70}$$

Introduce the function

$$K_\varepsilon(x,y) := \Big(\frac{1}{|x-y|}-\frac{1}{\varepsilon}\Big)\,\chi(x-y), \quad x\in\mathbb{R}^3, \;\; y\in\overline{\Omega}, \tag{71}$$

with χ defined in (9). It is easy to see that $K_\varepsilon(\,\cdot\,,y)$ belongs to the space $H^1_{comp}(\mathbb{R}^3\setminus B(y,\delta))$ for arbitrarily small $\delta>0$ and vanishes on the spherical surface $\Sigma(y,\varepsilon)$. Therefore

$$K_\varepsilon(x,y) = \frac{1}{|x-y|}-\frac{1}{\varepsilon} \text{ for } x\in S(y,\varepsilon),\;\; y\in\overline{\Omega},$$

and evidently

$$r_{S(y,\varepsilon)}K_\varepsilon(\,\cdot\,,y)\in\widetilde{H}^{\frac{1}{2}}\big(S(y,\varepsilon)\big) \text{ for } y\in\overline{\Omega}.$$

Consequently we can rewrite (70) as

$$-\frac{1}{4\pi}\int_{\partial\Omega(y,\varepsilon)} \frac{1}{|x-y|}\,T^{+}_{\partial\Omega(y,\varepsilon)}(x,\partial_x)u(x)\,dS_x =$$
$$= -\frac{1}{4\pi}\big\langle T^{+}_{S(y,\varepsilon)}u\,,\,K_\varepsilon(\,\cdot\,,y)\big\rangle_{S(y,\varepsilon)} - \frac{1}{4\pi\,\varepsilon}\int_{\Omega(y,\varepsilon)} A(x,\partial_x)u(x)\,dx, \tag{72}$$

where the first term in the right hand side of (72) is well defined duality pairing between $T^{+}_{S(y,\varepsilon)}u \in H^{-\frac{1}{2}}(S(y,\varepsilon))$ and $K_\varepsilon(\cdot\,,y) \in \widetilde{H}^{\frac{1}{2}}(S(y,\varepsilon))$.

Using (72) and the data of the Robin problem in relations (68)–(69) we arrive at the following system of boundary-domain integral equations with respect to u,

$$
\begin{aligned}
&a(y)\,u(y) + \int\limits_{\Omega(y,\varepsilon)} R(x,y)\,u(x)\,dx + \frac{1}{4\pi}\int\limits_{S(y,\varepsilon)} \Big(\frac{1}{|x-y|} - \frac{1}{\varepsilon}\Big)\,\varkappa\,\gamma^{+}_{S(y,\varepsilon)}u(x)\,dS_x\\
&+\frac{1}{4\pi}\int\limits_{\partial\Omega(y,\varepsilon)} \Big[T(x,\partial_x)\frac{1}{|x-y|}\Big]\,\gamma^{+}_{\partial\Omega(y,\varepsilon)}u(x)\,dS_x = -\frac{1}{4\pi}\int\limits_{\Omega(y,\varepsilon)} \Big(\frac{1}{|x-y|} - \frac{1}{\varepsilon}\Big)\,f(x)\,dx\\
&+\frac{1}{4\pi}\int\limits_{S(y,\varepsilon)} \Big(\frac{1}{|x-y|} - \frac{1}{\varepsilon}\Big)\psi_0(x)\,dS_x \qquad y \in \Omega,
\end{aligned} \tag{73}
$$

$$
\begin{aligned}
&\int\limits_{\Omega(y,\varepsilon)} R(x,y)\,u(x)\,dx + \frac{1}{4\pi}\int\limits_{S(y,\varepsilon)} \Big(\frac{1}{|x-y|} - \frac{1}{\varepsilon}\Big)\,\varkappa\,\gamma^{+}_{S(y,\varepsilon)}u(x)\,dS_x\\
&\quad+\frac{1}{2}\,a(y)\,\gamma^{+}_{S}u(y) + \frac{1}{4\pi}\int\limits_{\partial\Omega(y,\varepsilon)} \Big[T(x,\partial_x)\frac{1}{|x-y|}\Big]\,\gamma^{+}_{\partial\Omega(y,\varepsilon)}u(x)\,dS_x\\
&= -\frac{1}{4\pi}\int\limits_{\Omega(y,\varepsilon)} \Big(\frac{1}{|x-y|} - \frac{1}{\varepsilon}\Big)\,f(x)\,dx + \frac{1}{4\pi}\int\limits_{S(y,\varepsilon)} \Big(\frac{1}{|x-y|} - \frac{1}{\varepsilon}\Big)\psi_0(x)\,dS_x, \quad y \in S.
\end{aligned} \tag{74}
$$

Let us now prove the equivalence theorem.

Theorem 2 *Let $f \in H^0(\Omega)$, $\psi_0 \in H^{-\frac{1}{2}}(\partial\Omega)$, and $\varkappa \in C^{0,1}(S)$. The Robin problem* (65)–(66) *and the system of localized boundary-domain integral equations* (73)–(74) *are equivalent in the following sense:*

(i) If $u \in H^{1,0}(\Omega, A)$ solves the Robin problem (65)–(66)*, then u is a solution to the system of localized boundary-domain integral equations* (73)–(74)*, and vice versa,*

(ii) If $u \in H^{1,0}(\Omega, A)$ solves the system of localized boundary domain integral equations (73)–(74)*, then u is a solution to the Robin problem* (65)–(66).

Proof The proof of item (i) directly follows from the derivation of system (73)–(74). To prove the reverse item, let us assume that $u \in H^{1,0}(\Omega, A)$ is a solution to the system of localized boundary-domain integral equations (73)–(74). We can write Green's third formula (68) for u which, with the help of relation (70), can be rewritten as

$$a(y)\,u(y) + \int\limits_{\Omega(y,\varepsilon)} R(x,y)\,u(x)\,dx - \frac{1}{4\pi}\int\limits_{S(y,\varepsilon)}\Big(\frac{1}{|x-y|}-\frac{1}{\varepsilon}\Big)\,T^+_{S(y,\varepsilon)}(x,\partial_x)u(x)\,dS_x$$
$$+\frac{1}{4\pi}\int\limits_{\partial\Omega(y,\varepsilon)}\Big[T(x,\partial_x)\frac{1}{|x-y|}\Big]\,\gamma^+_{\partial\Omega(y,\varepsilon)}u\,dS_x$$
$$= -\frac{1}{4\pi}\int\limits_{\Omega(y,\varepsilon)}\Big(\frac{1}{|x-y|}-\frac{1}{\varepsilon}\Big)\,A(x,\partial_x)u(x)\,dx\,. \tag{75}$$

Subtracting (75) from (73) we arrive at the relation

$$\frac{1}{4\pi}\int\limits_{S(y,\varepsilon)}\Big(\frac{1}{|x-y|}-\frac{1}{\varepsilon}\Big)\big[T^+_{S(y,\varepsilon)}(x,\partial_x)u(x)+\varkappa\,\gamma^+_{S(y,\varepsilon)}u(x)-\psi_0(x)\big]\,dS_x$$
$$-\frac{1}{4\pi}\int\limits_{\Omega(y,\varepsilon)}\Big(\frac{1}{|x-y|}-\frac{1}{\varepsilon}\Big)\big[A(x,\partial_x)u(x)-f(x)\big]\,dx = 0, \qquad y\in\Omega. \tag{76}$$

Let us set

$$\widetilde{V}_{\widetilde{\chi}}\,\psi(y) := -\int\limits_{S}\widetilde{P}_{\widetilde{\chi}}(x-y)\psi(x)\,dS_x, \qquad \psi\in H^{-\frac{1}{2}}(\partial\Omega), \qquad y\in\mathbb{R}^3, \tag{77}$$

$$\widetilde{\mathscr{P}}_{\widetilde{\chi}}\,h(y) := \int\limits_{\Omega}\widetilde{P}_{\widetilde{\chi}}(x-y)\,h(x)\,dS_x, \qquad h\in H^0(\Omega), \qquad y\in\mathbb{R}^3, \tag{78}$$

where $\widetilde{P}_{\widetilde{\chi}}$ is defined by (64).

The Eq. (76) can be rewritten then as

$$\widetilde{V}_{\widetilde{\chi}}\big(T^+_S u + \varkappa\,\gamma^+_S u - \psi_0\big)(y) + \widetilde{\mathscr{P}}_{\widetilde{\chi}}\,\big(Au - f\big)(y) = 0, \qquad y\in\Omega, \tag{79}$$

where $T^+_S u + \varkappa\,\gamma^+_S u - \psi_0 \in H^{-\frac{1}{2}}(\partial\Omega)$ and $Au - f \in H^0(\partial\Omega)$.

By Lemma 6.3 in [2] then we conclude that the densities of both potentials vanish, i.e., $Au - f = 0$ in Ω and $T^+_S u + \varkappa\,\gamma^+_S u - \psi_0 = 0$ on S, which completes the proof.

Corollary 3 *Let $f\in H^0(\Omega)$, $\psi_0\in H^{-\frac{1}{2}}(\partial\Omega)$, and $\varkappa\in C^{0,1}(S)$ with $\varkappa\le 0$ and $\varkappa\not\equiv 0$. Then the system of localized boundary-domain integral equations* (73)–(74) *is uniquely solvable in the space $H^{1,0}(\Omega, A)$.*

Proof It is easy to show that the right hand side of system (73)–(74) vanishes if and only if $f = 0$ in Ω and $\psi_0 = 0$ on S. This can be proved by the arguments employed in the final part of the proof of Theorem 2. Indeed, the equality

$$-\frac{1}{4\pi}\int\limits_{\Omega(y,\varepsilon)}\Big(\frac{1}{|x-y|}-\frac{1}{\varepsilon}\Big)f(x)\,dx+\frac{1}{4\pi}\int\limits_{S(y,\varepsilon)}\Big(\frac{1}{|x-y|}-\frac{1}{\varepsilon}\Big)\psi_0(x)\,dS_x=0,\quad y\in\Omega \tag{80}$$

is equivalent to the relation

$$\widetilde{V}_{\widetilde{\chi}}\psi_0(y)+\widetilde{\mathscr{P}}_{\widetilde{\chi}}\,f(y)=0,\qquad y\in\Omega, \tag{81}$$

which implies $\psi_0=0$ on S and $f=0$ in Ω due to the above mentioned Lemma 6.3 in [2].

Therefore the proof of the corollary follows form equivalence Theorem 2 and unique solvability of the Robin problem (65)–(66).

Remark 2 If $\varkappa\equiv 0$, then the system of localized boundary-domain integro-differential equations corresponds to the Neumann problem and in view of equivalence Theorem 2 it is solvable in the space $H^{1,0}(\Omega,A)$ if the necessary and sufficient condition (67) holds. The general solution of the homogeneous system is a constant function.

4 Reduction to Segregated Systems of Boundary Domain Integral Equations for the Dirichlet Problem

We can rewrite system (59)–(60) as a segregated system with respect to $u\in H^{1,0}(\Omega,A)$ and $\psi=T_S^+u$. Indeed, let us consider the system of equations

$$\begin{aligned}a(y)\,u(y)&+\int\limits_{\Omega(y,\varepsilon)}R(x,y)\,u(x)\,dx-\frac{1}{4\pi}\int\limits_{S(y,\varepsilon)}\Big[\frac{1}{|x-y|}-\frac{1}{\varepsilon}\Big]\psi(x)\,dS_x\\&-\frac{1}{4\pi\varepsilon^2}\int\limits_{\Sigma_1(y,\varepsilon)}a(x)\,\gamma^+_{\Sigma_1(y,\varepsilon)}u(x)\,dS_x=-\frac{1}{4\pi}\int\limits_{\Omega(y,\varepsilon)}\Big[\frac{1}{|x-y|}-\frac{1}{\varepsilon}\Big]f(x)\,dx\\&-\frac{1}{4\pi}\int\limits_{S(y,\varepsilon)}\Big[T(x,\partial_x)\frac{1}{|x-y|}\Big]\varphi_0(x)\,dS_x,\qquad y\in\Omega,\end{aligned} \tag{82}$$

$$\begin{aligned}&\int\limits_{\Omega(y,\varepsilon)}R(x,y)\,u(x)\,dx-\frac{1}{4\pi}\int\limits_{S(y,\varepsilon)}\Big[\frac{1}{|x-y|}-\frac{1}{\varepsilon}\Big]\psi(x)\,dS_x\\&\quad-\frac{1}{4\pi\varepsilon^2}\int\limits_{\Sigma_1(y,\varepsilon)}a(x)\,\gamma^+_{\Sigma_1(y,\varepsilon)}u(x)\,dS_x=-\frac{1}{4\pi}\int\limits_{\Omega(y,\varepsilon)}\Big[\frac{1}{|x-y|}-\frac{1}{\varepsilon}\Big]f(x)\,dx\\&\quad-\frac{1}{2}a(y)\,\varphi_0(y)-\frac{1}{4\pi}\int\limits_{S(y,\varepsilon)}\Big[T(x,\partial_x)\frac{1}{|x-y|}\Big]\varphi_0(x)\,dS_x,\qquad y\in S,\end{aligned} \tag{83}$$

where f and φ_0 are as in (43)–(44).

Let us prove the following equivalence theorem.

Theorem 3 *The Dirichlet problem* (43)–(44) *and system* (82)–(83) *are equivalent in the following sense:*

(i) If $u \in H^{1,0}(\Omega)$ *solves the Dirichlet problem* (43)–(44), *then the pair* $(u, \psi) \in H^{1,0}(\Omega) \times H^{-\frac{1}{2}}(S)$ *with*

$$\psi = T_S^+ u \;\; on \;\; S \tag{84}$$

solves system (82)–(83) *and, vice versa,*

(ii) If a pair $(u, \psi) \in H^{1,0}(\Omega) \times H^{-\frac{1}{2}}(S)$ *solves system* (82)–(83), *then it is unique, u is a solution of the Dirichlet problem* (43)–(44), *and relation* (84) *holds true.*

Proof The item (i) directly follows from the derivation of the system of localized boundary-domain integro-differential equations (47)–(48) and Corollary 2. To prove the item (ii) we proceed as follows. Let $(u, \psi) \in H^{1,0}(\Omega) \times H^{-\frac{1}{2}}(S)$ be a solution of system (82)–(83). Take the trace of Eq. (82) on S and subtract it from Eq. (83) to obtain

$$\gamma^+ u = \varphi_0 \;\; \text{on} \;\; S. \tag{85}$$

By the same arguments as in Sect. 3.1 for the function $u \in H^{1,0}(\Omega)$ satisfying (85) we obtain formula (61), and subtracting this formula from (82) we find (cf. (76))

$$\frac{1}{4\pi} \int\limits_{S(y,\varepsilon)} \Big[\frac{1}{|x-y|} - \frac{1}{\varepsilon}\Big] \big[T_S^+(x, \partial_x)u(x) - \psi(x)\big]\, dS_x - \\ - \frac{1}{4\pi} \int\limits_{\Omega(y,\varepsilon)} \Big[\frac{1}{|x-y|} - \frac{1}{\varepsilon}\Big] \big[A(x, \partial_x)u(x) - f(x)\big]\, dx = 0 \;\; \text{in} \;\; \Omega. \tag{86}$$

As we have shown in the roof of Theorem 2, this relation can be rewritten in the form

$$\widetilde{V}_{\widetilde{\chi}}\big(T_S^+ u - \psi_0\big)(y) + \widetilde{\mathscr{P}}_{\widetilde{\chi}}\big(Au - f\big)(y) = 0, \qquad y \in \Omega, \tag{87}$$

where $\widetilde{V}_{\widetilde{\chi}}$ and $\widetilde{\mathscr{P}}_{\widetilde{\chi}}$ are defined by (77) and (78) respectively. Therefore, by Lemma 6.3 in [2] we conclude that $Au - f = 0$ in Ω and $T_S^+ u - \psi_0 = 0$ on S. Thus, u is a solution to the Dirichlet problems and relation (84) holds.

Further, let us show that the right hand side expressions of the system (82)–(83) vanish if and only if $f = 0$ in Ω and $\varphi_0 = 0$ on S. Indeed, let

$$\frac{1}{4\pi}\int\limits_{\Omega(y,\varepsilon)}\Big[\frac{1}{|x-y|}-\frac{1}{\varepsilon}\Big]f(x)\,dx+\frac{1}{4\pi}\int\limits_{S(y,\varepsilon)}\Big[T(x,\partial_x)\frac{1}{|x-y|}\Big]\varphi_0(x)\,dS_x=0,\quad y\in\Omega, \tag{88}$$

$$\frac{1}{4\pi}\int\limits_{\Omega(y,\varepsilon)}\Big[\frac{1}{|x-y|}-\frac{1}{\varepsilon}\Big]f(x)\,dx+\frac{1}{2}\,a(y)\,\varphi_0(y)$$
$$-\frac{1}{4\pi}\int\limits_{S(y,\varepsilon)}\Big[T(x,\partial_x)\frac{1}{|x-y|}\Big]\varphi_0(x)\,dS_x=0,\quad y\in S. \tag{89}$$

By comparison of (89) and the trace on S of (88) we deduce $\varphi_0=0$ on S. Consequently,

$$\frac{1}{4\pi}\int\limits_{\Omega(y,\varepsilon)}\Big[\frac{1}{|x-y|}-\frac{1}{\varepsilon}\Big]f(x)\,dx=0,\quad y\in\Omega$$

that is, $\widetilde{\mathscr{P}}_{\widetilde{\chi}}f=0$ in Ω, and by Lemma 6.3 in [2] we deduce $f=0$ in Ω.

Therefore the homogeneous system (82)–(83) possesses only the trivial solution since it is equivalent to the homogeneous Dirichlet problem due to the equivalence theorem which has been just proved. This completes the proof.

Acknowledgements This work was supported by Shota Rustaveli National Science Foundation of Georgia (SRNSF) (Grant number FR-18-126).

References

1. Chkadua, O., Mikhailov, S.E., Natroshvili, D.: Analysis of direct boundary-domain integral equations for a mixed BVP with variable coefficient, I: equivalence and invertibility. J. Integral Equ. Appl. **21**(4), 499–543 (2009)
2. Chkadua, O., Mikhailov, S.E., Natroshvili, D.: Analysis of some localized boundary-domain integral equations. J. Integral Equ. Appl. **21**(3), 405–445 (2009)
3. Chkadua, O., Mikhailov, S.E., Natroshvili, D.: Analysis of direct boundary-domain integral equations for a mixed BVP with variable coefficient, II: solution regularity and asymptotics. J. Integral Equ. Appl. **22**(1), 19–37 (2010)
4. Chkadua, O., Mikhailov, S.E., Natroshvili, D.: Localized boundary-domain singular integral equations based on harmonic parametrix for divergence-form elliptic PDEs with variable matrix coefficients. Integral Equ. Oper. Theory **76**(4), 509–547 (2013)
5. Costabel, M.: Boundary integral operators on Lipschitz domains: elementary results. SIAM J. Math. Anal. **19**, 613–626 (1988)
6. Dautray, R., Lions, J.: Mathematical Analysis and Numerical Methods for Science and Technology, Volume 4: Integral Equations and Numerical Methods. Springer, Berlin (1990)
7. Ding, Z.: A proof of the trace theorem of Sobolev spaces on Lipschitz domains. Proc. Am. Math. Soc. **124**, 591–600 (1996)
8. Eskin, G.: Boundary Value Problems for Elliptic Pseudodifferential Equations. Translations of Mathematical Monographs, vol. 52. American Mathematical Society, Providence (1981)

9. Jerison, D., Kenig, C.: The inhomogeneous Dirichlet problem in Lipschitz domains. J. Funct. Anal. **130**, 161–219 (1995)
10. Lions, J.-L., Magenes, E.: Non-homogeneous Boundary Value Problems and Applications, vol. 1. Springer, Berlin (1972)
11. McLean, W.: Strongly Elliptic Systems and Boundary Integral Equations. Cambridge University Press, Cambridge (2000)
12. Mikhailov, S.E.: Traces, extensions and co-normal derivatives for elliptic systems on Lipschitz domains. J. Math. Anal. Appl. **378**, 324–342 (2011)
13. Miranda, C.: Partial Differential Equations of Elliptic Type, 2nd edn. Springer, Berlin (1970)
14. Pomp, A.: The Boundary-Domain Integral Method for Elliptic Systems. With Applications in Shells. Lecture Notes in Mathematics, vol. 1683. Springer, Berlin (1998)

Boundary Value Problems of the Plane Theory of Elasticity for Materials with Voids

Bakur Gulua and Roman Janjgava

Abstract The present paper deals with plane strain problem for linear elastic materials with voids. In the spirit of N.I. Muskhelishvili the governing system of equations of the plane strain is rewritten in the complex form and its general solution is represented by means of two analytic functions of the complex variable and a solution of the Helmholtz equation. The constructed general solution enables us to solve analytically a problem for a circle and a problem for the plane with a circular hole.

Keywords Materials with voids · The plane strain · Kolosov-Muskhelishvili formulas

1 Introduction

The present paper deals with plane strain problem for linear elastic materials with voids. The theory for granular materials with interstitial voids was presented by Goodman and Cowin [10]. The nonlinear [19] and linear [9] theories for the behaviour of porous solids, in which the skeletal or matrix material is elastic and the interstices are voids of the material, was develop by Nunziato and Cowin. Within the framework of the theory of isotropic materials we using the different methods the plane strain problem with voids of stresses around a circular hole has been studied by Cowin [7] and by Ieşan [15, 16]. The aim of this paper is to solve this problem and a

B. Gulua (✉)
Sokhumi State University, 61 Politkovskaya Street, 0186 Tbilisi, Georgia

Faculty of Exact and Natural Sciences of I. Javakhishvili Tbilisi State University, 2 University Street, 0186 Tbilisi, Georgia
e-mail: bak.gulua@gmail.com

R. Janjgava
I. Vekua Institute of Applied Mathematics of Iv. Javakhishvili Tbilisi State University, 2 University Street, 0186 Tbilisi, Georgia
e-mail: roman.janjgava@gmail.com

Georgian National University SEU, 9 Tsinandali Str, 0144 Tbilisi, Georgia

G. Jaiani and D. Natroshvili (eds.), *Applications of Mathematics and Informatics in Natural Sciences and Engineering*, Springer Proceedings in Mathematics & Statistics 334, https://doi.org/10.1007/978-3-030-56356-1_13

problem for a circle in the spirit of Muskhelishvili [18]. Ieşan [14] describes the linear theory of thermoelastic bodies with voids, with some principal theorems (uniqueness, reciprocal and variation). The linear theory of thermoelasticity of porous bodies is presented, the uniqueness and existence theorems are proved, and Galerkin-type solutions are constructed by Ciarletta and Scalia [3, 4]. In Puri and Cowin [20] the behavior of plane harmonics waves and their properties are studied.

In [1, 2, 5, 6, 21, 22] (see also references therein), some results of the 2D and 3D theories of elasticity for materials with voids are given.

The paper is organized as follows. In Sects. 2 and 3, for readers convenience we remind basic equations in 3D and plane strain cases, respectively. In Sects. 4, 5 and 6 we state and prove our main results.

2 Basic Equations for Materials with Voids of the 3D Model

Let $x = (x_1; x_2; x_3)$ be a point of the Euclidean three dimensional space R^3. We assume that the subscripts preceded by a comma denote partial differentiation with respect to the corresponding Cartesian coordinate, repeated indices are summed over the range (1; 2; 3).

In what follows we consider an isotropic and homogeneous elastic solid with voids occupying a region of $\Omega \in R^3$. The governing equations of the theory of elastic materials with voids can be expressed in the following form [9, 17]:

• Equations of equilibrium

$$T_{ij,j} + \Phi_i = 0, \quad j = 1, 2, 3, \tag{1}$$

$$h_{i,i} + g + \Psi = 0, \tag{2}$$

where T_{ij} is the symmetric stress tensor, Φ_i are the volume force components, h_i is the equilibrated stress vector, g is the intrinsic equilibrated body force and Ψ is the extrinsic equilibrated body force. Equation (2) was first suggested by Goodman and Cowin [10]; it was derived from a variational arguments by Cowin and Goodman [8], and specific interpretations were given by Nunziato and Cowin [19] and by Jenkins [4].

• Constitutive equations

$$\begin{aligned} &T_{ij} = \lambda e_{kk}\delta_{ij} + 2\mu e_{ij} + \beta\phi\delta_{ij}, \quad i, j = 1, 2, 3, \\ &h_i = \alpha\phi_{,i}, \quad i = 1, 2, 3, \\ &g = -\xi\phi - \beta e_{kk}, \end{aligned} \tag{3}$$

where λ and μ are the Lamé constants; α, β and ξ are the constants characterizing the body porosity; δ_{ij} is the Kronecker delta; $\phi := \nu - \nu_0$ is the change of the volume fraction from the matrix reference volume fraction ν_0 (clearly, the bulk density $\rho = \nu\gamma$, $0 < \nu \leq 1$, here γ is the matrix density and ρ is the mass density); e_{ij} is the strain tensor and

$$e_{ij} = \frac{1}{2}\left(u_{i,j} + u_{j,i}\right), \tag{4}$$

where $u_i,\ i = 1, 2, 3$ are the components of the displacement vector.

The constitutive equations also meet some other conditions, following from physical considerations

$$\begin{aligned} &\mu > 0, \quad \alpha > 0, \quad \xi > 0, \\ &3\lambda + 2\mu > 0, \quad (3\lambda + 2\mu)\xi > 3\beta^2. \end{aligned} \tag{5}$$

Substituting (3) into (1) and (2), we obtain equations with respect to the components of the displacement and the function ϕ

$$\begin{aligned} &\mu\Delta u_j + (\lambda + \mu)\partial_j\Theta + \beta\partial_j\phi + \Phi_i = 0, \quad j = 1, 2, 3 \\ &(\alpha\Delta - \xi)\phi - \beta\Theta + \Psi = 0, \end{aligned}$$

where $\partial_i \equiv \frac{\partial}{\partial x_i}$, $\Theta = \partial_k u_k$, $\Delta \equiv \partial_{11} + \partial_{22} + \partial_{33}$ is the three-dimensional Laplace operator.

3 Basic (Governing) Equations of the Plane Strain

From the basic three-dimensional equations we obtain the basic equations for the case of plane strain. Let Ω be a sufficiently long cylindrical body with generatrix parallel to the Ox_3-axis. Denote by V the cross section of this cylindrical body, thus $V \subset R^2$. In the case of plane deformation $u_3 = 0$ while the functions u_1, u_2 and ϕ do not depend on the coordinate x_3 [14, 18].

As it follows from formulas (3) and (4), in the case of plane strain

$$T_{k3} = T_{3k} = 0, \quad h_3 = 0, \quad k = 1, 2.$$

Therefore, assuming $\Phi_i \equiv 0$ and $\Psi \equiv 0$, the system of equilibrium equations (1), (2) takes the form

$$\begin{aligned} &\partial_1 T_{11} + \partial_2 T_{21} = 0, \\ &\partial_1 T_{12} + \partial_2 T_{22} = 0, \\ &\partial_k h_k + g = 0. \end{aligned} \tag{6}$$

Note that $\Delta_2 \equiv \partial_{11} + \partial_{22}$ is the two-dimensional Laplace operator, $\theta = \partial_1 u_1 + \partial_2 u_2$.

Now, Relations (3) are rewritten as

$$
\begin{aligned}
&T_{11} = \lambda\theta + 2\mu\partial_1 u_1 + \beta\phi,\\
&T_{22} = \lambda\theta + 2\mu\partial_2 u_2 + \beta\phi,\\
&T_{12} = T_{21} = \mu(\partial_1 u_2 + \partial_2 u_1),\\
&T_{33} = \sigma(T_{11} + T_{22}),\\
&h_k = \alpha\partial_k\phi, \quad k = 1, 2,\\
&g = -\xi\phi - \beta\theta,
\end{aligned}
\tag{7}
$$

where σ is the Poisson ratio.

If relations (7) are substituted into system (6) then we obtain the following system of governing equations of statics with respect to the functions u_1, u_2 and ϕ

$$
\begin{aligned}
&\mu\Delta u_k + (\lambda + \mu)\partial_k\theta + \beta\partial_k\phi = 0, \quad k = 1, 2\\
&(\alpha\Delta - \xi)\Delta\phi - \beta\theta = 0.
\end{aligned}
\tag{8}
$$

On the plane Ox_1x_2, we introduce the complex variable $z = x_1 + ix_2 = re^{i\vartheta}$, $(^2 = -1)$ and the operators $\partial_z = 0.5(\partial_1 - i\partial_2)$, $\partial_{\bar z} = 0.5(\partial_1 + i\partial_2)$, $\bar z = x_1 - ix_2$, and $\Delta_2 = 4\partial_z\partial_{\bar z}$.

To write system (6) in the complex form, the second equation of this system we multiplied by i and sum up with the first equation

$$
\begin{aligned}
&\partial_z(T_{11} - T_{22} + 2iT_{12}) + \partial_{\bar z}(T_{11} + T_{22}) = 0,\\
&\partial_z h_+ + \partial_{\bar z}\bar h_+ + g = 0,
\end{aligned}
\tag{9}
$$

where $h_+ = h_1 + ih_2$ and formulas (7) we rewrite as follows

$$
\begin{aligned}
&T_{11} - T_{22} + 2iT_{12} = 4\mu\partial_{\bar z}u_+,\\
&T_{11} + T_{22} = 2(\lambda + \mu)\theta + 2\beta\phi,\\
&h_+ = 2\alpha\partial_{\bar z}\phi,\\
&g = -\xi\phi - \beta\theta,
\end{aligned}
\tag{10}
$$

$$
\theta = \partial_z u_+ + \partial_{\bar z}\bar u_+, \quad u_+ = u_1 + iu_2.
$$

Substituting relations (10) into system (9), we rewrite system (8) in the complex form

$$
\begin{aligned}
&2\mu\partial_{\bar z}\partial_z u_+ + (\lambda + \mu)\partial_{\bar z}\theta + \beta\partial_{\bar z}\phi = 0,\\
&(\alpha\Delta - \xi)\phi - \beta\theta = 0.
\end{aligned}
\tag{11}
$$

4 Kolosov–Muskhelishvili Formulas for (11) System

In this section, we construct the analogues to the Kolosov–Muskhelishvili formulas [18] (see also [11–13]) for system (11).

We take the operator $\partial_{\bar z}$ out of the brackets in the left-hand part of the first equation of system (11)

$$\partial_{\bar{z}}(2\mu\partial_z u_+ + (\lambda+\mu)\theta + \beta\phi) = 0. \tag{12}$$

Since (12) is a system of Cauchy–Riemann equations, we have

$$2\mu\partial_z u_+ + (\lambda+\mu)\theta + \beta\phi = A\varphi'(z), \tag{13}$$

where $\varphi(z)$ is an arbitrary analytic function of z and A an arbitrary constant.

A conjugate equation to (13) has the form

$$2\mu\partial_{\bar{z}}\bar{u}_+ + (\lambda+\mu)\theta + \beta\phi = A\overline{\varphi'(z)}, \tag{14}$$

Summing up Eqs. (13) and (14) and taking into account that

$$\theta = \partial_z u_+ + \partial_{\bar{z}}\bar{u}_+$$

we obtain

$$\theta = \frac{A}{2(\lambda+\mu)}(\varphi'(z) + \overline{\varphi'(z)}) - \frac{\beta}{\lambda+2\mu}\phi. \tag{15}$$

Substituting formula (15) into the second equation of system (11), we have

$$\Delta\phi - \frac{\xi(\lambda+2\mu)-\beta^2}{\alpha(\lambda+2\mu)}\phi = \frac{\beta A}{2\alpha(\lambda+2\mu)}(\varphi'(z) + \overline{\varphi'(z)}). \tag{16}$$

The general solution of Eq. (16) we may write in the form

$$\phi = \chi(z,\bar{z}) - \frac{\beta A}{2(\xi(\lambda+2\mu)-\beta^2)}(\varphi'(z) + \overline{\varphi'(z)}), \tag{17}$$

where $\chi(z,\bar{z})$ is a general solution of the Helmholtz equation

$$\Delta\chi(z,\bar{z}) - \gamma^2\chi(z,\bar{z}) = 0,$$

where $\gamma^2 = \frac{\xi(\lambda+2\mu)-\beta^2}{\alpha(\lambda+2\mu)}$ and from (5) $\gamma > 0$.

Substituting formulas (15) and (17) into Eq. (13), we obtain

$$2\mu\partial_z u_+ = \frac{\xi(\lambda+3\mu)-\beta^2}{2(\xi(\lambda+2\mu)-\beta^2)}A\varphi'(z) - \frac{\xi(\lambda+\mu)-\beta^2}{2(\xi(\lambda+2\mu)-\beta^2)}A\overline{\varphi'(z)} - \frac{\beta\mu}{\lambda+2\mu}\chi(z,\bar{z}).$$

Now, let $A := \frac{2(\xi(\lambda+2\mu)-\beta^2)}{\xi(\lambda+\mu)-\beta^2}$ then we get

$$2\mu u_+ = \kappa\varphi(z) - z\overline{\varphi'(z)} - \overline{\psi(z)} - \frac{4\alpha\beta\mu}{\xi(\lambda+2\mu)-\beta^2}\partial_{\bar{z}}\chi(z,\bar{z}),$$

where $\kappa = \frac{\xi(\lambda+3\mu)-\beta^2}{\xi(\lambda+\mu)-\beta^2}$, $\psi(z)$ is an arbitrary analytic function of z.

Thus, we have proved

Theorem 1 *The general solution of the system (11) is represented as follows:*

$$2\mu u_+ = \kappa\varphi(z) - z\overline{\varphi'(z)} - \overline{\psi(z)} - \frac{4\alpha\beta\mu}{\xi(\lambda+2\mu)-\beta^2}\partial_{\bar{z}}\chi(z,\bar{z}),$$

$$\phi = \chi(z,\bar{z}) - \frac{\beta}{\xi(\lambda+\mu)-\beta^2}(\varphi'(z)+\overline{\varphi'(z)}),$$

where $\kappa = \frac{\xi(\lambda+3\mu)-\beta^2}{\xi(\lambda+\mu)-\beta^2}$, $\varphi(z)$ *and* $\psi(z)$ *are arbitrary analytic functions of a complex variable* z *in the domain* V, $\chi(z,\bar{z})$ *is an arbitrary solution of the Helmholtz equation*

$$\Delta\chi(z,\bar{z}) - \gamma^2\chi(z,\bar{z}) = 0.$$

From (10) we have

$$T_{11} - T_{22} + 2iT_{12} = -2z\overline{\varphi''(z)} - 2\overline{\psi'(z)} - \frac{8\alpha\beta\mu}{\xi(\lambda+2\mu)-\beta^2}\partial_{\bar{z}}\partial_{\bar{z}}\chi(z,\bar{z}),$$

$$T_{11} + T_{22} = \frac{2\xi(\lambda+2\mu)^2 - 2(\lambda+3\mu)\beta^2}{(\lambda+2\mu)(\xi(\lambda+2\mu)-\beta^2)}\left(\varphi'(z)+\overline{\varphi'(z)}\right) + \frac{2\mu\beta}{\lambda+2\mu}\chi(z,\bar{z}),$$

$$h_+ = 2\alpha\partial_{\bar{z}}\chi(z,\bar{z}) - \frac{2\alpha\beta}{\xi(\lambda+\mu)-\beta^2}\overline{\varphi''(z)},$$

$$g = \left(\frac{\beta^2}{\lambda+2\mu} - \xi\right)\chi(z,\bar{z}) - \frac{\beta\mu(\xi(\lambda+2\mu)-\beta^2)}{(\lambda+\mu)(\lambda+2\mu)(\xi(\lambda+\mu)-\beta^2)}\left(\varphi'(z)+\overline{\varphi'(z)}\right).$$

5 A Problem for a Circle

Let the origin of coordinates be at the centre of the circle with radius R. On the boundary of the considered domain the values of ϕ and the displacement vector are given.

We consider the following problem

$$\begin{aligned} 2\mu u_+|_{r=R} &= 2\mu(G_1 + iG_2) = \sum_{-\infty}^{+\infty} A_n e^{in\vartheta}, \\ \phi|_{r=R} &= G_3 = \sum_{-\infty}^{+\infty} B_n e^{in\vartheta}. \end{aligned} \qquad (18)$$

The analytic functions $\varphi(z)$, $\psi(z)$ and the metaharmonic function $\chi(z,\bar{z})$ are represented as the series [18]

$$\varphi(z)=\sum_{n=1}^{\infty}a_nz^n,\quad \psi(z)=\sum_{n=0}^{\infty}b_nz^n,\quad \chi(z,\bar z)=\sum_{-\infty}^{+\infty}\alpha_nI_n(\gamma r)e^{in\vartheta},$$

where $I_n(\gamma r)$ is the modified Bessel function of the n-th order, after substituting into the boundary conditions (18) we have

$$\kappa\sum_{n=1}^{\infty}a_nR^ne^{in\vartheta}-\bar a_1Re^{i\vartheta}-\sum_{n=0}^{\infty}(n+2)\bar a_{n+2}R^{n+2}e^{-in\vartheta}-\sum_{n=0}^{\infty}\bar b_nR^ne^{-in\vartheta}$$

$$-\frac{\delta a}{2}\sum_{-\infty}^{+\infty}\alpha_nI_{n+1}(\gamma R)e^{i(n+1)\vartheta}=\sum_{-\infty}^{+\infty}A_ne^{in\vartheta},$$

$$\sum_{-\infty}^{+\infty}\alpha_nI_n(\gamma R)e^{in\vartheta}-\eta\sum_{n=1}^{\infty}\left(na_nR^{n-1}e^{i(n-1)\vartheta}+n\bar a_nR^{n-1}e^{-i(n-1)\vartheta}\right)=\sum_{-\infty}^{+\infty}B_ne^{in\vartheta},$$

where $\delta=\frac{4\alpha\beta\mu}{\xi(\lambda+2\mu)-\beta^2}$, $\eta=\frac{\beta}{\xi(\lambda+\mu)-\beta^2}$.

Comparing the coefficients of members with equal degrees, we obtain the following system

$$\begin{aligned}
&\kappa Ra_1-R\bar a_1-\frac{\delta a}{2}I_1(\gamma R)\alpha_0=A_1,\\
&\kappa R^na_n-\frac{\delta a}{2}I_n(\gamma R)\alpha_{n-1}=A_n,\quad n>1,\\
&-(n+2)R^{n+2}\bar a_{n+2}-R^n\bar b_n-\frac{\delta a}{2}I_n(aR)\alpha_{-n-1}=A_{-n},\quad n\geq 0,\\
&I_0(\gamma R)\alpha_0-\eta(a_1+\bar a_1)=B_0,\\
&I_n(\gamma R)\alpha_n-\eta(n+1)R^na_{n+1}=B_n,\quad n>0.
\end{aligned}$$

All coefficients are determined by these formulas.

It is easily seen the absolute and uniform convergence of the series obtained in the circle (including the contours) when the functions prescribed on the boundary are sufficiently smooth.

6 A Problem for the Plane with a Circular Hole

Let the origin of coordinates be at the centre of the hole of radius R.

We consider the following problem

$$2\mu u_+|_{r=R}=2\mu(H_1+iH_2)=\sum_{-\infty}^{+\infty}C_ne^{in\vartheta},\quad \phi|_{r=R}=H_3=\sum_{-\infty}^{+\infty}D_ne^{in\vartheta}.\qquad(19)$$

The analytic function $\varphi(z)$, $\psi(z)$ and the metaharmonic function $\chi(z,\bar z)$ we represent as series

$$\varphi'(z)=\sum_{n=0}^{+\infty}c_n z^{-n},\quad \psi'(z)=\sum_{n=0}^{+\infty}d_n z^{-n},\quad \chi(z,\bar z)=\sum_{-\infty}^{+\infty}\beta_n K_n(ar)e^{in\vartheta},$$

where $K_n(ar)$ is the modified Bessel function of the n-th order, and are substituted in the boundary conditions (19) we have

$$\kappa\left(Rc_0e^{i\vartheta}+\ln Rc_1+c_1\vartheta i-\sum_{n=2}^{+\infty}\frac{c_ne^{i(n-1)\vartheta}}{(n-1)R^{n-1}}\right)-\sum_{n=0}^{+\infty}\frac{\bar c_n}{R^{n-1}}e^{i(n+1)\vartheta}-R\bar d_0e^{-i\vartheta}$$

$$-\ln R\bar d_1+\bar d_1\vartheta i+\sum_{n=2}^{+\infty}\frac{\bar d_ne^{i(n-1)\vartheta}}{(n-1)R^{n-1}}-\frac{\delta\gamma}{2}\sum_{-\infty}^{+\infty}\beta_nK_{n+1}(aR)e^{i(n+1)\vartheta}=\sum_{-\infty}^{+\infty}C_ne^{in\vartheta},$$

$$\sum_{-\infty}^{+\infty}\beta_nK_n(\gamma R)e^{in\vartheta}-\eta\sum_{n=0}^{+\infty}\left(\frac{b_n}{R^n}e^{-in\vartheta}+\frac{\bar b_n}{R^n}e^{in\vartheta}\right)=\sum_{-\infty}^{+\infty}D_ne^{in\vartheta}.$$

Comparing the coefficients of members with equal degrees, we obtain from the constant term and from those involving $e^{i\vartheta}$, $e^{-i\vartheta}$, and $e^{2i\vartheta}$, respectively,

$$\begin{cases}\kappa\ln R\bar c_1-\ln Rd_1-\dfrac{\delta\gamma}{2}K_0(\gamma R)\beta_1=C_0,\\ K_1(\gamma R)\beta_1-\dfrac{1}{R}\bar c_1=D_1,\end{cases}\tag{20}$$

$$\begin{cases}\kappa R\bar c_0-R\bar c_0+\dfrac{1}{R}\bar d_2-\dfrac{\delta\gamma}{2}K_1(\gamma R)\beta_0=C_1,\\ K_0(\gamma R)\beta_0-\eta(c_0+\bar c_0)=D_0,\end{cases}\tag{21}$$

$$\begin{cases}-\dfrac{\kappa}{R}c_2-R\bar d_0-\dfrac{\delta\gamma}{2}K_{-1}(\gamma R)\beta_{-2}=C_{-1},\\ K_2(\gamma R)\beta_2-\dfrac{\eta}{R^2}\bar c_2=D_2.\end{cases}\tag{22}$$

For $e^{in\vartheta}$ $(n=\pm2,\pm3,...)$ gives

$$\begin{aligned}&-\frac{\kappa}{(n-1)R^{n-1}}\bar c_n-\frac{\delta\gamma}{2}K_{n-1}(\gamma R)\beta_n=\bar C_{-n+1},\quad n\ge 3,\\ &K_n(\gamma R)\beta_n-\tfrac{\eta}{R^n}\bar c_n=D_n,\quad n\ge 3,\\ &\tfrac{1}{(n-1)R^{n-1}}\bar d_n-\tfrac{1}{R^{n-3}}\bar c_{n-2}-\tfrac{\delta\gamma}{2}K_{n-1}(\gamma R)\beta_{n-2}=C_{n-1}.\end{aligned}\tag{23}$$

It is known that

$$c_0=\Gamma,\quad d_0=\Gamma',\tag{24}$$

where Γ, Γ' are known quantities, specifying the stress distribution at infinity (It is also assumed that b_0 has a real value).

In order to find expressions for b_1 and c_1, it is necessary to refer to the condition of single-valuedness of the displacements

$$\kappa c_1 + d_1 = 0. \tag{25}$$

From (20)–(25) we may find coefficients $c_n,\ d_c,\ \beta_n$.

It is easily seen the absolute and uniform convergence of the series obtained in the circle (including the contours) when the functions prescribed on the boundary are sufficiently smooth.

7 Conclusion

We consider plane strain of elastic materials with voids, a general solution of the governing system of differential equations is constructed by means of two analytic functions of a complex variable and a solution of the Helmholtz equation. Applying the constructed solutions we have obtained explicit solutions for the boundary value problems for a circle and the plane with a circular hole.

References

1. Bitsadze, L.: About some solutions of system of equations of steady vibrations for thermoelastic materials with voids. Semin. I. Vekua Inst. Appl. Math. Rep. **44**, 3–14 (2018)
2. Bitsadze, L.: The Neumann BVP of the linear theory of thermoelasticity for the sphere with voids and microtemperatures. Semin. I. Vekua Inst. Appl. Math. Rep. **44**, 15–29 (2018)
3. Ciarletta, M., Scalia, A.: On uniqueness and reciprocity in linear thermoelasticity of materials with voids. J. Elast. **32**, 1–17 (1993). https://doi.org/10.1007/BF00042245
4. Ciarletta, M., Scalia, A.: Results and applications in thermoelasticity of materials with voids. Le Matematiche **XLVI**, 85–96 (1991)
5. Ciarletta, M., Scalia, A., Svanadze, M.: Fundamental solution in the theory of micropolar thermoelasticity for materials with voids. J. Therm. Stress. **30**(3), 213–229 (2007)
6. Ciarletta, M., Svanadze, M., Buonanno, L.: Plane waves and vibrations in the theory of micropolar thermoelasticity for materials with voids. Eur. J. Mech. A Solids **28**(4), 897–903 (2009)
7. Cowin, S.C.: The stresses around a hole in a linear elastic material with voids. Q. J. Mech. Appl. Math. **37**(3), 441–465 (1984)
8. Cowin, S.C., Goodman, M.A.: A variational principle for granular materials. ZAMP **56**, 281–286 (1976)
9. Cowin, S.C., Nunziato, J.W.: Linear elastic materials with voids. J. Elast. **13**, 125–147 (1983)
10. Goodman, M.A., Cowin, S.C.: A continuum theory for granular materials. Arch. Rat. Mech. Anal. **44**, 249–266 (1972)
11. Gulua, B., Janjgava, R.: On construction of general solutions of equations of elastostatic problems for the elastic bodies with voids. PAMM **18**(1), e201800306 (2018). https://doi.org/10.1002/pamm.201800306
12. Gulua, B.: On one boundary value problems for a circular ring with double porosity. Proc. I. Vekua Inst. Appl. Math. **67**, 34–40 (2017)
13. Gulua, B., Janjgava, R., Kasrashvili, T.: About one problem of porous Cosserat media for solids with triple-porosity. Appl. Math. Inf. Mech. **22**(2), 3–15 (2017)
14. Ieşan, D.: A theory of thermoelastic materials with voids. Acta Mech. **60**, 67–89 (1986). https://doi.org/10.1007/BF01302942

15. Ieşan, D.: Classical and Generalized Models of Elastic Rods. CRC Series - Modern Mechanics and Mathematics. A. CHAPMAN & HALL BOOK, London (2009)
16. Ieşan, D.: On a theory of thermoviscoelastic materials with voids. J. Elast. **104**, 369–384 (2011). https://doi.org/10.1007/s10659-010-9300-7
17. Jenkins, J.T.: Static equilibrium of granular materials. J. Appl. Mech. **42**, 603–606 (1975)
18. Muskhelishvili, N.I.: Some Basic Problems of the Mathematical Theory of Elasticity. "Nauka", Moscow (1966)
19. Nunziato, J.W., Cowin, S.C.: A nonlinear theory of elastic materials with voids. Arch. Rat. Mech. Anal. **72**, 175–201 (1979)
20. Puri, P., Cowin, S.C.: Plane waves in linear elastic materials with voids. J. Elast. **15**, 167–183 (1985). https://doi.org/10.1007/BF00041991
21. Tsagareli, I.: Explicit solution of elastostatic boundary value problems for the elastic circle with voids. Adv. Math. Phys. 6 p. (2018). Art. ID 6275432
22. Tsagareli, I.: Solution of the boundary value problems of elastostatics for a plane with circular hole with voids. Semin. I. Vekua Inst. Appl. Math. Rep. **44**, 54–62 (2018)

Objective and Subjective Consistent Criteria for Hypotheses Testing

Omar Purtukhia and Zurab Zerakidze

Abstract In this work, we define objective and subjective consistent criteria for hypotheses testing. Sufficient conditions for the existence of such criteria are given in the case of Borel probability measures. At the same time, we construct the subjective consistent criterion for hypotheses testing, as well as the statistical structure, which admits the objective consistent criterion for hypotheses testing.

Keywords Statistical structure · Hypothesis testing · Consistent criterion · Objective criterion · Subjective criterion

1 Auxiliary Notions and Results

At first, we give some definitions (see [1–6]). Let V be a complete metric linear space, and let $B(V)$ be the σ-algebra of Borel sets. The symbol S_{+v} denotes the set obtained by displacing the set $S \subseteq V$ by the vector $v \in V$.

Definition 1 A mesure μ is called to be transverse to a Borel set $S \subset V$ if the following two conditions hold:

(1) there is a compact $U \subset V$ for which $0 < \mu(U) < 1$;
(2) $\mu(S_{+v}) = 0, \quad \forall v \in V$.

Definition 2 A Borel set $S \subset V$ is called shy if there exists a measure transverse to S. A subset of V is also called a shy-set if it is a subset of a Borel shy-set.

O. Purtukhia (✉)
Department of Mathematics, A. Razmadze Mathematical Institute, Ivane Javakhishvili Tbilisi State University, 13 University st, Tbilisi, Georgia
e-mail: o.purtukhia@gmail.com

Z. Zerakidze
Gori State Teaching University, 53 Chavchavadze Ave, Gori, Georgia
e-mail: zura.zerakidze@mail.ru

G. Jaiani and D. Natroshvili (eds.), *Applications of Mathematics and Informatics in Natural Sciences and Engineering*, Springer Proceedings in Mathematics & Statistics 334, https://doi.org/10.1007/978-3-030-56356-1_14

Definition 3 The complement of a shy set is called a prevalent set.

Definition 4 A set that is neither a shy-set nor a prevalent one is called ambivalent.

Definition 5 We will say that "almost all" elements of the set V satisfy some condition $P(V)$ if the set of elements for which $P(V)$ is true is prevalent.

Definition 6 A sequence of real numbers $(x_k)_{k\in N}$ is called uniformly distributed over the interval (a, b) if for any subinterval $[c, d]$ in (a, b) the following equality holds:

$$\lim_{n\to\infty} \frac{1}{n}\#(\{x_1, x_2, ..., x_n\} \cap [c, d]) = \frac{d-c}{b-a},$$

where # denotes a measure of counting.

Let $X \subset V$ be a compact from the Polish space, and let μ be a Borel probability measure on X. Let $C_b(X)$ be the space of all continuous bounded functions on X.

Definition 7 We will say that the sequence $(x_k)_{k\in N}$ of elements of X is μ-uniformly distributed on X, if $\forall f \in C_b(X)$ the following equality holds:

$$\lim_{n\to\infty} \frac{1}{n}\sum_{k=1}^{n} f(x_k) = \int_X f(x)\mu(dx).$$

Definition 8 Let F be the distribution function of the Borel probability measure μ on R. We will say that the sequence of real numbers $(x_k)_{k\in N}$ is μ-uniformly distributed on R if for any interval $[a, b]$ $(-\infty \le a < b \le +\infty)$ the following equality holds:

$$\lim_{n\to\infty} \frac{1}{n}\#(\{x_1, x_2, ..., x_n\} \cap [a, b]) = F(b) - F(a).$$

Theorem 1 (see Corollary 2.4 from [3]) *Let F be a strictly increasing continuous distribution function on R, and let μ_F be a Borel probability measure on R, defined by F. If we denote by D_F the set of all μ_F-uniformly distributed Borel probability measures on R, then the following statements are true:*

(1) $D_F = \{(F^{-1}(x_k))_{k\in N} : (x_k)_{k\in N} \in D\}$,
where D is the set of all equidistributed sequences on $(0, 1)$ *under the Lebesgue measure;*

(2) $\mu_F^N(D_F) = 1$,
where μ_F^N is an infinite (countable) product of the measure μ_F on itself: $\mu_F^N = \mu_F \times \mu_F \times \cdots$.

Let $\{\mu_h, h \in H\}$ be a family of probability measures, defined on the measurable space $(X, B(X))$. Let the class $S(X)$ be defined as follows:

$$S(X) = \cap_{h\in H} dom(\overline{\mu}_h),$$

where $\overline{\mu}_h$ denotes the completion of the measure μ_h, and $dom(\overline{\mu}_h)$ is the σ-algebra of all $\overline{\mu}_h$-measurable subsets, $h \in H$.

Definition 9 (*see* [1]) A statistical structure $\{X, B(X), \mu_h, h \in H\}$ is called strongly separable if there exists a partition $\{C_h, h \in H\}$ of the set X by elements of σ-algebra $S(X)$ such that

$$\overline{\mu}_h(C_h) = 1, \quad \forall h \in H.$$

Let H be a set of hypotheses, and let $B(H)$ be the minimal σ-algebra, generated by all subsets with finitely many elements in H.

Definition 10 We will say that the statistical structure $\{X, S(X), \mu_h, h \in H\}$ admits a consistent criterion for hypotheses testing if there exists at least one measurable mapping $T : (X, S(X)) \longrightarrow (H, B(H))$, such that

$$\mu_h(\{x : T(x) = h\}) = 1, \quad \forall\, h \in H.$$

Theorem 2 (see Theorem 2.2 from [1]) *Let* $\{X, S(X), \mu_h, h \in H\}$ *be a statistical structure. Then the following two statement are equivalent:*

(1) the statistical structure $\{X, S(X), \mu_h, h \in H\}$ *is strongly separable;*

(2) the statistical structure $\{X, S(X), \mu_h, h \in H\}$ *admits a consistent criterion for hypotheses testing.*

2 Subjective Consistent Criterion for Hypotheses Testing

Let $X_1, X_2, \ldots$ be an infinite sample, obtained by observing a sequence of independent random variables with an unknown distribution function F. We only know that F belongs to the family of distribution functions $\{F_h, h \in H\}$, where H is not an empty set. Using this infinite sample, we want to estimate the unknown distribution function F. Let μ_h denote the Borel probability measure on R, defined by the distribution function F, $\forall h \in H$, and let μ_h^N be an infinite (countable) product of the measureμ_h on itself. i.e. $\mu_h^N = \mu_h \times \mu_h \times \cdots$. Therefore, we have the statistical structure $\{R^N, B(R^N), \mu_h^N, h \in H\}$. It admits a consistent criterion for hypotheses testing if there exists at least one measurable mapping $T : (R^N, B(R^N)) \longrightarrow (H, B(H))$, such that

$$\mu_h^N(\{(x_k)_{k\in N} : (x_k)_{k\in N} \in R^N \ \&\ T((x_k)_{k\in N}) = h\}) = 1, \quad \forall\, h \in H.$$

Definition 11 A consistent criterion for hypotheses testing T is called objective if $T^{-1}(h)$ is the Haar ambivalent $\forall h \in H$. Otherwise, a consistent criterion for hypotheses testing T is called subjective.

Definition 12 Let H be a nonempty set, and let $B(H)$ be the minimal σ-algebra on H generated by singleton sets. Let the statistical structure $\{R^N, B(R^N), \mu_h^N, h \in H\}$

admit a consistent criterion for hypotheses testing $T : R^N \longrightarrow H$. We will say that the statistical structure $\{R^N, B(R^N), \mu_h^N, h \in H\}$ admits a strong objective consistent criterion for hypotheses testing if the following two conditions are true:

(1) $\forall h \in H : T^{-1}(h)$ is the Haar ambivalent;

(2) $\forall h_1, h_2 \in H$: there exists an isometrical transform $A_{h_1,h_2} : R^n \longrightarrow R^N$ such that $A_{h_1,h_2}(T^{-1}(h_1) \,\Delta\, T^{-1}(h_2))$ is a shy-set.

Theorem 3 *Let μ_h be a probability measure on R generated by a random variable with zero mathematical expectation and standard square deviation h ($h > 0$). For $(x_k)_{k\in N} \in R^N$ we denote*

$$T_1((x_k)_{k\in N}) = \lim_{n\longrightarrow\infty} \sup \frac{|\sum_{k=1}^n x_k|}{\sqrt{2nloglogn}},$$

if

$$\lim_{n\longrightarrow\infty} \sup \frac{|\sum_{k=1}^n x_k|}{\sqrt{2nloglogn}} \text{ exists and finite}$$

and

$$T_1((x_k)_{k\in N}) = 1, \quad otherwise.$$

Then T_1 is the subjective consistent criterion for hypotheses testing $h \in (0, \infty)$.

Proof Since $(x_k)_{k\in N}$ is a sequence of independent identically distributed random variables with zero mathematical expectation and a variation h^2, it is easy to prove that with probability 1 we have:

$$\lim_{n\longrightarrow\infty} \sup \frac{|\sum_{k=1}^n x_k|}{\sqrt{2h^2nloglogn}} = h, \quad i.\,e.$$

$$\mu_h^N(\{(x_k)_{k\in N} : (x_k)_{k\in N} \in R^N \;\&\; \lim_{n\longrightarrow\infty} \sup \frac{|\sum_{k=1}^n x_k|}{\sqrt{2h^2nloglogn}} = h\}) = 1, \quad h \in (0, \infty).$$

Therefore, the following also holds:

$$\mu_h^N(\{(x_k)_{k\in N} : (x_k)_{k\in N} \in R^N \;\&\; T_1((x_k)_{k\in N}) = h\}) = 1, \quad h \in (0, \infty).$$

Thus, T_1 is a consistent criterion. Now we prove now that in fact it is a subjective consistent criterion.

Let's define S as follows:

$$S = \{(x_k)_{k\in N} : (x_k)_{k\in N} \in R^N \;\&\; \lim_{n\longrightarrow\infty} \sup \frac{|\sum_{k=1}^n x_k|}{\sqrt{2nloglogn}} < \infty\}$$

and prove that it is a Borel shy set in R^N.

First, we verify that S is a vector space. Indeed, if $(x_k)_{k\in N}$ and $(y_k)_{k\in N}$ are alements of S and $\alpha,\ \beta \in R$, we can easily see that

$$\limsup_{n\longrightarrow\infty} \frac{|\sum_{k=1}^{n}(\alpha x_k+\beta y_k)|}{\sqrt{2nloglogn}} \leq \limsup_{n\longrightarrow\infty} \frac{|\sum_{k=1}^{n}\alpha x_k|}{\sqrt{2nloglogn}}+$$

$$+\limsup_{n\longrightarrow\infty} \frac{|\sum_{k=1}^{n}\beta y_k|}{\sqrt{2nloglogn}} = |\alpha| \limsup_{n\longrightarrow\infty} \frac{|\sum_{k=1}^{n}x_k|}{\sqrt{2nloglogn}}+$$

$$+|\beta| \limsup_{n\longrightarrow\infty} \frac{|\sum_{k=1}^{n}y_k|}{\sqrt{2nloglogn}} < \infty.$$

Hence, S is a vector subspace of R^N. Next we verify that S is a Borel set in R^N. For any $i \in N$ we consider the projections P_{r_i} into R^N, defined as follows

$$P_{r_i}((x_k)_{k\in N}) = x_i, \quad (x_k)_{k\in N} \in R^N.$$

Since

$$\{(x_k)_{k\in N} : (x_k)_{k\in N} \in R^N \ \&\ \limsup_{n\longrightarrow\infty} \frac{|\sum_{k=1}^{n}x_k|}{\sqrt{2nloglogn}} < \infty\} =$$

$$= \cup_{r=1}^{\infty}\{(x_k)_{k\in N} : (x_k)_{k\in N} \in R^N \ \&\ \limsup_{n\longrightarrow\infty} \frac{|\sum_{k=1}^{n}x_k|}{\sqrt{2nloglogn}} < r\}$$

it is evident that $\overline{T}_n = \frac{|\sum_{i=1}^{n}P_{r_i}|}{\sqrt{2nloglogn}}$ $(n \in N)$ is a Borel measurable function in R^N. Therefore, we conclude that S is a Borel measurable subset of R^N.

Let us prove now that S is a Borel shy set. For this purpose, we define the vector $v = (v_n)_{n\in N}$ as follows:

$$v_n = 0, \quad if\ 1 \leq n \leq 10;$$

$$v_n = 11\sqrt{22loglog11}, \quad if\ n = 11;$$

$$v_n = n\sqrt{2nloglogn} - (n-1)\sqrt{2(n-1)loglog(n-1)}, \quad if\ n > 11$$

and show that it defines a line L, all displacing of which can have at most one common point with the set S. This will prove that the line L will be a probe [1] of the complement of S.

Assume the opposite, then for $(z_k)_{k\in N} \in R^N$ there are two different parameter $t_1,\ t_2 \in R$ such that $(z_k)_{k\in N} + t_1 v \in S$ and $(z_k)_{k\in N} + t_2 v \in S$. Since S is a vector

[1] We call a finite-dimensional subspace $P \subset V$ a probe for a set $T \subset V$ if Lebesgue measure supported on P is transverse to a Borel set which contains the complement of T (see Definition 6 from [4]).

space, we have $(t_2 - t_1)v \in S$, and because $t_2 - t_1 \neq 0S$ we conclude that $v \in S$. But this leads to a contradiction:

$$\lim_{n \longrightarrow \infty} \sup \frac{|\sum_{k=1}^{n} v_k|}{\sqrt{2nloglogn}} \geq$$

$$\geq \lim_{n \longrightarrow \infty} \sup \frac{|\sum_{k=12}^{n}(k\sqrt{2kloglogk} - (k-1)\sqrt{2(k-1)loglog(k-1)})|}{\sqrt{2nloglogn}} \geq$$

$$\geq \lim_{n \longrightarrow \infty} \sup \frac{|n\sqrt{2nloglogn}|}{\sqrt{2nloglogn}} = \lim_{n \longrightarrow \infty} \sup n = +\infty.$$

Hence, S is a Borel shyset. We now note that T_1 is a subjective consistent criterion for hypotheses testing, since the set

$$\{(x_k)_{k \in N} : (x_k)_{k \in N} \in R^N \;\&\; T_1((x_k)_{k \in N}) = 1\}$$

is a completion of the shy set $S \setminus S_1$, where

$$S_1 = \{(x_k)_{k \in N} : (x_k)_{k \in N} \in R^N \;\&\; \lim_{n \longrightarrow \infty} \sup \frac{|\sum_{k=1}^{n} x_k|}{\sqrt{2nloglogn}} = 1\}.$$

Thus, the theorem is proved. □

3 Objective Consistent Criterion for Hypotheses Testing

Let J be a subset of N, and denote by A_J the following Borel subset of R^N :

$$A_J = \{(x_i)_{i \in N} : x_i \geq 0, \; if \; i \in J \;\&\; x_I < 0, \; if \; i \in N \setminus J\}.$$

It is clear that the A_J is Haar's ambivalent.

Let $P(N)$ be a Boolean of natural numbers, denote by Φ a one-to-one mapping from $R^+ = (0, +\infty)$ to $P(N)$, and suppose that

$$S_h = \{(x_k)_{k \in N} : (x_k)_{k \in N} \in R^N \;\&\; \lim_{n \longrightarrow \infty} \sup \frac{|\sum_{k=1}^{n} x_k|}{\sqrt{2nloglogn}} = h\}, \quad h \in R^+.$$

Since $S = \cup_{h \in R^+} S_h$ and $S_{h_1} \cap S_{h_2} = \emptyset$, if $h_1 \neq h_2$, therefore, it is clear that S_h $(\forall h \in R^+)$ is a shy set.

Let's denote

$$D_h = (A_{\Phi(h)} \setminus S) \cup S_h, \quad h \in R^+.$$

It is easy to see that $(D_h)_{h \in R^+}$ is a partition of R^N, where each D_h is a Borel subset and Haar's ambivalent. Moreover, for any pair of hypotheses $h_1, h_2 \in R^+$ there exists an isometric (according to Tychonoff metric) transformation $A_{(h_1,h_2)}$ of the space R^N such that $A_{(h_1,h_2)}(D_{h_1})\Delta(D_{H_2})$ is a shy set.

Indeed, as the transformation $A_{(h_1,h_2)}$ we consider the transformation, defined for all $(x_k)_{k \in N} \in R^N$ as follows:

$$A_{(h_1,h_2)}(x_k)_{k \in N} = (x_k)_{k \in N}, \quad if \ k \in \Phi(h_1) \cap \Phi(h_2) \ or \ k \in N \setminus \Phi(h_1) \cup \Phi(h_2)$$

and

$$A_{(h_1,h_2)}(x_k)_{k \in N} = (-x_k)_{k \in N}, \quad otherwise.$$

If we denote now

$$T^0((x_k)_{k \in N}) = h, \ if \ (x_k)_{k \in N} \in D_h,$$

then, is is not difficult to conclude that the statistical structure $\{R^N, B(R^N), \mu_h^N, h \in R^+\}$ admits the objective consistent criterion T^0 for hypotheses testing.

References

1. Zerakidze, Z.S., Purtukhia, O.G.: Consistent criteria for hypotheses testing. Ukr. Math. J. **71**(4), 554–571 (2019). https://doi.org/10.1007/s11253-019-01663-2
2. Pantsulaia, G., Kintsurashvili, M.: Why is null hypothesis rejected for "almost every" infinite sample by some hypothesis testing of maximal reliability. Adv. Theory Appl. **11**(1), 45–70 (2014)
3. Zerakidze, Z., Pantsulaia, G., Saatashvili, G.: On the separation problem for a family of Borel and Baire G-powers of shift-measures on R. Ukr. Math. J. **65**(4), 470–485 (2013)
4. Hunt, B.R., Sauer, T., Yorke, J.A.: Prevalence: A translation-invariant "almost every" on infinite-dimensional spaces. Bull. (New Series) Am. Math. Soc. **27**(2), 217–238 (1992)
5. Kuipers, L., Niederreiter, H.: Uniform Distribution of Sequences. A Wiley Interscience Publication, New York (1974)
6. Halmos, P.R.: Measure Theory. Springer, Heidelberg (1950)

Review of Rational Electrodynamics: Deformation and Force Models for Polarizable and Magnetizable Matter

Wilhelm Rickert and Wolfgang H. Müller

Abstract In this paper a rational derivation of MAXWELL's equations is presented in a purely spatial description. On a macroscopic scale this can be done by means of localization of global balance laws. The mathematical tools for the localization in a spatial description are presented. Subsequently the balance laws of electric charge and magnetic flux are discussed and localized in order to obtain MAXWELL's equations. Furthermore, short historical remarks on the origin of the governing laws in mechanics are presented. In order to illustrate the coupling of electrodynamics in matter to mechanics, two exemplary problems are analyzed. The procedure and the arising difficulties are discussed.

Keywords Rational electrodynamics · Electromechanical coupling · Force model analysis · spatial description

1 Introduction

The governing equations of electromagnetism are MAXWELL's equations. They come in different forms depending on the scale of the considered system. On a continuous macroscopic level they can be derived from balance laws. As customary in continuum mechanics, material balances are often considered. However, electromagnetic fields are not always bound to matter and it is thus reasonable to consider non-material balances for open systems. Surprisingly this is rarely seen in the pertinent literature.

Classical continuum mechanics relies on balance equations of additive quantities, which represent global formulations of physical laws. Because integral equations are difficult to deal with mathematically one usually localizes these global balances laws.

W. Rickert (✉) · W. H. Müller
Institute of Mechanics, Chair of Continuum Mechanics and Constitutive Theory, Technische Universität Berlin, Einsteinufer 5, 10587 Berlin, Germany
e-mail: rickert@tu-berlin.de

W. H. Müller
e-mail: wolfgang.h.mueller@tu-berlin.de

G. Jaiani and D. Natroshvili (eds.), *Applications of Mathematics and Informatics in Natural Sciences and Engineering*, Springer Proceedings in Mathematics & Statistics 334, https://doi.org/10.1007/978-3-030-56356-1_15

This is done in classical textbooks on continuum mechanics and thermodynamics, e.g., [17, Chap. 2], [21, Chap. 3], [11, Chap. 2]. In most references, the balance laws and their mathematical treatment are based on material representations. However, for electromagnetism we consider electromagnetic fields that are not bound to matter and thus a material description should be replaced by a spatial one. By considering a spatial description one also has to deal with open systems, which renders the formulation of balance laws as well as their localization more complicated. Finally note that in literature the (material) EULERean description if often confused with the spatial one, see the discussion in [13]. On the other hand, more general formulations as in [26] are often not appropriate for engineers. Therefore, the mathematical tools necessary for the localization of balance laws for open systems in a spatial description are presented in this section.

First, deformation geometry is introduced in material and spatial description in order to point out the differences. Then, some integral theorems are presented, which are partly derived in Appendix 8. Finally, the local forms of special balance laws used in this article are derived.

In continuum mechanics the two different descriptions commonly introduced are the material and spatial ones. The corresponding variables are labeled as

$$\boldsymbol{x}_{\mathrm{m}} = \boldsymbol{\chi}_{\mathrm{m}}(\boldsymbol{X}_{\mathrm{m}}, t) \quad \text{and} \quad \boldsymbol{x}_{\mathrm{s}} = \boldsymbol{\chi}_{\mathrm{s}}(\boldsymbol{X}_{\mathrm{s}}, t)\,, \tag{1}$$

Both position vectors $\boldsymbol{x}_{\mathrm{m}}$ and $\boldsymbol{x}_{\mathrm{s}}$ are obtained by one-to-one (bijective) mappings $\boldsymbol{\chi}_{\mathrm{m}}$ and $\boldsymbol{\chi}_{\mathrm{s}}$, respectively, which are referring to their reference systems. These reference systems are actual material points $\boldsymbol{X}_{\mathrm{m}}$ of the body at some reference time or the reference position of the non-material control point, $\boldsymbol{X}_{\mathrm{s}}$. A material representation using the current particle position $\boldsymbol{x}_{\mathrm{m}}$ is called EULERean whereas a description relying on $\boldsymbol{X}_{\mathrm{m}}$ is called LAGRANGEian.

From these, different velocities may be considered. First, the material velocity $\boldsymbol{v}_{\mathrm{m}} = \boldsymbol{v}_{\mathrm{m}}(\boldsymbol{X}_{\mathrm{m}}, t)$ as a function of the considered particle, which is identified by its original position, $\boldsymbol{X}_{\mathrm{m}}$. Second, the mapping velocity, $\boldsymbol{w}(\boldsymbol{X}_{\mathrm{s}}, t)$, which describes the movement of non-material points, which originally have been at the position $\boldsymbol{X}_{\mathrm{s}}$. In addition, the material velocity may also be expressed in terms of a spatial position such that $\boldsymbol{V}_{\mathrm{m}}(\boldsymbol{x}_{\mathrm{s}}, t)$. In contrast to the first two velocities $\boldsymbol{v}_{\mathrm{m}}$ and $\boldsymbol{w}$, which arise from time derivatives of the functions of motion in Eq. (1), i.e.,

$$\boldsymbol{v}_{\mathrm{m}} = \frac{\partial \boldsymbol{\chi}_{\mathrm{m}}}{\partial t}\,, \quad \boldsymbol{w} = \frac{\partial \boldsymbol{\chi}_{\mathrm{s}}}{\partial t}\,, \tag{2}$$

the material velocity $\boldsymbol{V}_{\mathrm{m}}(\boldsymbol{x}_{\mathrm{s}}, t)$ cannot be represented as a time derivative, but can be expressed in terms of the material velocity if the particle currently at the position $\boldsymbol{x}_{\mathrm{s}}$ is considered

$$\boldsymbol{V}_{\mathrm{m}} = \boldsymbol{V}_{\mathrm{m}}(\boldsymbol{x}_{\mathrm{s}}, t) \stackrel{!}{=} \boldsymbol{v}_{\mathrm{m}}\big(\boldsymbol{\chi}_{\mathrm{m}}^{-1}(\boldsymbol{x}_{\mathrm{s}}, t), t\big)\,. \tag{3}$$

Note that this identification is only valid for spatial points $\boldsymbol{x}_{\mathrm{s}}$ which are currently occupied by the material body. In the following, the material velocity in a spatial description is denoted as $\boldsymbol{v} = \boldsymbol{V}_{\mathrm{m}}$.

With the material and spatial variables, different control volumes may be considered, namely material and spatial ones. Theorems such as NANSON's formula look similar for both surface elements, i.e.,

$$\boldsymbol{n}\,\mathrm{d}A = J_{\mathrm{m}}\boldsymbol{F}_{\mathrm{m}}^{-\mathrm{T}} \cdot \boldsymbol{n}_0\,\mathrm{d}A_0 \,, \quad \boldsymbol{n}\,\mathrm{d}A = J_{\mathrm{s}}\boldsymbol{F}_{\mathrm{s}}^{-\mathrm{T}} \cdot \boldsymbol{n}_0\,\mathrm{d}A_0 \,, \tag{4}$$

with the deformation gradients

$$\boldsymbol{F}_{\mathrm{m}} = \frac{\partial \boldsymbol{\chi}_{\mathrm{m}}}{\partial \boldsymbol{X}_{\mathrm{m}}} \,, \quad \boldsymbol{F}_{\mathrm{s}} = \frac{\partial \boldsymbol{\chi}_{\mathrm{s}}}{\partial \boldsymbol{X}_{\mathrm{s}}} \,, \quad J_{\mathrm{m}} = \det(\boldsymbol{F}_{\mathrm{m}}) \,, \quad J_{\mathrm{s}} = \det(\boldsymbol{F}_{\mathrm{s}}) \,. \tag{5}$$

If time derivatives of vectorial surface elements are considered, different velocities are involved, because a material surface moves with the particle velocity, $\boldsymbol{v}$, and a non-material surface with $\boldsymbol{w}$. As a result, the time derivatives are given by:

$$\begin{aligned} \frac{\mathrm{d}}{\mathrm{d}t}\big(\boldsymbol{n}\,\mathrm{d}A\big) &= (\nabla_{\mathrm{m}} \cdot \boldsymbol{v}_{\mathrm{m}})\boldsymbol{n}\,\mathrm{d}A - (\nabla_{\mathrm{m}} \otimes \boldsymbol{v}_{\mathrm{m}}) \cdot \boldsymbol{n}\,\mathrm{d}A \,, \\ \frac{\mathrm{d}}{\mathrm{d}t}\big(\boldsymbol{n}\,\mathrm{d}A\big) &= (\nabla_{\mathrm{s}} \cdot \boldsymbol{w})\boldsymbol{n}\,\mathrm{d}A - (\nabla_{\mathrm{s}} \otimes \boldsymbol{w}) \cdot \boldsymbol{n}\,\mathrm{d}A \,, \end{aligned} \tag{6}$$

with

$$\nabla_{\mathrm{m}} = \frac{\partial}{\partial \boldsymbol{x}_{\mathrm{m}}} \,, \quad \nabla_{\mathrm{s}} = \frac{\partial}{\partial \boldsymbol{x}_{\mathrm{s}}} \,. \tag{7}$$

One should note, that the choice of the variables, $\boldsymbol{x}_{\mathrm{m}}$ or $\boldsymbol{x}_{\mathrm{s}}$, affects the type of nabla operator that is considered. The same applies to the time derivatives of the volume elements in a material and a spatial description, respectively:

$$\frac{\mathrm{d}}{\mathrm{d}t}(\mathrm{d}V) = (\nabla_{\mathrm{m}} \cdot \boldsymbol{v}_{\mathrm{m}})\,\mathrm{d}V \,, \quad \frac{\mathrm{d}}{\mathrm{d}t}(\mathrm{d}V) = (\nabla_{\mathrm{s}} \cdot \boldsymbol{w})\,\mathrm{d}V \,. \tag{8}$$

Following this argument, two functions corresponding to a field quantity $\boldsymbol{\psi}$ should be distinguished:

$$\boldsymbol{\psi} = \boldsymbol{\psi}_{\mathrm{m}}(\boldsymbol{x}_{\mathrm{m}}, t) \,, \quad \boldsymbol{\psi} = \boldsymbol{\psi}_{\mathrm{s}}(\boldsymbol{x}_{\mathrm{s}}, t) \,, \tag{9}$$

where $\boldsymbol{\psi}_{\mathrm{m}}$ and $\boldsymbol{\psi}_{\mathrm{s}}$ are different functions and the same identification problem arises as in Eq. (3). However, for fields other than the velocity, the distinction between the two functions will not be denoted explicitly, because we are interested in a purely spatial description. Therefore, all fields are considered in spatial description if not stated otherwise.

2 Balance Laws and Integral Theorems

Consider an open control surface as in Fig. 1. The non-material surface S is cut into two parts, S^- and S^+, by a singular line ℓ_I. While the surface itself moves with $\boldsymbol{w}$, the singular line is allowed to move separately with $\boldsymbol{w}_I$. Integral theorems such as GAUSS' theorem and the REYNOLDS transport theorem require the integrands to be continuous in the domain of integration. At a singular surface however, continuum field quantities are allowed to be discontinuous and hence the integral theorems need to be modified.

Consider the differentiable field quantity $\boldsymbol{f}$ that is continuous everywhere in S except for $S \setminus \ell_I$. The limits $\boldsymbol{f}^+$ and $\boldsymbol{f}^-$ from the respective domains S^+ and S^- at the singular surface are allowed to be different. This difference is captured by the jump evaluated at the interface $\boldsymbol{x}_s = \boldsymbol{x}_I$

$$[\![\boldsymbol{f}]\!] = \boldsymbol{f}^+ - \boldsymbol{f}^- \quad \boldsymbol{f}^{\pm}(\boldsymbol{x}_I, t) = \lim_{\boldsymbol{x}_s \to \boldsymbol{x}_I} \boldsymbol{f}(\boldsymbol{x}_s, t) \,, \quad \boldsymbol{x}_s \in \Omega^{\pm} \,, \quad \boldsymbol{x}_I \in I \,. \tag{10}$$

We are free to choose material or non-material control volumes and in the end the localized equations look essentially the same except for different functional dependencies, i.e., $\boldsymbol{x}_m$ or $\boldsymbol{x}_s$, and different velocities. It turns out that the non-material description is more general and the material one can be deduced as a special case. Therefore, in this section the domain of integration will be non-material and in a spatial description with $\nabla = \nabla_s$.

One is now interested in a general surface flux balance for the domain depicted in Fig. 1. The temporal change of the total flux due a flux density $\boldsymbol{f}$ through the surface S is given by

$$\frac{\mathrm{d}}{\mathrm{d}t} \int_S \boldsymbol{e} \cdot \boldsymbol{f} \, \mathrm{d}A = - \int_{\ell^- \cup \ell^+} \boldsymbol{\tau} \cdot \big(\boldsymbol{\phi} + \boldsymbol{f} \times (\boldsymbol{v} - \boldsymbol{w}) \big) \, \mathrm{d}\ell + \int_{S^- \cup S^+} \boldsymbol{e} \cdot \boldsymbol{s} \, \mathrm{d}A + \int_{\ell_I} \boldsymbol{\nu}_I \cdot \boldsymbol{s}_I \, \mathrm{d}\ell \,. \tag{11}$$

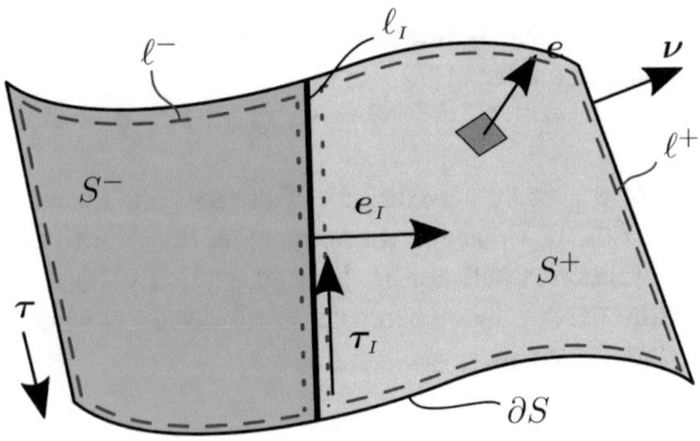

Fig. 1 Depiction of a surface S cut by a singular line ℓ_I such that its boundary is given by $\partial S = \ell^+ \cup \ell^- \cup \partial \ell_I$. Also the binormal vector to the singular line is given by $\boldsymbol{\nu}_I = \boldsymbol{e}_I \times \boldsymbol{\tau}_I$

Therein, $\boldsymbol{\phi}$ is the conductive (or non-convective) flux (of the flux $\boldsymbol{f}$) and $\boldsymbol{f} \times (\boldsymbol{v} - \boldsymbol{w})$ is the convective flux of the flux. Furthermore, $\boldsymbol{s}$ and $\boldsymbol{s}_I$ are regular and singular supply terms. Note that both, $\boldsymbol{s}$ and $\boldsymbol{s}_I$, may contain conductive as well as convective supply since the domain S is open. For the convective flux $\boldsymbol{f} \times (\boldsymbol{v} - \boldsymbol{w})$ it is assumed that $\boldsymbol{f}$ is transported with mass and hence with the material velocity $\boldsymbol{v}$. If this is not the case, i.e., $\boldsymbol{f}$ is a matter independent field, then $\boldsymbol{v} \equiv \boldsymbol{0}$ in the above equation.

In order to localize this balance law in regular and singular points, several integral theorems for domains with singular lines are required. They are derived in the Appendix 8 and the process of localization in regular and singular points is shown in Appendix 9. By means of these methods the following local variants of the balance law in regular and singular points, respectively, are found:

$$\frac{\partial \boldsymbol{f}}{\partial t} + (\nabla \cdot \boldsymbol{f})\boldsymbol{w} + \nabla \times \big(\boldsymbol{\phi} - \boldsymbol{v} \times \boldsymbol{f}\big) = \boldsymbol{s} \,,$$
$$\boldsymbol{e}_I \times [\![\boldsymbol{\phi} - (\boldsymbol{v} - \boldsymbol{w}_I) \times \boldsymbol{f}]\!] = \boldsymbol{s}_I \,. \tag{12}$$

If $\boldsymbol{f}$ is a field that is not convected with matter and has no supply $\boldsymbol{s}$ or $\boldsymbol{s}_I$, one simply has

$$\frac{\partial \boldsymbol{f}}{\partial t} + (\nabla \cdot \boldsymbol{f})\boldsymbol{w} + \nabla \times \boldsymbol{\phi} = \boldsymbol{0} \,,$$
$$\boldsymbol{e}_I \times [\![\boldsymbol{\phi}]\!] + \boldsymbol{e}_I \cdot [\![\boldsymbol{f} \otimes \boldsymbol{w}_I - \boldsymbol{w}_I \otimes \boldsymbol{f}]\!] = \boldsymbol{0} \,. \tag{13}$$

3 Rational Electrodynamics

There are several approaches to electrodynamics and, consequently, to the MAXWELL equations in literature. The rational derivation of MAXWELL's equations relies on four basic axioms:

- the conservation of charge,
- the decomposition of charges and currents into bound and free parts,
- the conservation of magnetic flux and
- the MAXWELL-LORENTZ-aether relations,

which are also considered in the derivations of [12, 14] or [16], to name a few. However, all of these authors restrict their presentation to material, i.e., closed systems. In [26] a more general approach is pursued. However, the mathematics involved are quite complicated. Before we start with a rational derivation of MAXWELL's equations in spatial formulation, a word of caution regarding the term "rational" is in order. In [26, p. 669] the truth about the rationality of modern electrodynamics is phrased as follows "[...] the integral equations (270.5) and (270.9) were deduced by BATEMAN[1], who took as a starting point the differential equations commonly referred to as MAXWELL's equations." Nonetheless, this approach is still more rational than a purely phenomenological one. The presentation in this paper is restricted to classical equations, i.e., no theory of relativity is considered. The interested reader is referred to [16, 26], where the conservation of charge is analyzed relativistically. Furthermore we adapt the notation of [26, p. 689].

3.1 Electromagnetic Fields

The electric and magnetic fields $\boldsymbol{E}$ and $\boldsymbol{B}$ are originally defined via the LORENTZ force,

$$\boldsymbol{F} = Q\boldsymbol{E} + \boldsymbol{J} \times \boldsymbol{B} \,, \tag{14}$$

that acts on a point-like particle carrying the charge Q moving at a velocity $\boldsymbol{v}$ through an electric field and appears as an electric current $\boldsymbol{J} = Q\boldsymbol{v}$ to a magnetic field. From this it can be seen that there is an intimate connection between electrodynamics and mechanics. The LORENTZ force expression contains two contributions experimentally found by COULOMB and AMPÉRE. While COULOMB performed experiments with localized non-moving distributions of charge to find a force per unit charge, $\boldsymbol{E}$, AMPÉRE investigated the force per unit current in a conducting wire, $\boldsymbol{B}$.

Later, in 1830, FARADAY found that there is a close relation between the electric field and the magnetic flux. It was thought that the voltage V due an electric field $\boldsymbol{E}$ around a closed material loop $\mathscr{C}$ is zero, i.e.,

$$V = \oint\limits_{\mathscr{C}} \boldsymbol{\tau} \cdot \boldsymbol{E} \, \mathrm{d}\ell = 0 \,. \tag{15}$$

where $\boldsymbol{\tau}$ is the tangent vector of $\mathscr{C}$. In view of the LORENTZ force expression, this voltage is sometimes called electromotive force. However, FARADAY found that this is not the case if a varying magnetic field is present, see [8],

$$\oint\limits_{\mathscr{C}} \boldsymbol{\tau} \cdot \boldsymbol{E} \, \mathrm{d}\ell = -\frac{\mathrm{d}}{\mathrm{d}t} \int\limits_{\mathscr{S}} \boldsymbol{n} \cdot \boldsymbol{B} \, \mathrm{d}A \,, \quad \mathscr{C} = \partial\mathscr{S} \,. \tag{16}$$

This is to be interpreted as the flow of an electric current due to a magnetic field and therefore this is also known as FARADAY's law of induction. Note that in this context the surface $\mathscr{S}$ spanned by the material loop of a conducting wire is arbitrary and thus non-material. Additionally, one assumes that this law is equally valid, i.e., the same electromotive force is produced, even if the loop $\mathscr{C}$ is non-material as well. Then, of course, there would be no resulting current. This motivates the last section where all presentations are given with respect to non-material control domains.

With this notion of the fields $\boldsymbol{E}$ and $\boldsymbol{B}$ one can start to derive the differential equations they are governed by, MAXWELL's equations. However, one should note that the transition from some free charges Q in space or in a conducting wire to a continuum description of matter is not trivial. Even the force expression in Eq. (14)

cannot simply be translated into a continuum force density in the presence of (magnetizable or polarizable) matter. This conundrum is part of the ABRAHAM–MINKOWSKI controversy and will be discussed in Sect. 4.

3.2 *Derivation of* MAXWELL*'s Equations in Matter*

A rational approach to MAXWELL's equations becomes possible through the balance laws for the total charge and the magnetic flux. Note that the latter is referred to as FARADAY's law.

Consider a volumetric region Ω cut by a singular surface I as depicted in Fig. 2. Then, the balance of charge reads,

$$\frac{\mathrm{d}}{\mathrm{d}t}\int\limits_{\Omega^-\cup\Omega^+} q\,\mathrm{d}V + \frac{\mathrm{d}}{\mathrm{d}t}\int\limits_{I} q_I\,\mathrm{d}A = -\int\limits_{\Gamma^-\cup\Gamma^+} \boldsymbol{n}\cdot\big(\boldsymbol{j}+q(\boldsymbol{v}-\boldsymbol{w})\big)\,\mathrm{d}A - \oint\limits_{\partial I} \boldsymbol{\nu}\cdot\big(\boldsymbol{j}_I + q_I(\boldsymbol{v}-\boldsymbol{w}_I)\big)\,\mathrm{d}\ell\,, \tag{17}$$

If one considers the definitions of the total currents, *cf.* [21],

$$\boldsymbol{J} = \boldsymbol{j} + q\boldsymbol{v}\,,\quad \boldsymbol{J}_I = \boldsymbol{j}_I + q_I\boldsymbol{v}\,, \tag{18}$$

the charge balance is denoted more compactly by

$$\frac{\mathrm{d}}{\mathrm{d}t}\int\limits_{\Omega^-\cup\Omega^+} q\,\mathrm{d}V + \frac{\mathrm{d}}{\mathrm{d}t}\int\limits_{I} q_I\,\mathrm{d}A = -\int\limits_{\Gamma^-\cup\Gamma^+} \boldsymbol{n}\cdot(\boldsymbol{J}-q\boldsymbol{w})\,\mathrm{d}A - \oint\limits_{\partial I} \boldsymbol{\nu}\cdot\big(\boldsymbol{J}_I - q_I\boldsymbol{w}_I\big)\,\mathrm{d}\ell\,. \tag{19a}$$

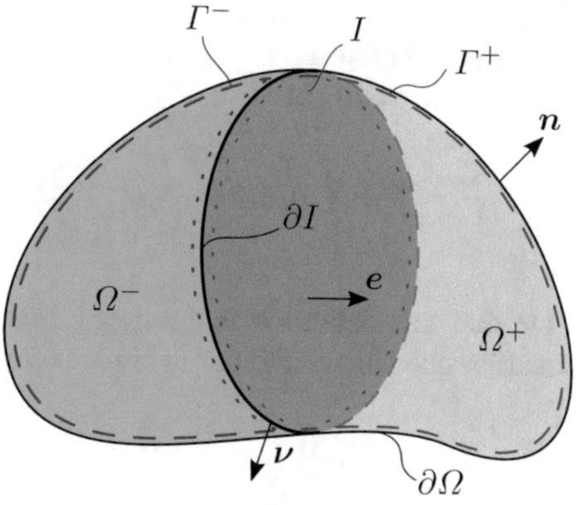

Fig. 2 Depiction of a volumetric domain Ω. The domain Ω is cut by a singular surface I such that the surface is given by $\partial\Omega = \Gamma^-\cup\Gamma^+\cup\partial I$, where $\Gamma^\pm = \partial\Omega^\pm \setminus I$ are non-closed surfaces

Furthermore, for some other non-material surface S the balance of magnetic flux is given by

$$\frac{\mathrm{d}}{\mathrm{d}t}\int_S \boldsymbol{e}\cdot\boldsymbol{B}\,\mathrm{d}A = -\oint_{\partial S} \boldsymbol{\tau}\cdot\big(\boldsymbol{E}+\boldsymbol{w}\times\boldsymbol{B}\big)\,\mathrm{d}\ell\,. \tag{19b}$$

Therein, q is the total charge density, $\boldsymbol{j}$ is the conductive electric current density, $\boldsymbol{B}$ is the magnetic flux density, $\boldsymbol{E}$ is the electric field, $\boldsymbol{v}$ is the material velocity in spatial description and $\boldsymbol{w}$ as well as $\boldsymbol{w}_I$ are the mapping velocities of the volume and the singular surface, respectively. The additional quantities q_I and $\boldsymbol{j}_I$ are the (singular) surface charge density and the (singular) surface conductive current density, respectively.

Note that the balance of the magnetic flux as presented here appears as a postulate. However, originally it was supposed to summarize the experimental observations due to FARADAY, see the discussion in Sect. 3.1. As a generalization to non-material control surfaces, a convective flux is $\boldsymbol{w}\times\boldsymbol{B}$ is added without a dependence upon the material velocity. Note that most authors restrict themselves to material domains and thus are required to introduce the material velocity, $\boldsymbol{v}\times\boldsymbol{B}$.

First, let us consider the charge balance. It can be derived from the mass balances utilizing a mixture theory, see [24]. Therefore, most authors tend to restrict themselves to material domains $\Omega = \Omega_{\mathrm{m}}$. However, we refrain from doing so. Since Ω is a volumetric region, its surface $\partial\Omega$ is closed and the boundary of the surface is empty, i.e., $\partial\partial\Omega = \emptyset$. Hence, we are free to add an arbitrary zero to the charge balance and write

$$\begin{aligned}\frac{\mathrm{d}}{\mathrm{d}t}\int_{\Omega^-\cup\Omega^+} q\,\mathrm{d}V + \frac{\mathrm{d}}{\mathrm{d}t}\int_I q_I\,\mathrm{d}A &= -\int_{\Gamma^-\cup\Gamma^+} \boldsymbol{n}\cdot\big(\boldsymbol{J}-q\boldsymbol{w}\big)\,\mathrm{d}A \\ &\quad -\oint_{\partial I} \boldsymbol{\nu}\cdot\big(\boldsymbol{J}_I - q_I\boldsymbol{w}_I\big)\,\mathrm{d}\ell + \oint_{\partial\partial\Omega} \boldsymbol{a}\cdot\boldsymbol{\tau}\,\mathrm{d}\ell\,,\end{aligned} \tag{20}$$

where $\boldsymbol{a}$ is an additional auxiliary field, which does not contribute to the balance of charge (19a). In order to apply the localization theorem, all integrals must agree in their type. Therefore, the so-called total charge potential, $\boldsymbol{D}$, is introduced and the generalized version of GAUSS' theorem from Eq. (117) is applied

$$\int_{\Omega^-\cup\Omega^+} q\,\mathrm{d}V + \int_I q_I\,\mathrm{d}A \equiv \oint_{\partial\Omega} \boldsymbol{D}\cdot\boldsymbol{n}\,\mathrm{d}A = \int_{\Omega^-\cup\Omega^+} \nabla\cdot\boldsymbol{D}\,\mathrm{d}V + \int_I \boldsymbol{e}\cdot\boldsymbol{D}\,\mathrm{d}A\,. \tag{21}$$

Note that this definition is a purely formal step. By comparing both sides of the equation one obtains the following local relations:

$$\nabla\cdot\boldsymbol{D} = q\,,\quad \boldsymbol{e}\cdot[\![\boldsymbol{D}]\!] = q_I\,. \tag{22}$$

Due to the definition of this potential, the total charge balance reduces to a surface balance with $S = \partial\Omega$, $S^{\pm} = \Gamma^{\pm}$ and $\ell_I = \partial I$

$$\frac{\mathrm{d}}{\mathrm{d}t}\oint\limits_{S} \boldsymbol{D}\cdot\boldsymbol{n}\,\mathrm{d}A = -\int\limits_{S^{+}\cup S^{-}} \boldsymbol{n}\cdot\left(\boldsymbol{J} - q\boldsymbol{w}\right)\mathrm{d}A - \oint\limits_{\ell_I} \boldsymbol{v}\cdot\left(\boldsymbol{J}_I - q_I\boldsymbol{w}_I\right)\mathrm{d}\ell + \int\limits_{\ell^{-}\cup\ell^{+}} \boldsymbol{a}\cdot\boldsymbol{\tau}\,\mathrm{d}\ell\,. \tag{23}$$

In view of a general surface balance for open control volumes in Eq. (11) it seems natural to introduce the auxiliary field as a conductive term plus a convective one

$$\boldsymbol{a} = \boldsymbol{H} + \boldsymbol{D}\times\boldsymbol{w}\,, \tag{24}$$

where $\boldsymbol{H}$ is another "auxiliary field." Since we do not know anything about the charge potential $\boldsymbol{D}$, it is assumed that it is not convective with respect to matter. However, the definition of $\boldsymbol{a}$ is completely arbitrary. Now the localization theorem of a generalized surface balance can be applied. With the given quantities

$$\boldsymbol{f} = \boldsymbol{D}\,,\quad \boldsymbol{s} = -\boldsymbol{J} + q\boldsymbol{w}\,,\quad \boldsymbol{s}_I = -\boldsymbol{J}_I + q_I\boldsymbol{w}_I\,, \tag{25}$$

one has from the localization theorem that

$$\begin{gathered} -\frac{\partial\boldsymbol{D}}{\partial t} + \nabla\times\boldsymbol{H} = \boldsymbol{J} = \boldsymbol{j} + q\boldsymbol{v}\,,\\ \boldsymbol{e}\times[\![\boldsymbol{H} + \boldsymbol{D}\times\boldsymbol{w}_I]\!] = \boldsymbol{J}_I - q_I\boldsymbol{w}_I\,. \end{gathered} \tag{26}$$

This is also called ØRSTED's law. From this equation the name current potential for the field $\boldsymbol{H}$ can be motivated. However, we will later discuss this auxiliary field in more detail. It should be noted that in [16, Chap. 3] the current potential is introduced via the components of a so-called charge current potential in a EUCLIDean space-time formulation of the charge conservation. Furthermore note that by observing

$$\boldsymbol{e}\times[\![\boldsymbol{D}\times\boldsymbol{w}_I]\!] = (\boldsymbol{e}\cdot\boldsymbol{w}_I)[\![\boldsymbol{D}]\!] - (\boldsymbol{e}\cdot[\![\boldsymbol{D}]\!])\boldsymbol{w}_I = (\boldsymbol{e}\cdot\boldsymbol{w}_I)[\![\boldsymbol{D}]\!] - q_I\boldsymbol{w}_I \tag{27}$$

the balance in singular points can be simplified to read

$$(\boldsymbol{e}\cdot\boldsymbol{w}_I)[\![\boldsymbol{D}]\!] + \boldsymbol{e}\times[\![\boldsymbol{H}]\!] = \boldsymbol{J}_I = \boldsymbol{j}_I + q_I\boldsymbol{v}\,. \tag{28}$$

Now we turn to FARADAY's law. From the localization theorem in Eq. (13) one obtains

$$\begin{gathered} \frac{\partial\boldsymbol{B}}{\partial t} + \nabla\times\boldsymbol{E} + (\nabla\cdot\boldsymbol{B})\boldsymbol{w} = \boldsymbol{0}\,,\\ \boldsymbol{e}\times[\![\boldsymbol{E}]\!] + \boldsymbol{e}\cdot[\![\boldsymbol{B}\otimes\boldsymbol{w}_I - \boldsymbol{w}_I\otimes\boldsymbol{B}]\!] = \boldsymbol{0}\,. \end{gathered} \tag{29}$$

Therein, the electric field $\boldsymbol{E}$ is also called MINKOWSKIan electric field strength, *cf.* [12, p. 14]. It is now very surprising that the local equation contains a dependence upon the mapping velocity $\boldsymbol{w}$. It turns out that this is an artifact from the arbitrary surface spanned by the line $\mathscr{C}$ in Sect. 3.1. In fact, there is an additional way of localizing the global form. Consider a surface of some volumetric region $S = \partial\Omega_s$, that may be cut by a singular surface I. Then the line integral vanishes and from FARADAY's law in global form it follows that

$$\frac{\mathrm{d}}{\mathrm{d}t}\int_{\partial\Omega_s} \boldsymbol{B}\cdot\boldsymbol{e}\,\mathrm{d}A = 0 \quad\Rightarrow\quad 0 = \int_{\partial\Omega_s} \boldsymbol{B}\cdot\boldsymbol{e}\,\mathrm{d}A = \int_{\Omega_s^-\cup\Omega_s^+} \nabla\cdot\boldsymbol{B}\,\mathrm{d}V + \int_I \boldsymbol{e}\cdot[\![\boldsymbol{B}]\!]\,\mathrm{d}A\,. \tag{30}$$

Via localization one obtains GAUSS' law

$$\nabla\cdot\boldsymbol{B} = 0\,,\quad \boldsymbol{e}\cdot[\![\boldsymbol{B}]\!] = 0\,. \tag{31}$$

Note that the constant due to the time integration is neglected here. Furthermore, with this result, Eq. (29) simplifies, which shows that FARADAY's law in regular points does actually not depend upon the mapping velocity,

$$\begin{aligned}\frac{\partial\boldsymbol{B}}{\partial t} + \nabla\times\boldsymbol{E} &= \boldsymbol{0}\,,\\ \boldsymbol{e}\times[\![\boldsymbol{E}]\!] - (\boldsymbol{e}\cdot\boldsymbol{w}_I)[\![\boldsymbol{B}]\!] &= \boldsymbol{0}\,.\end{aligned} \tag{32}$$

So far, the following set of equations in regular points is obtained from the conservation of charge

$$-\frac{\partial\boldsymbol{D}}{\partial t} + \nabla\times\boldsymbol{H} = \boldsymbol{j} + q\boldsymbol{v}\,,\quad \nabla\cdot\boldsymbol{D} = q \tag{33a}$$

and from the conservation of magnetic flux

$$\frac{\partial\boldsymbol{B}}{\partial t} + \nabla\times\boldsymbol{E} = \boldsymbol{0}\,,\quad \nabla\cdot\boldsymbol{B} = 0\,. \tag{33b}$$

Note that the (total) charge potential as well as the (total) current potential are not exclusively due to matter, rather matter contributes to the fields $\boldsymbol{D}$ and $\boldsymbol{H}$, as will be seen in the next section.

3.3 Charges in Materials

The equations up to now are valid for every material, but not very convenient if dielectrics or magnets are considered. Therefore, it is assumed that the charge densities as well as the current densities are additively decomposed into free and bound parts:

$$q = q^{\mathrm{f}} + q^{\mathrm{b}} \,, \quad q_I = q_I^{\mathrm{f}} + q_I^{\mathrm{b}} \,, \quad \boldsymbol{J} = \boldsymbol{J}^{\mathrm{f}} + \boldsymbol{J}^{\mathrm{b}} \,, \quad \boldsymbol{J}_I = \boldsymbol{J}_I^{\mathrm{f}} + \boldsymbol{J}_I^{\mathrm{b}} \,, \tag{34}$$

where the free currents are decomposed into conductive and convective parts

$$\boldsymbol{J}^{\mathrm{f}} = \boldsymbol{j}^{\mathrm{f}} + q^{\mathrm{f}}\boldsymbol{v} \,, \quad \boldsymbol{J}_I^{\mathrm{f}} = \boldsymbol{j}_I^{\mathrm{f}} + q_I^{\mathrm{f}}\boldsymbol{v} \,. \tag{35}$$

The bound charges account for pseudo charges due to electric dipoles. The free charges and currents are properly defined as they can be obtained from particle densities and partial mass flows in a mixture theory, see [24] for an extensive discussion of this topic. Hence, in the following a closed material domain Ω_{m} is considered. It can then be shown, that with $\boldsymbol{w} = \boldsymbol{v}$ the balance of free charges is given by

$$\frac{\mathrm{d}}{\mathrm{d}t}\int\limits_{\Omega_{\mathrm{m}}^{+}\cup\Omega_{\mathrm{m}}^{-}} q^{\mathrm{f}}\,\mathrm{d}V + \frac{\mathrm{d}}{\mathrm{d}t}\int\limits_{I} q_I^{\mathrm{f}}\,\mathrm{d}A = -\int\limits_{\Gamma^{-}\cup\Gamma^{+}} \boldsymbol{j}^{\mathrm{f}}\cdot\boldsymbol{n}\,\mathrm{d}A - \oint\limits_{I} \boldsymbol{j}_I^{\mathrm{f}}\cdot\boldsymbol{\nu}\,\mathrm{d}\ell \tag{36}$$

and by combining this with Eq. (19a) one obtains the balance of bound charges

$$\frac{\mathrm{d}}{\mathrm{d}t}\int\limits_{\Omega_{\mathrm{m}}^{+}\cup\Omega_{\mathrm{m}}^{-}} q^{\mathrm{b}}\,\mathrm{d}V + \frac{\mathrm{d}}{\mathrm{d}t}\int\limits_{I} q_I^{\mathrm{b}}\,\mathrm{d}A = -\int\limits_{\Gamma^{-}\cup\Gamma^{+}} \boldsymbol{j}^{\mathrm{b}}\cdot\boldsymbol{n}\,\mathrm{d}A - \oint\limits_{I} \boldsymbol{j}_I^{\mathrm{b}}\cdot\boldsymbol{\nu}\,\mathrm{d}\ell \,. \tag{37}$$

Both balances look identical and it is thus natural to decompose the charge and current potentials as well into free and bound parts:

$$\boldsymbol{D} = \boldsymbol{\mathfrak{D}} - \boldsymbol{P} \,, \quad \boldsymbol{H} = \boldsymbol{\mathfrak{H}} + \boldsymbol{M} \,, \tag{38}$$

where $\boldsymbol{\mathfrak{D}}$ and $\boldsymbol{\mathfrak{H}}$ are the free charge and current potentials. The fields $\boldsymbol{P}$ (the minus sign in Eq. (38) is based on convention) and $\boldsymbol{M}$ are analogously referred to as bound charge and bound current potentials. However, they allow for an intuitive interpretation, see [18, 21], which in turn lead to the names polarization and magnetization. Electric dipoles around a body introduce a current as they change their direction over time. This is what is considered as a bound current since there is no net movement. Similarly, with magnetic dipoles a magnetization current is introduced at the surface of a body. Of course one could add higher multipole moments to the consideration, which is usually done in microscopic theories, [14, p. 232]. However, the dipole moments are dominant and higher moments can be neglected on a continuum scale. Note that the magnetization vector $\boldsymbol{M}$ as introduced above is also called MINKOWSKI magnetization. It is also common to introduce the so-called LORENTZ magnetization $\mathscr{M}$, see [16], which is related to $\boldsymbol{M}$ via:

$$\boldsymbol{M} = \mathscr{M} - \boldsymbol{v}\times\boldsymbol{P} \,. \tag{39}$$

The LORENTZ magnetization emphasizes the fact, that for a medium in motion the effective magnetization is altered by the polarization, see [14, p. 234]

Then, by the same arguments as before, from Eqs. (36) and (37) the following localized versions arise:

$$
\begin{aligned}
-\frac{\partial \mathfrak{D}}{\partial t} + \nabla \times \mathfrak{H} &= \boldsymbol{J}^{\mathrm{f}} = \boldsymbol{j}^{\mathrm{f}} + q^{\mathrm{f}} \boldsymbol{v} \,, & \nabla \cdot \mathfrak{D} &= q^{\mathrm{f}} \,, \\
\frac{\partial \boldsymbol{P}}{\partial t} + \nabla \times \boldsymbol{M} &= \boldsymbol{J}^{\mathrm{b}} \,, & -\nabla \cdot \boldsymbol{P} &= q^{\mathrm{b}} \,.
\end{aligned} \tag{40}
$$

and in singular points:

$$
\begin{aligned}
(\boldsymbol{e} \cdot \boldsymbol{w}_I)[\![\mathfrak{D}]\!] + \boldsymbol{e} \times [\![\mathfrak{H}]\!] &= \boldsymbol{j}_I^{\mathrm{f}} + q_I^{\mathrm{f}} \boldsymbol{v} \,, & \boldsymbol{e} \cdot [\![\mathfrak{D}]\!] &= q_I^{\mathrm{f}} \,, \\
-(\boldsymbol{e} \cdot \boldsymbol{w}_I)[\![\boldsymbol{P}]\!] + \boldsymbol{e} \times [\![\boldsymbol{M}]\!] &= \boldsymbol{J}_I^{\mathrm{b}} \,, & -\boldsymbol{e} \cdot [\![\boldsymbol{P}]\!] &= q_I^{\mathrm{b}} \,.
\end{aligned} \tag{41}
$$

3.4 *Universal Connections: The* MAXWELL-LORENTZ-*æther Relations*

An illuminating discussion of the æther relations is given in [16, Chap. 5], and there is nothing to add. Hence, we may just denote the relations in SI-units and briefly comment on them:

$$
\boldsymbol{B} = \mu_0 \boldsymbol{H} \,, \quad \epsilon_0 \boldsymbol{E} = \boldsymbol{D} \,, \tag{42}
$$

with the vacuum permeability μ_0 and the vacuum permittivity ϵ_0. Together they define the speed of light *viz.*, $c^{-1} = \mu_0 \epsilon_0$. It is very interesting to note that these equations relate the physical fields $\boldsymbol{B}$ and $\boldsymbol{E}$ and the potentials $\boldsymbol{H}$ and $\boldsymbol{D}$. Therefore, the potentials, as such not uniquely defined, become unique quantities.

It is clear that these relations cannot hold for every frame of reference. However, those frames for which the æther relations do hold are called LORENTZ rest frames, see [14, 26].

Note that from the æther relations it follows that

$$
\frac{1}{\mu_0} \boldsymbol{B} = \mathfrak{H} + \boldsymbol{M} \,, \quad \epsilon_0 \boldsymbol{E} = \mathfrak{D} - \boldsymbol{P} \,, \tag{43}
$$

where the use of the MINKOWSKI magnetization is rather convenient.

4 Coupling of Electromagnetism and Mechanics

In the last section the governing equations of electromagnetism were derived. As it was pointed out, electromagnetism is originally strongly connected to mechanics. In continuum mechanics, additive quantities are balanced. In balances of conserved quantities, there may be supply terms (that can be deactivated, at least in principle) but no so-called production terms. A production term models some kind of reaction that cannot be controlled directly. Let us consider the balance of linear momentum with electromagnetic influences in regular and singular points, respectively,

$$\begin{aligned}\frac{\partial \rho \boldsymbol{v}}{\partial t} + \nabla \cdot (\rho \boldsymbol{v} \otimes \boldsymbol{v} - \boldsymbol{\sigma}) &= \rho \boldsymbol{f} + \boldsymbol{f}^{(\mathrm{EM})} , \\ \boldsymbol{n} \cdot [\![\boldsymbol{\sigma} + \rho(\boldsymbol{w}_I - \boldsymbol{v}) \otimes \boldsymbol{v}]\!] &= -\boldsymbol{f}_I^{(\mathrm{EM})} . \end{aligned} \tag{44}$$

Therein, ρ denotes the mass density, $\boldsymbol{\sigma}$ is the mechanical stress tensor, $\boldsymbol{f}$ is the mechanical volume force density and $\boldsymbol{f}^{(\mathrm{EM})}$ is the volumetric force density due to electromagnetic fields. Furthermore, $\boldsymbol{w}_I$ is the surface velocity and $\boldsymbol{f}_I^{(\mathrm{EM})}$ is the electromagnetic surface force density. The volumetric force density due to electromagnetic fields cannot be controlled directly and, hence, must be considered as a production of linear momentum. Therefore, the (mechanical) linear momentum is not conserved and the following electromagnetic momentum balance is postulated

$$\frac{\partial \boldsymbol{g}^{(\mathrm{EM})}}{\partial t} - \nabla \cdot \boldsymbol{\sigma}^{(\mathrm{EM})} = -\boldsymbol{f}^{(\mathrm{EM})} , \tag{45}$$

such that the total momentum is conserved

$$\frac{\partial}{\partial t}(\rho \boldsymbol{v} + \boldsymbol{g}^{(\mathrm{EM})}) + \nabla \cdot (\rho \boldsymbol{v} \otimes \boldsymbol{v} - \boldsymbol{\sigma} - \nabla \cdot \boldsymbol{\sigma}^{(\mathrm{EM})}) = \rho \boldsymbol{f} . \tag{46}$$

Therein, $\boldsymbol{\sigma}^{(\mathrm{EM})}$ is called electromagnetic stress tensor and $\boldsymbol{g}^{(\mathrm{EM})}$ is regarded as the electromagnetic momentum. Note that from Eq. (45) it follows that the surface force density is given by, see [23],

$$\boldsymbol{f}_I^{(\mathrm{EM})} = \boldsymbol{e} \cdot [\![\boldsymbol{w}_I \otimes \boldsymbol{g}^{(\mathrm{EM})} + \boldsymbol{\sigma}^{(\mathrm{EM})}]\!] . \tag{47}$$

Unfortunately, both of the newly introduced additional quantities $\boldsymbol{g}^{(\mathrm{EM})}$ and $\boldsymbol{\sigma}^{(\mathrm{EM})}$ are, in general, unknown. Which concrete form they have differs in the literature and this conundrum is referred to as the ABRAHAM–MINKOWSKI controversy and is still subject to discussion, see e.g., [2, 3].

In their famous publications on electromagntism of ponderable media, ABRAHAM [1] and MINKOWSKI [20] proposed the following different momentum densities, respectively:

$$\boldsymbol{g}^{\mathrm{A}} = \boldsymbol{D} \times \mu_0 \mathfrak{H} = (\mathfrak{D} - \boldsymbol{P}) \times \mu_0 \mathfrak{H} , \quad \boldsymbol{g}^{\mathrm{M}} = \mathfrak{D} \times \boldsymbol{B} . \tag{48}$$

Furthermore, in [26, p. 689] another momentum density is introduced

$$\boldsymbol{g}^{\mathrm{L}} = \boldsymbol{D} \times \boldsymbol{B} , \tag{49}$$

which we shall refer to as generalized LORENTZ momentum density. However, even if one of these is accepted, the expression for $\boldsymbol{f}^{(\mathrm{EM})}$ is not uniquely defined as $\boldsymbol{\sigma}^{(\mathrm{EM})}$ is not determined yet.

The usual procedure for the determination of a force model starts from accepting a momentum density. Then MAXWELL's equations are used to find an identity of the form as in Eq. (45). However, there are infinitely many possibilities to find a combination of a momentum density and an electromagnetic stress tensor such that the electromagnetic momentum balance in Eq. (45) is fulfilled. Furthermore, the problem of ambiguity extends to the unknown expression for the electromagnetic energy in the presence of matter, see [15, 23]. The so-called POYNTING's theorem is the balance of electromagnetic energy u

$$\frac{\partial u}{\partial t} = -\nabla \cdot \boldsymbol{S} + r \,, \tag{50}$$

in which $\boldsymbol{S}$ is the POYNTING vector and r is the electromagnetic power. This equation looks similar to the electromagnetic momentum balance and shares the fact that the quantities it contains are unknown in general. In particular, FEYNMAN, [6], wrote: "Before we take up some applications of the Poynting formulas [...], we would like to say that we have not really "proved" them. All we did was to find a possible "u" and a possible "$\boldsymbol{S}$." How do we know that by juggling the terms around some more we couldn't find another formula ... It can be done, but the forms that have been found always involve various derivatives of the field (and always with second-order terms like a second derivative or the square of a first derivative). There are, in fact, an infinite number of different possibilities ... and so far no one has thought of an experimental way to tell which one is right! People have guessed that the simplest one is probably the correct one, but we must say that we do not know for certain what is the actual location in space of the electromagnetic field energy."

In literature, several approaches to the ABRAHAM–MINKOWSKI controversy are proposed. In [19], for example, *gedankenexperiments* with light waves are investigated. Another approach relies on the solution of boundary value problems for different force models, because these can be compared with real experiments, see [5, 23].

Therefore, we need to first solve some boundary value problems, that can be compared to actual experiments. In the following sections two different problems will be considered. In order to analyze the effects of particular force models, we take some models derived in [23] and investigate their impact on mechanical responses. The following force models are considered:

$$\begin{aligned}
\boldsymbol{f}^{\mathrm{L}} &= q\boldsymbol{E} + \boldsymbol{J} \times \boldsymbol{B} \, , \\
\boldsymbol{f}^{\mathrm{A}} &= q\boldsymbol{E} + \boldsymbol{J} \times \mu_0 \mathfrak{H} - \nabla \cdot (\boldsymbol{M} \otimes \boldsymbol{B}) + \mu_0 \boldsymbol{D} \times \frac{\partial \boldsymbol{M}}{\partial t} \, , \\
\boldsymbol{f}^{\mathrm{M}} &= q^{\mathrm{f}}\boldsymbol{E} + \boldsymbol{J}^{\mathrm{f}} \times \boldsymbol{B} - (\nabla \otimes \boldsymbol{M}) \cdot \boldsymbol{B} + (\nabla \otimes \boldsymbol{E}) \cdot \boldsymbol{P} \, , \\
\boldsymbol{f}^{\mathrm{EL}} &= q^{\mathrm{f}}\boldsymbol{E} + \boldsymbol{J}^{\mathrm{f}} \times \mu_0 \mathfrak{H} + \boldsymbol{P} \cdot (\nabla \otimes \boldsymbol{E}) + \\
&\quad + \frac{\partial \boldsymbol{P}}{\partial t} \times \mu_0 \mathfrak{H} + \mu_0 \boldsymbol{M} \cdot (\nabla \otimes \mathfrak{H}) - \mu_0 \frac{\partial \boldsymbol{M}}{\partial t} \times \boldsymbol{D} \, ,
\end{aligned} \tag{51}$$

with the corresponding surface forces

$$\begin{aligned}
\boldsymbol{f}_I^{\mathrm{L}} &= q_I \langle \boldsymbol{E} \rangle + \boldsymbol{J}_I \times \langle \boldsymbol{B} \rangle \, , \\
\boldsymbol{f}_I^{\mathrm{A}} &= q_I \langle \boldsymbol{E} \rangle + \boldsymbol{J}_I \times \mu_0 \langle \mathfrak{H} \rangle - \mu_0 w_\perp \langle \boldsymbol{D} \rangle \times [\![\boldsymbol{M}]\!] + \\
&\qquad + (\boldsymbol{n} \times [\![\boldsymbol{B}]\!]) \times \langle \boldsymbol{M} \rangle - \boldsymbol{n} \cdot [\langle \boldsymbol{M} \rangle \otimes [\![\boldsymbol{B}]\!] + [\![\boldsymbol{M}]\!] \otimes \langle \boldsymbol{B} \rangle] \, , \\
\boldsymbol{f}_I^{\mathrm{M}} &= q_I^{\mathrm{f}} \langle \boldsymbol{E} \rangle + \boldsymbol{J}_I^{\mathrm{f}} \times \langle \boldsymbol{B} \rangle + \boldsymbol{n}(\langle \boldsymbol{P} \rangle \cdot [\![\boldsymbol{E}]\!] - \langle \boldsymbol{B} \rangle \cdot [\![\boldsymbol{M}]\!]) \, , \\
\boldsymbol{f}_I^{\mathrm{EL}} &= \boldsymbol{f}_I^{\mathrm{M}_2} + \boldsymbol{n}[\![\boldsymbol{B} \cdot \boldsymbol{M} - \tfrac{\mu_0}{2} \boldsymbol{M} \cdot \boldsymbol{M}]\!] - w_\perp [\![\mathfrak{D} \times \mu_0 \boldsymbol{M} + \boldsymbol{P} \times \mu_0 \mathfrak{H}]\!] \, .
\end{aligned} \tag{52}$$

5 Dielectrics

A dielectric is an electrical insulator with low conductivity that can be polarized by an external field $\boldsymbol{E}_0$ such that

$$\boldsymbol{P} = \hat{\boldsymbol{P}}(\boldsymbol{E}) \quad \text{with} \quad \hat{\boldsymbol{P}}(\boldsymbol{E} = \boldsymbol{0}) = \boldsymbol{0} \, , \tag{53}$$

where $\boldsymbol{E}$ is the resulting electric field, which is a composition of the external excitation $\boldsymbol{E}_0$ and the material response $\boldsymbol{P}$. For the special case of a linear (and isotropic) dielectric the polarization is a linear function of the electric field, $\boldsymbol{P} = \chi \epsilon_0 \boldsymbol{E}$, with the electric susceptibility χ. Some special dielectrics can possess a remanent polarization and are called electrets in analogy to a magnet, $\boldsymbol{P}_0 = \boldsymbol{P}(\boldsymbol{E} = \boldsymbol{0})$. Additionally, a dielectric may retain excess surface charges, e.g., a homogeneously distributed charge Q on the surface of the body with surface area A such that $q_I^{\mathrm{f}} = Q/A$. From these properties five different spherical dielectrics are analyzed in this paper:

(I) A linear dielectric, without any surface charge, in an external field $\boldsymbol{E}_0$:

$$q_I^{\mathrm{f}} = 0 \, , \quad \boldsymbol{P} = \chi \boldsymbol{D} \, , \quad \boldsymbol{E}_0 = E_0 \boldsymbol{e}_z \, . \tag{54}$$

Using the æther relations in Eq. (42) the polarization and the free charge potential can be rewritten as

$$\boldsymbol{P} = \epsilon_0 \chi \boldsymbol{E} \quad \Leftrightarrow \quad \mathfrak{D} = \epsilon_0 \epsilon_{\mathrm{r}} \boldsymbol{E} \, . \tag{55}$$

(II) A real-charge electret possesses a constant surface charge $q_I^f = Q/A_{sph}$ with the surface of the sphere $A_{sph} = 4\pi R^2$, see [4, p. 510]. It is assumed that the polarization vanishes, $\boldsymbol{P} = \boldsymbol{0}$ and that no external field is applied, $\boldsymbol{E}_0 = \boldsymbol{0}$.

(III) For the oriented dipole model no surface charge and no external field are considered, i.e., $q_I^f = 0$ and $\boldsymbol{E}_0 = \boldsymbol{0}$, but a constant and homogeneous polarization is used $\boldsymbol{P} = P_0\boldsymbol{e}_z$.

In addition, combinations of these three material models are possible, from which the following two are analyzed:

(IV) Real-charge electrets made of linear dielectric material are more complicated as they posses not only a surface charge but can also be excited by an external field $\boldsymbol{E}_0$:

$$q_I^{\mathrm{f}} = Q/A_{\mathrm{sph}} \,, \quad \mathfrak{D} = \epsilon_0\epsilon_{\mathrm{r}}\boldsymbol{E} \,. \tag{56}$$

(V) A mix of all of the previous models is a real-charge electret with affine linear polarization that is placed in an external field:

$$q_I^{\mathrm{f}} = Q/A_{\mathrm{sph}} \boldsymbol{P} = P_0\boldsymbol{e}_z + \epsilon_0\chi\boldsymbol{E} \quad \Leftrightarrow \quad \mathfrak{D} = P_0\boldsymbol{e}_z + \epsilon_0\epsilon_{\mathrm{r}}\boldsymbol{E}\boldsymbol{E}_0 = E_0\boldsymbol{e}_z \,. \tag{57}$$

5.1 Electric Fields

In stationary problems, the curl of the electric field vanishes, i.e., $\nabla \times \boldsymbol{E} = \boldsymbol{0}$. Therefore it is possible to replace the field with an electric potential. However, in some of the discussed problems there is an external field $\boldsymbol{E}_0$ that acts as a source. It is therefore convenient to decompose the total electric field $\boldsymbol{E}$ into the external field $\boldsymbol{E}_0$ and the (local) stray field $\hat{\boldsymbol{E}}$, i.e., $\boldsymbol{E} = \boldsymbol{E}_0 + \hat{\boldsymbol{E}}$. For simplification, a potential is introduced for the stray field by putting $\hat{\boldsymbol{E}} = -\nabla V$. In order to obtain a differential equation for this potential, the MAXWELL equation for the free charge potential, $\nabla \cdot \mathfrak{D} = q^{\mathrm{f}}$, is used. In all considered problems, the volumetric free electric charge density vanishes. Hence, the problems simplify to $\nabla \cdot \mathfrak{D} = 0$. In order to solve the considered problems generically, the free charge potential is connected to the electric field with the generalized relation

$$\mathfrak{D} = \epsilon_0\epsilon_{\mathrm{r}}\boldsymbol{E} + \boldsymbol{P}_0 \,, \quad \text{with} \quad \nabla \otimes \boldsymbol{P}_0 = \boldsymbol{0} \,. \tag{58}$$

To see that this relation is applicable, one can choose, e.g., $\epsilon_{\mathrm{r}} = 1$ and $\boldsymbol{P}_0 = \boldsymbol{0}$ to obtain the scenario (II) of a real-charge electret. As the free charge potential is solenoidal and any $\boldsymbol{P}_0$ is assumed constant and homogeneous, one obtains a LAPLACE equation for the stray potential, $\Delta V = 0$. This equation is also obtained outside of the spheres, as the relation $\boldsymbol{D} = \mathfrak{D} = \epsilon_0\boldsymbol{E}$ holds for vacuum. Every problem possesses (at least) azimuthal symmetry, hence, the solution can be written, w.r.t. spherical coordinates, as

$$V(\tilde{r}, x) = \sum_{n=0}^{\infty} \left(a_n \tilde{r}^n + b_n \tilde{r}^{-(n+1)} \right) \mathscr{P}_n(x) \,, \quad \tilde{r} := {}^r\!/\!{}_R \,, \quad x := \cos\vartheta \,. \tag{59}$$

This solution contains LEGENDRE polynomials $\mathscr{P}_n$. Note that the adapted structure of this solution must be different inside and outside of a considered sphere. Since the potential must be regular, it cannot be proportional to $\tilde{r}^{-n(n+1)}$ inside of a sphere for $n > 0$. Outside of it, the potential cannot be proportional to $\tilde{r}^n$ as the stray field must vanish as $\tilde{r} \to \infty$. At the interface, the total electric field must be continuous in tangential direction, i.e., $\boldsymbol{n} \times [\![\boldsymbol{E}]\!] = \boldsymbol{0}$. Because every external field $\boldsymbol{E}_0$ is constant, the relation must also hold for the stray field. Hence, the potential must satisfy $\boldsymbol{n} \times [\![\nabla V]\!] = \boldsymbol{0}$. By exploiting the gauge freedom, it can be seen that this relation is satisfied by choosing a continuous potential, i.e., $[\![V]\!] = 0$. Therefore, the inner and exterior solution can be denoted as

$$\begin{aligned} V^{(\mathrm{I})}(\tilde{r}, \vartheta) &= \sum_{n=0}^{\infty} a_n \tilde{r}^n \mathscr{P}_n(x) \,, && \tilde{r} < 1 \,, \\ V^{(\mathrm{O})}(\tilde{r}, \vartheta) &= \sum_{n=0}^{\infty} a_n \tilde{r}^{-(n+1)} \mathscr{P}_n(x) \,, && \tilde{r} > 1 \,. \end{aligned} \tag{60}$$

The coefficients a_n are determined by the remaining jump condition at $\tilde{r} = 1$, *viz.*,

$$\boldsymbol{n} \cdot [\![\mathfrak{D}]\!] = q_I^{\mathrm{f}} \quad \Leftrightarrow \quad \boldsymbol{n} \cdot (\epsilon_0 \boldsymbol{E}^{(\mathrm{O})} - \epsilon_0 \epsilon_{\mathrm{r}} \boldsymbol{E}^{(\mathrm{I})} - \boldsymbol{P}_0) = q_I^{\mathrm{f}} \,, \tag{61}$$

which results in the transmission problem for the potential

$$\boldsymbol{n} \cdot (\epsilon_{\mathrm{r}} \nabla V^{(\mathrm{I})} - \nabla V^{(\mathrm{O})}) = \tfrac{1}{\epsilon_0} q_I^{\mathrm{f}} + \tfrac{1}{\epsilon_0} \boldsymbol{n} \cdot \boldsymbol{P}_0 + (\epsilon_{\mathrm{r}} - 1) \boldsymbol{n} \cdot \boldsymbol{E}_0 \,. \tag{62}$$

To solve this, the right-hand side is rewritten by noting that $\boldsymbol{n} = \boldsymbol{e}_r$, and that $\boldsymbol{e}_r \cdot \boldsymbol{e}_z = \cos\vartheta$; one obtains

$$\sum_{n=0}^{\infty} a_n [n(\epsilon_{\mathrm{r}} + 1) + \epsilon_{\mathrm{r}}] \mathscr{P}_n(x) = \frac{q_I^{\mathrm{f}}}{\epsilon_0} \mathscr{P}_0(x) + \left[\frac{P_0}{\epsilon_0} + (\epsilon_{\mathrm{r}} - 1) E_0 \right] \mathscr{P}_1(x) \,. \tag{63}$$

Due to the orthogonality relations of the LEGENDRE polynomials, an uncoupled algebraic system for the coefficients a_n is obtained. Its solution reads

$$a_0 = \frac{q_I^{\mathrm{f}}}{\epsilon_0 \epsilon_{\mathrm{r}}} \,, \quad a_1 = \frac{1}{2\epsilon_{\mathrm{r}} + 1} \frac{P_0}{\epsilon_0} + \frac{\epsilon_{\mathrm{r}} - 1}{2\epsilon_{\mathrm{r}} + 1} E_0 \,, \quad a_n = 0 \;\forall\, n \in \mathbb{N} \backslash \{0, 1\} \,. \tag{64}$$

In the specialization of this general solution, the values of q_I^{f}, ϵ_{r}, P_0, and E_0, are respectively inserted in a_0 and a_1. To reduce complexity, a reference field $\mathscr{E}$ is introduced, paired with the scaling factors α, β, and γ, such that

$$\frac{q_I^{\mathrm{f}}}{\epsilon_0} = \alpha\mathscr{E}\,, \quad \frac{P_0}{\epsilon_0} = \beta\mathscr{E}\,, \quad E_0 = \gamma\mathscr{E} \quad \Rightarrow \quad a_0 = \frac{\alpha}{\epsilon_{\mathrm{r}}}\mathscr{E}\,, \quad a_1 = \frac{\beta + (\epsilon_{\mathrm{r}} - 1)\gamma}{2\epsilon_{\mathrm{r}} + 1}\mathscr{E}\,. \tag{65}$$

In the following, the relations $a_0 = \tilde{a}_0\mathscr{E}$, and $a_1 = \tilde{a}_1\mathscr{E}$, are also used. For the considered cases, the involved scales and factors can be chosen as

$$\begin{array}{llllll}
\text{(I)} & \mathscr{E} = E_0\,, & \alpha = 0\,, & \beta = 0\,, & \gamma = 1\,, & (\epsilon_{\mathrm{r}} \neq 1) \\
\text{(II)} & \mathscr{E} = \dfrac{q_I^{\mathrm{f}}}{\epsilon_0}\,, & \alpha = 1\,, & \beta = 0\,, & \gamma = 0\,, & (\epsilon_{\mathrm{r}} = 1) \\
\text{(III)} & \mathscr{E} = \dfrac{P_0}{\epsilon_0}\,, & \alpha = 0\,, & \beta = 1\,, & \gamma = 0\,, & (\epsilon_{\mathrm{r}} = 1) \\
\text{(IV)} & \mathscr{E} = \dfrac{q_I^{\mathrm{f}}}{\epsilon_0}\,, & \alpha = 1\,, & \beta = 0\,, & \gamma = 0\,, & (\epsilon_{\mathrm{r}} \neq 1) \\
\text{(V)} & \mathscr{E} = E_0\,, & \alpha = \dfrac{q_I^{\mathrm{f}}}{\epsilon_0 E_0}\,, & \beta = \dfrac{P_0}{\epsilon_0 E_0}\,, & \gamma = 1\,. & (\epsilon_{\mathrm{r}} \neq 1)
\end{array} \tag{66}$$

From Eq. (65) it can be seen that γ can only affect the solution if $\epsilon_{\mathrm{r}} \neq 1$. With these (potentially) non-vanishing coefficients, the electric stray fields read:

$$\begin{aligned}
\hat{\boldsymbol{E}}^{(\mathrm{I})} &= -a_1[\cos\vartheta\,\boldsymbol{e}_r - \sin\vartheta\,\boldsymbol{e}_\vartheta] = -a_1\boldsymbol{e}_z = \text{const.}\,, \\
\hat{\boldsymbol{E}}^{(\mathrm{O})} &= \tilde{r}^{-3}[(a_0\tilde{r} + 2a_1\cos\vartheta)\boldsymbol{e}_r + a_1\sin\vartheta\,\boldsymbol{e}_\vartheta]\,.
\end{aligned} \tag{67}$$

The total electric field, in which $\hat{\boldsymbol{E}}$ may be superposed by an external field $\boldsymbol{E}_0$, reads

$$\begin{aligned}
\boldsymbol{E}^{(\mathrm{I})} &= (\gamma - \tilde{a}_1)\mathscr{E}[\cos\vartheta\,\boldsymbol{e}_r - \sin\vartheta\,\boldsymbol{e}_\vartheta] = (\gamma - \tilde{a}_1)\mathscr{E}\boldsymbol{e}_z = \text{const.}\,, \\
\boldsymbol{E}^{(\mathrm{O})} &= \{[\tilde{a}_0\tilde{r}^{-2} + (2\tilde{a}_1\tilde{r}^{-3} + \gamma)\cos\vartheta]\boldsymbol{e}_r + (\tilde{a}_1\tilde{r}^{-3} - \gamma)\sin\vartheta\,\boldsymbol{e}_\vartheta\}\mathscr{E}\,.
\end{aligned} \tag{68}$$

For the computation of the electrostatic forces that act in the individual problems, the polarization must also be denoted in an explicit manner. By using Eq. (58), one finds for the general case

$$\boldsymbol{P}^{(\mathrm{I})} = \mathfrak{D}^{(\mathrm{I})} - \epsilon_0\boldsymbol{E}^{(\mathrm{I})} = \epsilon_0(\epsilon_{\mathrm{r}} - 1)\boldsymbol{E}^{(\mathrm{I})} + \boldsymbol{P}_0\,, \qquad \boldsymbol{P}^{(\mathrm{O})} = \boldsymbol{0}\,. \tag{69}$$

With Eq. $(5.1)_1$ and the definition

$$\kappa := \beta + (\gamma - \tilde{a}_1)(\epsilon_{\mathrm{r}} - 1) = \frac{\epsilon_{\mathrm{r}} + 2}{2\epsilon_{\mathrm{r}} + 1}\big[\beta + (\epsilon_{\mathrm{r}} - 1)\gamma\big]\,, \tag{70}$$

the polarization within the sphere is obtained as

$$\boldsymbol{P}^{(\mathrm{I})} = \kappa\epsilon_0\mathscr{E}[\cos\vartheta\,\boldsymbol{e}_r - \sin\vartheta\,\boldsymbol{e}_\vartheta] = \kappa\epsilon_0\mathscr{E}\boldsymbol{e}_z = \text{const.} \tag{71}$$

It can be seen that both $\boldsymbol{E}^{(\mathrm{I})}$ and $\boldsymbol{P}^{(\mathrm{I})}$ are constant and homogeneous fields within any considered sphere. Hence, there are no electrostatic volume force densities resulting from Eq. (51). However, due to the discontinues at the interface, surface force densities arise. To compute them for a specific electromagnetic force model, the jump and mean value fields of $\boldsymbol{E}$ and $\boldsymbol{P}$ are required. These read at $\tilde{r} = 1$:

$$
\begin{aligned}
[\![\boldsymbol{E}]\!] &= \boldsymbol{E}^{(\mathrm{O})} - \boldsymbol{E}^{(\mathrm{I})} = (\tilde{a}_0 + 3\tilde{a}_1 \cos\vartheta)\mathscr{E}\boldsymbol{e}_r \, , \\
\langle \boldsymbol{E} \rangle &= \tfrac{1}{2}(\boldsymbol{E}^{(\mathrm{O})} + \boldsymbol{E}^{(\mathrm{I})}) = \{\tfrac{1}{2}(\tilde{a}_0 + [\tilde{a}_1 + 2\gamma]\cos\vartheta)\boldsymbol{e}_r + (\tilde{a}_1 - \gamma)\sin\vartheta\,\boldsymbol{e}_\vartheta\}\mathscr{E} \, , \\
[\![\boldsymbol{P}]\!] &= \boldsymbol{0} - \boldsymbol{P}^{(\mathrm{I})} = -\kappa\epsilon_0\mathscr{E}[\cos\vartheta\,\boldsymbol{e}_r - \sin\vartheta\,\boldsymbol{e}_\vartheta] \, , \\
\langle \boldsymbol{P} \rangle &= \tfrac{1}{2}\boldsymbol{P}^{(\mathrm{I})} = \tfrac{1}{2}\kappa\epsilon_0\mathscr{E}[\cos\vartheta\,\boldsymbol{e}_r - \sin\vartheta\,\boldsymbol{e}_\vartheta] \, .
\end{aligned}
\tag{72}
$$

Also needed are the total densities of the surface charge and of the surface current. These follow as

$$
\begin{aligned}
q_I &= q_I^{\mathrm{f}} + q_I^{\mathrm{b}} = \epsilon_0\Big(\frac{q_I^{\mathrm{f}}}{\epsilon_0} - \boldsymbol{n}\cdot\frac{[\![\boldsymbol{P}]\!]}{\epsilon_0}\Big) = \epsilon_0\mathscr{E}[\alpha + \kappa\cos\vartheta] \, , \\
\boldsymbol{J}_I &= \boldsymbol{J}_I^{\mathrm{f}} + \boldsymbol{J}_I^{\mathrm{b}} = \boldsymbol{J}_I^{\mathrm{b}} = -[\![\boldsymbol{P}]\!]w_\perp + \boldsymbol{n}\times[\![\boldsymbol{M}]\!] = \boldsymbol{0} \, ,
\end{aligned}
\tag{73}
$$

due to the fact that there is no surface normal velocity $w_\perp$, and there is also no magnetization, $\boldsymbol{M} = \boldsymbol{0}$.

5.2 *Electrostriction*

Since the electric field as well as the polarization are constant in the interior, no electrostatic volume force densities results from the force models given in Eq. (51), i.e., $\boldsymbol{f}^{(\mathrm{EM})} = \boldsymbol{0}$. Hence, only surface forces $\boldsymbol{f}_I^{(\mathrm{EM})}$ need to be considered. In the given spherical setting and with the solution of the electric field, the simplification of the surface force densities in Eq. (52) reveals that only two distinct surface force densities (electrostatics) are to be considered:

$$
\boldsymbol{f}_I^{(1)} := \boldsymbol{f}_I^{\mathrm{L}} = \boldsymbol{f}_I^{\mathrm{A}} = q_I\langle\boldsymbol{E}\rangle \, , \qquad \boldsymbol{f}_I^{(2)} := \boldsymbol{f}_I^{\mathrm{M}} = \boldsymbol{f}_I^{\mathrm{EL}} = q_I^{\mathrm{f}}\langle\boldsymbol{E}\rangle + \boldsymbol{n}\big(\langle\boldsymbol{P}\rangle\cdot[\![\boldsymbol{E}]\!]\big) \, .
\tag{74}
$$

Similar to the analysis in [23], the solution of the elastic problem found by HIRAMATSU–OKA, see [10], is used. In this setup, gravitational effects are ignored. The balance of linear momentum in regular points is therefore homogeneous and reads $\nabla\cdot\boldsymbol{\sigma} = \boldsymbol{0}$. It is assumed that the material response is simple ("non-piezo"), such that the *mechanical* stress tensor $\boldsymbol{\sigma}$ is solely determined by mechanical strain. Potential stresses due to electromagnetic fields are neglected. For the isotropic linear-elastic material and the assumption of small strains, HOOKE's law

$$
\boldsymbol{\sigma} = \lambda(\nabla\cdot\boldsymbol{u})\boldsymbol{1} + \mu(\nabla\otimes\boldsymbol{u} + \boldsymbol{u}\otimes\nabla)
\tag{75}
$$

is employed. In this law, $\boldsymbol{u}$ is the displacement field and λ, μ denote LAMÉ's parameters. As mentioned before, this form of the elastic law assumes that the mechanical stress has no direct dependency on the electromagnetic field. In classical piezoelectricity the stress tensor is directly influenced by the electric field. However, in our case, the stresses may be influenced indirectly due to the electromagnetic forces. It can be noted that if a material possesses a direct dependency upon the electromagnetic fields, then there would be too many unknowns and experiments could not shed insight as to the validity of an electromagnetic force model. By inserting HOOKE's law into the homogeneous balance of linear momentum, the homogeneous LAMÉ–NAVIER equations are obtained,

$$(\lambda + \mu)\nabla(\nabla \cdot \boldsymbol{u}) + \mu \Delta \boldsymbol{u} = \boldsymbol{0} \,. \tag{76}$$

Since the setup of the problem bears azimuthal symmetry and a spherical body is considered, the generic solution of HIRAMATSU and OKA can be used, *cf.* [10]. They found a series solution for the displacement field that contains powers of the spherical radial coordinate and LEGENDRE polynomials,

$$u_r(\tilde{r}, \vartheta) = R \sum_{n=0}^{\infty} \left[-\frac{n\frac{\lambda}{\mu} + n - 2}{2(2n+3)} A_n \tilde{r}^{n+1} + n B_n \tilde{r}^{n-1} \right] \mathscr{P}_n(x) \,, \tag{77a}$$

$$u_\vartheta(\tilde{r}, \vartheta) = R \sum_{n=1}^{\infty} \left[-\frac{(n+3)\frac{\lambda}{\mu} + n + 5}{2(n+1)(2n+3)} A_n \tilde{r}^{n+1} + B_n \tilde{r}^{n-1} \right] \frac{\mathrm{d}\mathscr{P}_n(x)}{\mathrm{d}\vartheta} \,. \tag{77b}$$

Note that $B_1 = 0$ in order to ensure that the center of the sphere is not shifted. The stresses follow as

$$\sigma_{rr}(\tilde{r}, \vartheta) = \mu \sum_{n=0}^{\infty} \left[-\frac{(n^2 - n - 3)\frac{\lambda}{\mu} + (n+1)(n-2)}{2n+3} A_n \tilde{r}^{n} + 2n(n-1) B_n \tilde{r}^{n-2} \right] \mathscr{P}_n(x) \,, \tag{77c}$$

$$\sigma_{\vartheta\vartheta}(\tilde{r}, \vartheta) = \mu \sum_{n=0}^{\infty} \left[\frac{(n+3)\frac{\lambda}{\mu} - n + 2}{2n+3} A_n \tilde{r}^{n} + 2n B_n \tilde{r}^{n-2} \right] \mathscr{P}_n(x) + {} + \mu \sum_{n=2}^{\infty} \left[-\frac{(n+3)\frac{\lambda}{\mu} + n + 5}{(n+1)(2n+3)} A_n \tilde{r}^{n} + 2 B_n \tilde{r}^{n-2} \right] \frac{\mathrm{d}^2\mathscr{P}_n(x)}{\mathrm{d}\vartheta^2} \,, \tag{77d}$$

$$\sigma_{\varphi\varphi}(\tilde{r}, \vartheta) = \mu \sum_{n=0}^{\infty} \left[\frac{(n+3)\frac{\lambda}{\mu} - n + 2}{2n+3} A_n \tilde{r}^{n} + 2n B_n \tilde{r}^{n-2} \right] \mathscr{P}_n(x) + {} + \mu \sum_{n=1}^{\infty} \left[-\frac{(n+3)\frac{\lambda}{\mu} + n + 5}{(n+1)(2n+3)} A_n \tilde{r}^{n} + 2 B_n \tilde{r}^{n-2} \right] \frac{\mathrm{d}\mathscr{P}_n(x)}{\mathrm{d}\vartheta} \cot(\vartheta) \,, \tag{77e}$$

$$\sigma_{r\vartheta}(\tilde{r},\vartheta) = \mu \sum_{n=1}^{\infty} \left[-\frac{n(n+2)\frac{\lambda}{\mu} + n^2 + 2n - 1}{(n+1)(2n+3)} A_n \tilde{r}^n + 2(n-1) B_n \tilde{r}^{n-2} \right] \frac{\mathrm{d}\mathscr{P}_n(x)}{\mathrm{d}\vartheta} , \tag{77f}$$

$$\sigma_{r\varphi}(\tilde{r},\vartheta) = \sigma_{\vartheta\varphi}(\tilde{r},\vartheta) = 0 . \tag{77g}$$

The mathematical stress dependencies on the polar angle motivate the chosen expansion of the forces in Eq. (80). The usefulness can be seen by regarding the jump condition of linear momentum and neglecting any mechanical pressure outside of the sphere. One obtains

$$\boldsymbol{n} \cdot [\![\boldsymbol{\sigma}]\!] = -\boldsymbol{f}_I^{(\mathrm{EM})} \quad \Rightarrow \quad \boldsymbol{n} \cdot \boldsymbol{\sigma}(\tilde{r} = 1) = \boldsymbol{f}_I^{(\mathrm{EM})} . \tag{78}$$

Because the obtained surface forces have radial and polar components, the jump equation yields two non-trivial relations. With $\sigma_{rr} = \boldsymbol{e}_r \cdot \boldsymbol{\sigma} \cdot \boldsymbol{e}_r$ and $\sigma_{r\vartheta} = \boldsymbol{e}_r \cdot \boldsymbol{\sigma} \cdot \boldsymbol{e}_\vartheta$ they read:

$$\sigma_{rr}(\tilde{r} = 1, \vartheta) = f_r^{(\mathrm{EM})}(\vartheta) , \quad \sigma_{r\vartheta}(\tilde{r} = 1, \vartheta) = f_\vartheta^{(\mathrm{EM})}(\vartheta) . \tag{79}$$

Hence it is reasonable to expand the radial force component in terms of LEGENDRE polynomials and analogously the polar component in terms of derivatives of LEGENDRE polynomials. For the given force expressions the following expansion is sufficient

$$\boldsymbol{f}_I^{(i)} = \epsilon_0 \mathscr{E}^2 \sum_{k=0}^{2} c_k^{(i)} \mathscr{P}_k(x) \boldsymbol{e}_r + \epsilon_0 \mathscr{E}^2 \sum_{k=1}^{2} d_k^{(i)} \frac{\mathrm{d}\mathscr{P}_k(x)}{\mathrm{d}\vartheta} \boldsymbol{e}_\vartheta \tag{80a}$$

Therein, $d_1^{(2)} = 0$ and the other coefficients are given by:

$$\begin{aligned}
&c_0^{(1)} = \tfrac{1}{2}\alpha\tilde{a}_0 + \tfrac{1}{6}\kappa(\tilde{a}_1 + 2\gamma), && c_1^{(1)} = \tfrac{1}{2}\alpha(\tilde{a}_1 + 2\gamma) + \tfrac{1}{2}\kappa\tilde{a}_0, && c_2^{(1)} = \tfrac{1}{3}\kappa(\tilde{a}_1 + 2\gamma),\\
&d_1^{(1)} = \alpha(\gamma - \tilde{a}_1), && d_2^{(1)} = \tfrac{1}{3}\kappa(\gamma - \tilde{a}_1), && d_1^{(2)} = \alpha(\gamma - \tilde{a}_1),\\
&c_0^{(2)} = \tfrac{1}{2}\alpha\tilde{a}_0 + \tfrac{1}{2}\kappa\tilde{a}_1 , && c_1^{(2)} = \tfrac{1}{2}\alpha(\tilde{a}_1 + 2\gamma) + \tfrac{1}{2}\kappa\tilde{a}_0 , && c_2^{(2)} = \kappa\tilde{a}_1 .
\end{aligned} \tag{81}$$

The boundary conditions in Eq. (79) are used to find the unknown coefficients A_n, and B_n. As the expansions of the force components in Eq. (81) are finite, only few of these differ from zero. The last two relations read explicitly:

$$\mu\sum_{n=0}^{\infty}\left[-\frac{(n^2-n-3)\frac{\lambda}{\mu}+(n+1)(n-2)}{2n+3}A_n^{(i)}+2n(n-1)B_n^{(i)}\right]\mathscr{P}_n(x)=$$
$$=\epsilon_0\mathscr{E}^2\left[c_0^{(i)}\mathscr{P}_0(x)+c_1^{(i)}\mathscr{P}_1(x)+c_2^{(i)}\mathscr{P}_2(x)\right], \tag{82}$$

and

$$\mu\sum_{n=1}^{\infty}\left[-\frac{n(n+2)\frac{\lambda}{\mu}+n^2+2n-1}{(n+1)(2n+3)}A_n^{(i)}+2(n-1)B_n^{(i)}\right]\frac{\mathrm{d}\mathscr{P}_n(x)}{\mathrm{d}\vartheta}=$$
$$=\epsilon_0\mathscr{E}^2\left[d_1^{(i)}\frac{\mathrm{d}\mathscr{P}_1(x)}{\mathrm{d}\vartheta}+d_2^{(i)}\frac{\mathrm{d}\mathscr{P}_2(x)}{\mathrm{d}\vartheta}\right]. \tag{83}$$

It can be seen that it is convenient to introduce the rescaled coefficients:

$$\tilde{A}_n^{(i)}=\frac{\mu}{\epsilon_0\mathscr{E}^2}A_n^{(i)}\,,\quad \tilde{B}_n^{(i)}=\frac{\mu}{\epsilon_0\mathscr{E}^2}B_n^{(i)}\,. \tag{84}$$

For these, one obtains the following non-homogeneous system of equations:

$$\begin{gathered}\tfrac{1}{3}(3\tfrac{\lambda}{\mu}+2)\tilde{A}_0^{(i)}=c_0^{(i)},\ \tfrac{1}{5}(3\tfrac{\lambda}{\mu}+2)\tilde{A}_1^{(i)}=c_1^{(i)},\ \tfrac{1}{7}\tfrac{\lambda}{\mu}\tilde{A}_2^{(i)}+4\tilde{B}_2^{(i)}=c_2^{(i)}\,,\\ -\tfrac{1}{10}(3\tfrac{\lambda}{\mu}+2)\tilde{A}_1^{(i)}=d_1^{(i)}\,,-\tfrac{1}{21}(8\tfrac{\lambda}{\mu}+7)\tilde{A}_2^{(i)}+2\tilde{B}_2^{(i)}=d_2^{(i)}\,.\end{gathered} \tag{85}$$

Since there is no index shift in these equations, it follows that all coefficients $\tilde{A}_n^{(i)}$, and $\tilde{B}_n^{(i)}$, are zero for $n\in\mathbb{N}\backslash\{0,1,2\}$. This algebraic system of equations has no solution because Eqs. $(85)_2$ and $(85)_4$ are inconsistent. In order for the system to be solvable the first coefficients in the force expansion need to satisfy

$$c_1=-2d_1\,. \tag{86}$$

It turns out that this condition is necessary for the boundary value problem to be well posed, see the discussion in the next section. Note that in the cases (I)–(IV) condition (86) is fulfilled. Then, the linear system can be solved and the solution reads

$$\begin{aligned}\tilde{A}_0^{(i)}&=\frac{3c_0^{(i)}}{3\frac{\lambda}{\mu}+2}\,, & \tilde{A}_1^{(i)}&=\frac{5c_1^{(i)}}{3\frac{\lambda}{\mu}+2}\,,\\ \tilde{A}_2^{(i)}&=\frac{21(c_2^{(i)}-2d_2^{(i)})}{19\frac{\lambda}{\mu}+14}\,, & \tilde{B}_2^{(i)}&=\frac{(7+8\frac{\lambda}{\mu})c_2^{(i)}+3\frac{\lambda}{\mu}d_2^{(i)}}{38\frac{\lambda}{\mu}+28}\,.\end{aligned} \tag{87}$$

The question as to why the system cannot be solved in the mixed case (V) and how to address this problem is discussed in the next section.

5.3 Total Force Analysis

The total force on the spheres is solely given due to surface force densities. For the sake of the following argument, consider an expansion of the surface force density similar to that of Eq. (80):

$$\boldsymbol{f}_I^{(\mathrm{EM})} = \sum_{k=0}^{\infty} c_k \mathscr{P}_k(x) \boldsymbol{e}_r + \sum_{k=1}^{\infty} d_k \frac{\mathrm{d}\mathscr{P}_k(x)}{\mathrm{d}\vartheta} \boldsymbol{e}_\vartheta \,. \tag{88}$$

Then, since the radial and polar base vector are given by

$$\begin{aligned} \boldsymbol{e}_r &= \sin(\vartheta)\big(\cos(\varphi)\boldsymbol{e}_x + \sin(\varphi)\boldsymbol{e}_y\big) + \cos(\vartheta)\boldsymbol{e}_z \,, \\ \boldsymbol{e}_\vartheta &= \cos(\vartheta)\big(\cos(\varphi)\boldsymbol{e}_x + \sin(\varphi)\boldsymbol{e}_y\big) - \sin(\vartheta)\boldsymbol{e}_z \,, \end{aligned} \tag{89}$$

it follows that the total force is given by

$$\boldsymbol{F}^{(\mathrm{EM})} = \int_A \boldsymbol{f}_I^{(\mathrm{EM})} \,\mathrm{d}A = R^2 \int_{\varphi=0}^{2\pi} \int_{\vartheta=0}^{\pi} \boldsymbol{f}_I^{(\mathrm{EM})} \sin\vartheta \,\mathrm{d}\vartheta \,\mathrm{d}\varphi = \tfrac{4}{3}\pi R^2 (c_1 + 2d_1)\boldsymbol{e}_z \,. \tag{90}$$

This is due to the fact that the only φ-dependence in the x and y-component is given by the cosine and sine function, respectively. However, it is very surprising that also for the z-component from a general expansion as in Eq. (88) only two terms, namely $c_1 \mathscr{P}_1(x)\boldsymbol{e}_r$ and $d_1 \frac{\mathrm{d}\mathscr{P}_1(x)}{\mathrm{d}\vartheta}\boldsymbol{e}_\vartheta$, survive the integration. For static equilibrium one requires the total force on a body to vanish. Hence, one recovers the solvability condition in Eq. (86). Another possibility would be to introduce a more realistic bearing of the sphere in order to obtain a well-posed problem. The simplest way was to introduce a bearing in the center of the sphere. From the free body diagram one would then find the bearing force to be equal but opposite in sing to the force in Eq. (90). However, a point force on an elastic continuum body introduces singular stresses, which were neglected in the elastic solution in Eq. (77). Therefore, we refrain from investigating the mixed case (V) and proceed to analyze the solution for the cases (I)–(IV).

5.4 Displacement Solutions

By inserting the coefficients from Eq. (87) into the solution in Eqs. (77a) and (77b) one obtains the displacement field for the different materials. In Fig. 3 the displacement solutions are visualized by plotting the original spherical shape, the surface displacement and the deformed shaped. It turns out that the fourth mixed case (IV) looks

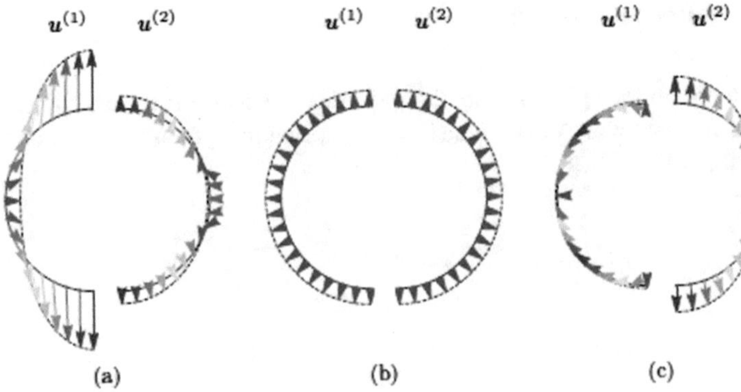

Fig. 3 Displacement figures for different dielectrics in **a**, **b** and **a** as well as different force models (1) and (2) in every plot. **a** Linear dielectric $\boldsymbol{P} = \epsilon_0 \chi \boldsymbol{E}$; **b** Real charge electret $q_I^{\mathrm{f}} = \frac{Q}{A_{\mathrm{sph}}}$; **c** Oriented dipole $\boldsymbol{P} = P_0 \boldsymbol{e}_z$

similar to the second case in Fig. 3b and is therefore suppressed. Due to azimuthal symmetry only one half of a slice through the sphere is shown. Furthermore, for each of the three cases two the displacement solutions $\boldsymbol{u}^{(1)}$ and $\boldsymbol{u}^{(2)}$, corresponding to the two different force expression $\boldsymbol{f}^{(1)}$ and $\boldsymbol{f}^{(2)}$ from Eq. (74), are shown.

From the figure it can be seen that the two different force expressions yield similar deformation figures for a linear dielectric, Fig. 3a, but the displacements have different magnitudes. Note that in both plots the same scaling was applied. In contrast to that the displacement predictions are identical for the real charge electret in Fig. 3b. The most significant difference arises for the oriented dipole model in Fig. 3c, in which not only the quantitative behavior is different but also the qualitative shape. The deformation due to the force expression $\boldsymbol{f}^{(1)}$ results in a mildly oblate spheroid whereas the $\boldsymbol{f}^{(2)}$ yields a more distinct prolate shape.

Note that in general, the mechanical response i.e., the deformation of the sphere influences the electric field as well. The boundary conditions change as the shape deforms. However, for the current situation the deformations are considered to be small and therefore this effect is neglected.

This analysis shows that the choice of a particular force model can have significant influence on the mechanical behavior of deformable bodies. However, a difference in the mechanical response may also arise for rigid bodies, as will be shown in the next section.

6 Total Forces and Moments Between Spherical Magnets

We consider the interaction between two spherical rigid permanent magnets, homogeneously magnetized by $M_0^{(\mathrm{I})}$ and $M_0^{(\mathrm{II})}$ of radii $R_{(\mathrm{I})}$ and $R_{(\mathrm{II})}$, respectively: Fig. 4. The magnetic field of the i-th magnet is given by, see [7, 14],

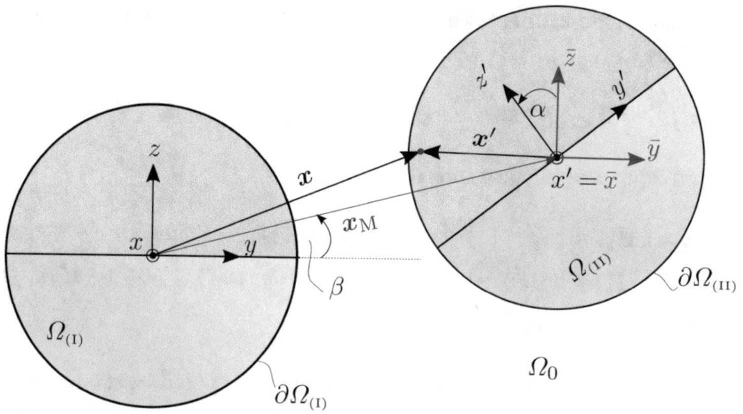

Fig. 4 Interacting spherical magnets with magnetizations $\boldsymbol{M}_{(\mathrm{I})} = M_0^{(\mathrm{I})}\boldsymbol{e}_z$ and $\boldsymbol{M}_{(\mathrm{II})} = M_0^{(\mathrm{II})}\boldsymbol{e}'_z$. The bar coordinate system is rotated about the x-axis by an angle of α, which represents the orientation angle

$$\boldsymbol{B}_i = \frac{2}{3}\frac{\mu_0 M_0^i R_i^3}{r_i^3}\big(\cos(\vartheta_i)\boldsymbol{e}_r^i + \tfrac{1}{2}\sin(\vartheta_i)\boldsymbol{e}_\vartheta^i\big)\,. \tag{91}$$

The total magnetic field is then a superposition of these. The mechanical interaction due to these magnetic fields is given by the total force and total moment, which are defined as

$$\begin{aligned}\boldsymbol{F}^{(\mathrm{EM})} &= \int \boldsymbol{f}^{(\mathrm{EM})}\,\mathrm{d}V + \int \boldsymbol{f}_I^{(\mathrm{EM})}\,\mathrm{d}A\,,\\ \boldsymbol{M}^{(\mathrm{EM})} &= \int \boldsymbol{x}\times\boldsymbol{f}^{(\mathrm{EM})}\,\mathrm{d}V + \int \boldsymbol{x}\times\boldsymbol{f}_I^{(\mathrm{EM})}\,\mathrm{d}A\,.\end{aligned} \tag{92}$$

Note that the electromagnetic moment is only due to force densities, *i.e.*, possible origin-independent moments are neglected. This of course weakens the significance of the following analysis. In order to perform a complete investigation, one needs to introduce another balance, namely the balance of electromagnetic angular momentum, $\boldsymbol{s}$,

$$\frac{\partial \boldsymbol{s}}{\partial t} - \nabla\cdot\boldsymbol{\pi} = -\boldsymbol{m}^{(\mathrm{EM})}\,, \tag{93}$$

where $\boldsymbol{\pi}$ are couple stresses and $\boldsymbol{m}^{(\mathrm{EM})}$ is the electromagnetic moment density. However, as $\boldsymbol{\pi}$ and $\boldsymbol{m}^{(\mathrm{EM})}$ are additional unknowns, we neglect them in order to simplify the analysis. Furthermore note that the existence of a moment density $\boldsymbol{m}^{(\mathrm{EM})}$ independent of $\boldsymbol{f}^{(\mathrm{EM})}$ implies that matter that interacts with electromagnetic fields is always polar from a mechanical point of view.

In a magnetostatic setting with constant magnetization the force expressions in Eq. (51) reduce to

$$\boldsymbol{f}^{\mathrm{L}} = \boldsymbol{f}^{\mathrm{M}} = \boldsymbol{0}\,, \quad \boldsymbol{f}^{\mathrm{A}} = -\nabla \cdot (\boldsymbol{M} \otimes \boldsymbol{B})\,, \quad \boldsymbol{f}^{\mathrm{EL}} = \mu_0 \boldsymbol{M} \cdot (\nabla \otimes \boldsymbol{\mathfrak{H}})\,, \tag{94}$$

with the corresponding surface forces

$$\begin{aligned}
\boldsymbol{f}_I^{\mathrm{L}} &= (\boldsymbol{n} \times [\![\boldsymbol{M}]\!]) \times \langle \boldsymbol{B} \rangle\,, \qquad \boldsymbol{f}_I^{\mathrm{M}} = -\boldsymbol{n}(\langle \boldsymbol{B} \rangle \cdot [\![\boldsymbol{M}]\!])\,,\\
\boldsymbol{f}_I^{\mathrm{A}} &= (\boldsymbol{n} \times [\![\boldsymbol{M}]\!]) \times \mu_0 \langle \boldsymbol{\mathfrak{H}} \rangle + (\boldsymbol{n} \times [\![\boldsymbol{B}]\!]) \times \langle \boldsymbol{M} \rangle - \boldsymbol{n} \cdot [\langle \boldsymbol{M} \rangle \otimes [\![\boldsymbol{B}]\!] + [\![\boldsymbol{M}]\!] \otimes \langle \boldsymbol{B} \rangle]\,,\\
\boldsymbol{f}_I^{\mathrm{EL}} &= \boldsymbol{n}([\![\boldsymbol{B}]\!] \cdot \langle \boldsymbol{M} \rangle) - \boldsymbol{n}[\![\tfrac{\mu_0}{2} \boldsymbol{M} \cdot \boldsymbol{M}]\!]\,.
\end{aligned} \tag{95}$$

In order to obtain the total force and total moment acting on the second magnet (II) due to the first one, the force expression are best represented in the coordinates of the second magnet. Note that since a magnet cannot be self-accelerated without any external influences. Therefore, only products of $\boldsymbol{M}_{(\mathrm{II})}$ and $\boldsymbol{B}_{(\mathrm{I})}$ survive the integration. For example, for the LORENTZ force density and $\boldsymbol{x} \in \Omega_{(\mathrm{II})}$

$$\boldsymbol{f}_I^{\mathrm{L}} = (\boldsymbol{n} \times [\![\boldsymbol{M}]\!]) \times \langle \boldsymbol{B} \rangle = (\boldsymbol{n} \times [\![\boldsymbol{M}_{(\mathrm{II})}]\!]) \times \left(\boldsymbol{B}_{(\mathrm{I})} + \langle \boldsymbol{B}_{(\mathrm{II})} \rangle\right) \tag{96}$$

where $\boldsymbol{B}_{(\mathrm{I})}$ is continuous, and the total force is given by

$$\boldsymbol{F}^{\mathrm{L}} = \int\limits_{\partial\Omega_{(\mathrm{II})}} (\boldsymbol{n} \times [\![\boldsymbol{M}_{(\mathrm{II})}]\!]) \times \boldsymbol{B}_{(\mathrm{I})}\, \mathrm{d}A\,. \tag{97}$$

Therefore, the force densities are replaced by effective ones, neglecting products of $\boldsymbol{M}_{(\mathrm{II})}$ and $\boldsymbol{B}_{(\mathrm{II})}$. For the coordinate systems in Fig. 4 the (effective) dimensionless surface force densities $\tilde{\boldsymbol{f}}_I^{(\mathrm{EM})} = \boldsymbol{f}_I^{(\mathrm{EM})} / \hat{f}$ are given by:

$$\begin{aligned}
\tilde{\boldsymbol{f}}_I^{\mathrm{L}} &= \sin\vartheta' \boldsymbol{e}'_\varphi \times \tilde{\boldsymbol{B}}_{(\mathrm{I})}\,, \quad \tilde{\boldsymbol{f}}_I^{\mathrm{M}} = \boldsymbol{n}(\tilde{\boldsymbol{B}}_{(\mathrm{I})} \cdot \boldsymbol{e}'_z)\,, \quad \tilde{\boldsymbol{f}}_I^{\mathrm{EL}} = \boldsymbol{0}\,,\\
\tilde{\boldsymbol{f}}_I^{\mathrm{A}} &= \sin\vartheta' \boldsymbol{e}'_\varphi \times \tilde{\boldsymbol{B}}_{(\mathrm{I})} + \cos\vartheta' \tilde{\boldsymbol{B}}_{(\mathrm{I})}\,,
\end{aligned} \tag{98}$$

where $\hat{f} = \mu_0 M_0^{(\mathrm{I})} M_0^{(\mathrm{II})}$ and $\boldsymbol{B}_{(\mathrm{I})} = \mu_0 M_0^{(\mathrm{I})} \tilde{\boldsymbol{B}}_{(\mathrm{I})}$. The volumetric force densities normalized with $\hat{f} R_{(\mathrm{II})}^{-1}$ read:

$$\tilde{\boldsymbol{f}}^{\mathrm{L}} = \tilde{\boldsymbol{f}}^{\mathrm{M}} = \boldsymbol{0}\,, \quad \tilde{\boldsymbol{f}}^{\mathrm{A}} = -\tilde{\boldsymbol{f}}^{\mathrm{EL}} = -\boldsymbol{e}'_z \cdot (\tilde{\nabla} \otimes \tilde{\boldsymbol{B}}_{(\mathrm{I})})\,. \tag{99}$$

In order to obtain the total force on the second magnet, the surface force densities are integrated across the surface of the second magnet and the volumetric forces across its volume. There the magnetic field $\tilde{\boldsymbol{B}}_{(\mathrm{I})}$ is expressed in terms of the $\{\bar{x}, \bar{y}, \bar{z}\}$-coordinates. This is done by noting that with

$$\boldsymbol{x}' = \boldsymbol{x} - \boldsymbol{x}_{\mathrm{M}}\,, \quad d = \|\boldsymbol{x}_{\mathrm{M}}\|\,, \tag{100}$$

it follows that:

$$
\begin{aligned}
\bar{r}\sin(\bar{\vartheta})\cos(\bar{\varphi}) &= r\sin(\vartheta)\cos(\varphi) \ , \\
\bar{r}\sin(\bar{\vartheta})\sin(\bar{\varphi}) &= r\sin(\vartheta)\sin(\varphi) - d\cos(\beta) \ , \\
\bar{r}\cos(\bar{\vartheta}) &= r\cos(\vartheta) - d\sin(\beta) \ .
\end{aligned} \tag{101}
$$

Furthermore, the bar system is connected to the dashed one via

$$
\boldsymbol{e}'_y = \cos(\alpha)\bar{\boldsymbol{e}}_y + \sin(\alpha)\bar{\boldsymbol{e}}_z \ , \ \boldsymbol{e}'_z \qquad = -\sin(\alpha)\bar{\boldsymbol{e}}_y + \cos(\alpha)\bar{\boldsymbol{e}}_z \ , \tag{102}
$$

Moreover, it can be shown with some effort that the resulting forces are all equal independently of the model:

$$
\boldsymbol{F}^{\mathrm{L}} = \boldsymbol{F}^{\mathrm{M}} = \boldsymbol{F}^{\mathrm{A}} = \boldsymbol{F}^{\mathrm{EL}} \ . \tag{103}
$$

It is suspected in [23] that all force models yield the same total force and later on proved in [22]. From the experimental point of view this is bad news because a measurement of the force would not allow us to identify the most realistic force density model. However, for the moments the situation is different. If two different functions yield the same value after integration, their weighted integrals are in general not equal. We find for the moments:

$$
\begin{aligned}
\boldsymbol{M}^{\mathrm{L}} &= \hat{f} R^3_{\text{(II)}} \int\limits_{\partial\Omega_{\text{(II)}}} \sin\vartheta' \boldsymbol{e}'_{\varphi} (\boldsymbol{e}'_r \cdot \tilde{\boldsymbol{B}}_{\text{(I)}}) \, \mathrm{d}\tilde{A} \ , \\
\boldsymbol{M}^{\mathrm{A}} &= -\hat{f} R^3_{\text{(II)}} \int\limits_{\Omega_{\text{(II)}}} \boldsymbol{e}'_r \times \Big[\boldsymbol{e}'_z \cdot (\tilde{\nabla} \otimes \tilde{\boldsymbol{B}}_{\text{(I)}}) \Big] \, \mathrm{d}\tilde{V} + \\
&\quad + \hat{f} R^3_{\text{(II)}} \int\limits_{\partial\Omega_{\text{(II)}}} \boldsymbol{e}'_r \times \Big[(\sin\vartheta' \boldsymbol{e}'_{\varphi} \times \tilde{\boldsymbol{B}}_{\text{(I)}} + \cos\vartheta' \tilde{\boldsymbol{B}}_{\text{(I)}}) \Big] \, \mathrm{d}\tilde{A} \ , \\
\boldsymbol{M}^{\mathrm{M}} &= \hat{f} R^3_{\text{(II)}} \int\limits_{\partial\Omega_{\text{(II)}}} \boldsymbol{e}'_r \times \Big[\boldsymbol{e}'_r (\tilde{\boldsymbol{B}}_{\text{(I)}} \cdot \boldsymbol{e}'_z) \Big] \, \mathrm{d}\tilde{A} = \boldsymbol{0} \ , \\
\boldsymbol{M}^{\mathrm{EL}} &= \hat{f} R^3_{\text{(II)}} \int\limits_{\Omega_{\text{(II)}}} \boldsymbol{e}'_r \times \Big[\boldsymbol{e}'_z \cdot (\tilde{\nabla} \otimes \tilde{\boldsymbol{B}}_{\text{(I)}}) \Big] \, \mathrm{d}\tilde{V} .
\end{aligned} \tag{104}
$$

Most interestingly, the version of the MINKOWSKI model yields no moment in any configuration of the two magnets. Therefore, we may conclude that this version of the model is unrealistic. The other integrals, unfortunately, need to be calculated numerically, because the coordinate transformations in Eq. (101) yield complicated expressions for $\tilde{\boldsymbol{B}}_{\text{(I)}}$. As to be expected, all non-vanishing moments point in x-direction

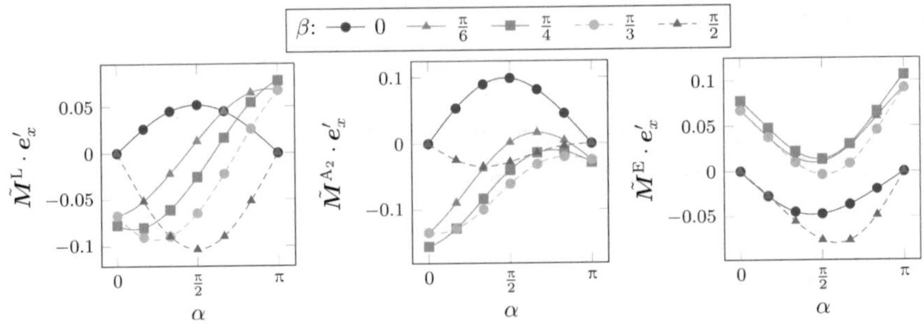

Fig. 5 The x-components of the moments exerted on the magnet (II) due to the magnetic field of the first magnet (I) for different models and for different orientation angles α. Additionally the position angle β is varied

perpendicular to the plane depicted in Fig. 4. The x-components of the resulting moments are shown in Fig. 5.

From the figure it is clear, that the total moment on the second magnet is different for the distinct models. All three models show completely different characteristics. Not only do they predict different scales of the dimensionless moments, but they also predict different qualitative behavior. Hence, by measurements the correct force model for this situation can be found in principle. Of course, performing measurements with spherical magnets is difficult. Therefore, one should analyze cylindrical magnets as in [23]. However, cylindrical magnets render the mathematical analysis more complicated.

7 Summary

In this paper the mathematical tools required for the analysis of open systems for electromagnetic phenomena were presented. Based on these and starting from basic principles MAXWELL's equations were derived for open systems in a spatial description in a rational manner. Since electromagnetism is deeply connected with mechanics, the coupling of these two branches of continuum physics was considered. While in pure electromagnetism there is a consensus on the theory, the situation is different for the coupling of mechanics and electromagnetism, because the forces and energies are not known for ponderable matter. This conundrum is called ABRAHAM–MINKOWKSI controversy and was discussed in detail. Furthermore, one electrostatic and one magnetostatic exemplary problem were analyzed, namely the deformation of a dielectric sphere and the forces and moments between spherical permanent magnets. For both problems it was shown that the prediction of the mechanical behavior strongly depends on the force model used. This in turn has influences on the electromagnetic fields and, in principle, allows by experiment to decide which force model is correct. The answer might depend on the specific material that is used.

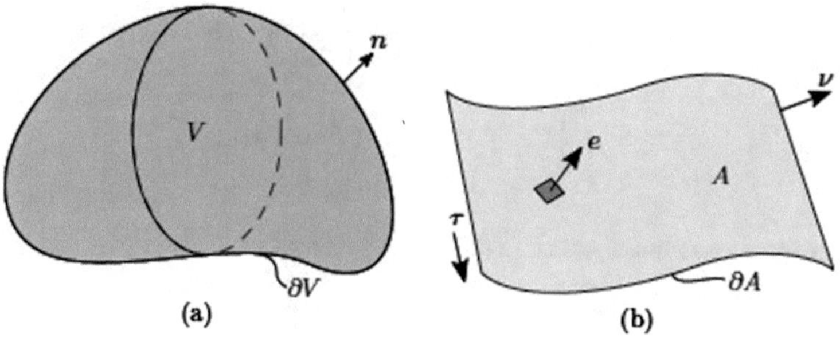

Fig. 6 Depiction of a regular V and a regular surface A

8 Derivation of Integral Theorems with Singular Surfaces

The derivation of generalized integral theorems relies on the classical integral theorems. Consider a regular volume V and a regular surface A as in Fig. 6.

Then, for a smooth field $\boldsymbol{\psi}$ and the domain V in Fig. 6a one has the theorem of GAUSS,

$$\int_V \nabla \oplus \boldsymbol{\psi} \, \mathrm{d}V = \oint_{\partial V} \boldsymbol{n} \oplus \boldsymbol{\psi} \, \mathrm{d}A \,, \tag{105}$$

with arbitrary product $\oplus \in \{\cdot, \times, \otimes\}$, and the REYNOLDS transport theorem

$$\frac{\mathrm{d}}{\mathrm{d}t} \int_V \boldsymbol{\psi} \, \mathrm{d}V = \int_V \frac{\mathrm{d}\boldsymbol{\psi}}{\mathrm{d}t} \, \mathrm{d}V + \int_V (\nabla \cdot \boldsymbol{w}) \boldsymbol{\psi} \, \mathrm{d}V = \int_V \frac{\partial \boldsymbol{\psi}}{\partial t} \, \mathrm{d}V + \int_{\partial V} (\boldsymbol{n} \cdot \boldsymbol{w}) \boldsymbol{\psi} \, \mathrm{d}A \,, \tag{106}$$

where the total time derivative in spatial coordinates is given by

$$\frac{\mathrm{d}\boldsymbol{\psi}(\boldsymbol{x}_{\mathrm{s}}, t)}{\mathrm{d}t} = \frac{\partial \boldsymbol{\psi}}{\partial t} + \boldsymbol{w} \cdot (\nabla \otimes \boldsymbol{\psi}) \,. \tag{107}$$

For a smooth field $\boldsymbol{f}$ and the surface A in Fig. 6b the theorem of KELVIN–STOKES holds,

$$\int_A \boldsymbol{e} \cdot (\nabla \times \boldsymbol{f}) \, \mathrm{d}A = \oint_{\partial A} \boldsymbol{\tau} \cdot \boldsymbol{f} \, \mathrm{d}\ell \,. \tag{108}$$

By noting that $\boldsymbol{\nu} = \boldsymbol{\tau} \times \boldsymbol{e}$ the so-called surface divergence theorem can be derived:

$$
\begin{aligned}
\oint\limits_{\partial A} \boldsymbol{\nu} \cdot \boldsymbol{f} \, \mathrm{d}\ell = \oint\limits_{\partial A} \boldsymbol{\tau} \cdot (\boldsymbol{e} \times \boldsymbol{f}) \, \mathrm{d}\ell = \int\limits_{A} \boldsymbol{e} \cdot (\nabla \times [\boldsymbol{e} \times \boldsymbol{f}]) \, \mathrm{d}A \\
= \int\limits_{A} \big[\nabla_I \cdot \boldsymbol{f} - (\boldsymbol{e} \cdot \boldsymbol{f})(\nabla \cdot \boldsymbol{e}) \big] \, \mathrm{d}A \, ,
\end{aligned} \tag{109}
$$

where the surface nabla operator is defined as

$$
\nabla_I \cdot \boldsymbol{f} = (\mathbf{1} - \boldsymbol{e} \otimes \boldsymbol{e}) \cdot\cdot (\nabla \otimes \boldsymbol{f}) \, . \tag{110}
$$

In there **1** denotes the unit tensor and the double contraction is defined as:

$$
(\boldsymbol{a} \otimes \boldsymbol{b}) \cdot\cdot (\boldsymbol{c} \otimes \boldsymbol{d}) = (\boldsymbol{a} \cdot \boldsymbol{c})(\boldsymbol{b} \cdot \boldsymbol{d}) \, . \tag{111}
$$

Note that $H = -\frac{1}{2}\nabla \cdot \boldsymbol{e}$ the curvature of the surface.

A similar theorem as the REYNOLDS transport theorem can be derived for the temporal change of a flux through a surface A. First, by use of NANSON's formula one has

$$
\begin{aligned}
\frac{\mathrm{d}}{\mathrm{d}t} \int\limits_{A} \boldsymbol{e} \cdot \boldsymbol{f} \, \mathrm{d}A = \int\limits_{A} \boldsymbol{e} \cdot \left[\frac{\mathrm{d}\boldsymbol{f}}{\mathrm{d}t} + (\nabla \cdot \boldsymbol{w}) \boldsymbol{f} - \boldsymbol{f} \cdot (\nabla \otimes \boldsymbol{v}_{\mathrm{s}}) \right] \mathrm{d}A \\
= \int\limits_{A} \boldsymbol{e} \cdot \left[\frac{\partial \boldsymbol{f}}{\partial t} + \boldsymbol{w} \cdot (\nabla \otimes \boldsymbol{f}) + (\nabla \cdot \boldsymbol{w}) \boldsymbol{f} - \boldsymbol{f} \cdot (\nabla \otimes \boldsymbol{w}) \right] \mathrm{d}A \, .
\end{aligned} \tag{112}
$$

Noting that

$$
\nabla \times (\boldsymbol{f} \times \boldsymbol{w}) = (\nabla \cdot \boldsymbol{w}) \boldsymbol{f} + \boldsymbol{w} \cdot (\nabla \otimes \boldsymbol{f}) - (\nabla \cdot \boldsymbol{f}) \boldsymbol{w} - \boldsymbol{f} \cdot (\nabla \otimes \boldsymbol{w}) \tag{113}
$$

the surface transport theorem is compactly written as

$$
\frac{\mathrm{d}}{\mathrm{d}t} \int\limits_{A} \boldsymbol{e} \cdot \boldsymbol{f} \, \mathrm{d}A = \int\limits_{A} \boldsymbol{e} \cdot \left[\frac{\partial \boldsymbol{f}}{\partial t} + (\nabla \cdot \boldsymbol{f}) \boldsymbol{w} + \nabla \times (\boldsymbol{f} \times \boldsymbol{w}) \right] \mathrm{d}A \, . \tag{114}
$$

If the theorem of KELVIN–STOKES is applied, another convenient representation is obtained

$$
\frac{\mathrm{d}}{\mathrm{d}t} \int\limits_{A} \boldsymbol{e} \cdot \boldsymbol{f} \, \mathrm{d}A = \int\limits_{A} \boldsymbol{e} \cdot \left[\frac{\partial \boldsymbol{f}}{\partial t} + (\nabla \cdot \boldsymbol{f}) \boldsymbol{w} \right] \mathrm{d}A + \oint\limits_{\partial S} \boldsymbol{\tau} \cdot (\boldsymbol{f} \times \boldsymbol{w}) \, \mathrm{d}\ell \, . \tag{115}
$$

In order to generalize these theorems to domains of integration cut by singular surfaces or singular lines, the classical theorems are applied to the regular subregions.

When the results are summed up the jump at the singular surface appears. All of the following theorems can be found in [25] or [21], but they are presented here again for the sake of convenience.

8.1 Generalized GAUSS Theorem

Consider a volumetric region Ω cut by a singular surface I as in Fig. 2. Then Eq. (105) can be applied to both regular regions $\Omega^{\pm}$ with boundaries $\partial\Omega^{\pm} = \Gamma^{\pm} \cup I$

$$\int_{\Omega^{\pm}} \nabla \oplus \boldsymbol{\psi}\, \mathrm{d}V = \oint_{\partial\Omega^{\pm}} \boldsymbol{n} \oplus \boldsymbol{\psi}\, \mathrm{d}A = \int_{\Gamma^{\pm}} \boldsymbol{n} \oplus \boldsymbol{\psi}\, \mathrm{d}A + \int_{I} (\mp\boldsymbol{e}) \oplus \boldsymbol{\psi}\, \mathrm{d}A\,. \tag{116}$$

The sum of these equations yields the generalized GAUSS theorem:

$$\int_{\Omega^{-}\cup\Omega^{+}} \nabla \oplus \boldsymbol{\psi}\, \mathrm{d}V = \int_{\Gamma^{-}\cup\Gamma^{+}} \boldsymbol{n} \oplus \boldsymbol{\psi}\, \mathrm{d}A - \int_{I} \boldsymbol{e} \oplus [\![\boldsymbol{\psi}]\!]\, \mathrm{d}A\,, \tag{117}$$

where the jump is defined as $[\![\boldsymbol{\psi}]\!] = \boldsymbol{\psi}^{+} - \boldsymbol{\psi}^{-}$ and $\boldsymbol{\psi}^{\pm}$ are the limits from the respective side.

8.2 Generalized STOKES Theorem

Analogously to the generalized GAUSS theorem, the regular version in Eq. (108) can be applied to the regional subdomains $\Gamma^{\pm}$ in order to obtain the generalized STOKES theorem

$$\int_{\Gamma^{-}\cup\Gamma^{+}} \boldsymbol{e} \cdot (\nabla \times \boldsymbol{\psi})\, \mathrm{d}A = \int_{\ell^{-}\cup\ell^{+}} \boldsymbol{\tau} \cdot \boldsymbol{\psi}\, \mathrm{d}\ell - \int_{\ell_I} \boldsymbol{\tau} \cdot [\![\boldsymbol{\psi}]\!]\, \mathrm{d}\ell\,. \tag{118}$$

8.3 Generalized Volumetric Transport Theorem

Now consider REYNOLDS' transport theorem for volumetric regions. Again, the classical theorem from Eq. (106) can be applied to the regular subsets $\Omega^{\pm}$:

$$\frac{\mathrm{d}}{\mathrm{d}t} \int_{\Omega^{\pm}} \boldsymbol{\psi}\, \mathrm{d}V = \int_{\Omega^{\pm}} \frac{\partial \boldsymbol{\psi}}{\partial t}\, \mathrm{d}V + \int_{\Gamma^{\pm}} (\boldsymbol{w} \cdot \boldsymbol{n})\boldsymbol{\psi}\, \mathrm{d}A + \int_{I} (\mp\boldsymbol{w}_I \cdot \boldsymbol{e})\boldsymbol{\psi}\, \mathrm{d}A \tag{119}$$

and hence for the total region

$$\frac{\mathrm{d}}{\mathrm{d}t}\int\limits_{\Omega^-\cup\Omega^+}\boldsymbol{\psi}\,\mathrm{d}V=\int\limits_{\Omega^-\cup\Omega^+}\frac{\partial\boldsymbol{\psi}}{\partial t}\,\mathrm{d}V+\int\limits_{\Gamma^-\cup\Gamma^+}(\boldsymbol{w}\cdot\boldsymbol{n})\boldsymbol{\psi}\,\mathrm{d}A-\int\limits_{I}(\boldsymbol{w}_I\cdot\boldsymbol{e})[\![\boldsymbol{\psi}]\!]\,\mathrm{d}A\,. \tag{120}$$

Furthermore, by means of the generalized GAUSS theorem one has

$$\int\limits_{\Gamma^-\cup\Gamma^+}(\boldsymbol{w}\cdot\boldsymbol{n})\boldsymbol{\psi}\,\mathrm{d}A=\int\limits_{\Omega^-\cup\Omega^+}\nabla\cdot(\boldsymbol{w}\otimes\boldsymbol{\psi})\,\mathrm{d}V+\int\limits_{I}\boldsymbol{e}\cdot[\![\boldsymbol{w}\otimes\boldsymbol{\psi}]\!]\,\mathrm{d}A \tag{121}$$

and therefore

$$\frac{\mathrm{d}}{\mathrm{d}t}\int\limits_{\Omega^-\cup\Omega^+}\boldsymbol{\psi}\,\mathrm{d}V=\int\limits_{\Omega^-\cup\Omega^+}\frac{\partial\boldsymbol{\psi}}{\partial t}\,\mathrm{d}V+\int\limits_{\Omega^-\cup\Omega^+}\nabla\cdot(\boldsymbol{w}\otimes\boldsymbol{\psi})\,\mathrm{d}V+\int\limits_{I}\boldsymbol{e}\cdot[\![(\boldsymbol{w}-\boldsymbol{w}_I)\otimes\boldsymbol{\psi}]\!]\,\mathrm{d}A\,. \tag{122}$$

8.4 *Generalized Surface Transport Theorem*

Analogously to the volumetric version one obtains the generalized surface transport theorem. For a singular surface I moving at the velocity $\boldsymbol{w}_I$ and a scalar density ψ_I the theorem follows as

$$\frac{\mathrm{d}}{\mathrm{d}t}\int\limits_{I}\psi_I\,\mathrm{d}A=\int\limits_{I}\left(\frac{\mathrm{d}_I\psi_I}{\mathrm{d}t}+(\nabla_I\cdot\boldsymbol{w}_I)\psi_I\right)\mathrm{d}A\,, \tag{123}$$

where ${}^{\mathrm{d}_I}/_t$ is the total surface derivative. It is important to note that ψ_I lives on the possibly moving surface, see for example [9]. The total surface derivatives need to be treated with care if material surfaces are considered and additionally material points are allowed to move relatively on the surface. If, however, a general non-material surface is considered or a material one without relative motion of the particles on the surface, then the surface total time derivative is given by

$$\frac{\mathrm{d}_I\psi_I}{\mathrm{d}t}=\frac{\partial\psi_I}{\partial t}+\boldsymbol{w}_I\cdot(\nabla_I\psi_I)\,. \tag{124}$$

This is to say that the surface has no intrinsic particles that are allowed to move freely.

8.5 Generalized Surface Flux Theorem

Consider the surface S is cut by a singular line ℓ_I from Fig. 1. Then, the surface flux theorem from Eq. (115) can be applied to both regular subsets $S^\pm$

$$\frac{\mathrm{d}}{\mathrm{d}t}\int_{S^\pm} \boldsymbol{\psi}\cdot\boldsymbol{e}\,\mathrm{d}A = \int_{S^\pm} \boldsymbol{e}\cdot\left[\frac{\partial\boldsymbol{\psi}}{\partial t} + (\nabla\cdot\boldsymbol{\psi})\boldsymbol{w}\right]\mathrm{d}A + \\ + \int_{\ell^\pm} \boldsymbol{\tau}\cdot(\boldsymbol{\psi}\times\boldsymbol{w})\,\mathrm{d}\ell + \int_{\ell_I} (\mp\boldsymbol{\tau}_I)\cdot(\boldsymbol{\psi}\times\boldsymbol{w}_I)\,\mathrm{d}\ell\,. \tag{125}$$

Thus the surface flux transport theorem incorporating singular surfaces reads

$$\frac{\mathrm{d}}{\mathrm{d}t}\int_{S^-\cup S^+} \boldsymbol{e}\cdot\boldsymbol{\psi}\,\mathrm{d}A = \int_{S^-\cup S^+} \boldsymbol{e}\cdot\left[\frac{\partial\boldsymbol{\psi}}{\partial t} + (\nabla\cdot\boldsymbol{\psi})\boldsymbol{w}\right]\mathrm{d}A + \\ + \int_{\ell^-\cup\ell^+} \boldsymbol{\tau}\cdot(\boldsymbol{\psi}\times\boldsymbol{w})\,\mathrm{d}\ell - \int_{\ell_I} \boldsymbol{\tau}_I\cdot([\![\boldsymbol{\psi}]\!]\times\boldsymbol{w}_I)\,\mathrm{d}\ell\,. \tag{126}$$

9 Localization Theorem

In oder to find the localized balance law in regular points, the global balance is considered of a regular surface region in which all singular contributions vanish. Then, the total domain is given by $S = S^-\cup S^+$ and $\partial S = \ell^-\cup\ell^+$. Hence, the surface transport theorem together with STOKES's theorem reduce to

$$\frac{\mathrm{d}}{\mathrm{d}t}\int_S \boldsymbol{e}\cdot\boldsymbol{f}\,\mathrm{d}A = \int_S \boldsymbol{e}\cdot\left[\frac{\partial\boldsymbol{f}}{\partial t} + (\nabla\cdot\boldsymbol{f})\boldsymbol{w} - \nabla\times(\boldsymbol{w}\times\boldsymbol{f})\right]\mathrm{d}A\,. \tag{127}$$

Furthermore, the right hand side of the balance follows as

$$\frac{\mathrm{d}}{\mathrm{d}t}\int_S \boldsymbol{e}\cdot\boldsymbol{f}\,\mathrm{d}A = \int_S \boldsymbol{e}\cdot\left[\boldsymbol{s} - \nabla\times\big(\boldsymbol{\phi} - (\boldsymbol{v}-\boldsymbol{w})\times\boldsymbol{f}\big)\right]\mathrm{d}A\,. \tag{128}$$

By combining both equations one has

$$0 = \int_S \boldsymbol{e}\cdot\left[\frac{\partial\boldsymbol{f}}{\partial t} + (\nabla\cdot\boldsymbol{f})\boldsymbol{w} + \nabla\times\big(\boldsymbol{\phi} - \boldsymbol{v}\times\boldsymbol{f}\big) - \boldsymbol{s}\right]\mathrm{d}A\,. \tag{129}$$

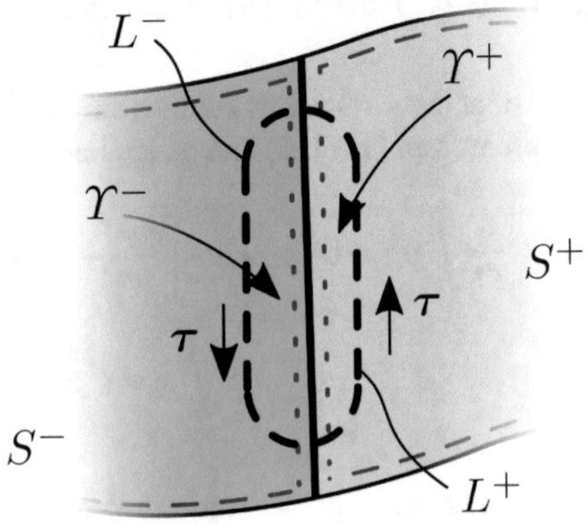

Fig. 7 Special control domain constructed for the localization in singular points on a line

Since the integration domain is arbitrary and the integrand is continuous, the local equation follows as

$$\frac{\partial \boldsymbol{f}}{\partial t} + (\nabla \cdot \boldsymbol{f})\boldsymbol{w} + \nabla \times \left(\boldsymbol{\phi} - \boldsymbol{v} \times \boldsymbol{f}\right) = \boldsymbol{s} \,. \tag{130}$$

It is interesting to note, that this equation contains both the mapping velocity, $\boldsymbol{w}$, as well as the material velocity $\boldsymbol{v}$. For fields $\boldsymbol{f}$ that are not transported by matter and have no supply $\boldsymbol{s}$ one simply has

$$\frac{\partial \boldsymbol{f}}{\partial t} + (\nabla \cdot \boldsymbol{f})\boldsymbol{w} + \nabla \times \boldsymbol{\phi} = \boldsymbol{0} \,. \tag{131}$$

For singular points the situation is more complicated. Consider a tight noose on S around a section of ℓ_I with the area $\Upsilon = \Upsilon^- \cup \Upsilon^+$ and the closed boundary line $\partial\Upsilon = L^- \cup L^+$, which is depicted in Fig. 7.

For this control surface the application of the transport theorem reads

$$\begin{aligned} \frac{\mathrm{d}}{\mathrm{d}t} \int\limits_{\Upsilon^- \cup \Upsilon^+} \boldsymbol{e} \cdot \boldsymbol{f} \,\mathrm{d}A = \int\limits_{\Upsilon^- \cup \Upsilon^+} \boldsymbol{e} \cdot \left[\frac{\partial \boldsymbol{f}}{\partial t} + (\nabla \cdot \boldsymbol{f})\boldsymbol{w} \right] \mathrm{d}A \\ - \int\limits_{L^- \cup L^+} \boldsymbol{\tau} \cdot (\boldsymbol{w} \times \boldsymbol{f}) \,\mathrm{d}\ell + \int\limits_{\ell_I} \boldsymbol{\tau}_I \cdot (\boldsymbol{w}_I \times [\![\boldsymbol{f}]\!]) \,\mathrm{d}\ell \,. \end{aligned} \tag{132}$$

Furthermore, if the generalized version of STOKES' theorem is applied to the right hand side of the balance one has

$$\frac{\mathrm{d}}{\mathrm{d}t}\int_{\Upsilon^-\cup\Upsilon^+} \boldsymbol{e}\cdot\boldsymbol{f}\,\mathrm{d}A = \int_{\Upsilon^-\cup\Upsilon^+} \boldsymbol{e}\cdot\boldsymbol{s}\,\mathrm{d}A - \int_{L^-\cup L^+} \boldsymbol{\tau}\cdot\big(\boldsymbol{\phi}-(\boldsymbol{v}-\boldsymbol{w})\times\boldsymbol{f}\big)\,\mathrm{d}\ell + \int_{\ell_I} \boldsymbol{v}_I\cdot\boldsymbol{s}_I\,\mathrm{d}\ell\,. \tag{133}$$

The combination of Eq. (132) and (133) gives

$$0 = \int_{\Upsilon^-\cup\Upsilon^+} \boldsymbol{e}\cdot\left[\frac{\partial\boldsymbol{f}}{\partial t}+(\nabla\cdot\boldsymbol{f})\boldsymbol{w}-\boldsymbol{s}\right]\mathrm{d}A + \int_{L^-\cup L^+} \boldsymbol{\tau}\cdot\big(\boldsymbol{\phi}-\boldsymbol{v}\times\boldsymbol{f}\big)\,\mathrm{d}\ell + \int_{\ell_I}\big[\boldsymbol{\tau}_I\cdot(\boldsymbol{w}_I\times[\![\boldsymbol{f}]\!])-\boldsymbol{v}_I\cdot\boldsymbol{s}_I\big]\,\mathrm{d}\ell\,. \tag{134}$$

In the limit, the tight noose contains no area, i.e., $\Upsilon^\pm \to 0$, the boundary parts $L^\pm$ reduce to I and the tangent is given by $\boldsymbol{\tau} = \pm\boldsymbol{\tau}_I$

$$\int_{L^-\cup L^+} \boldsymbol{\tau}\cdot(\boldsymbol{w}\times\boldsymbol{f})\,\mathrm{d}\ell \to \int_{\ell_I} \boldsymbol{\tau}_I\cdot[\![\boldsymbol{w}\times\boldsymbol{f}]\!]\,\mathrm{d}\ell\,. \tag{135}$$

Therefore, one has

$$0 = \int_{\ell_I}\big[\boldsymbol{\tau}_I\cdot([\![\boldsymbol{\phi}+\boldsymbol{f}\times(\boldsymbol{v}-\boldsymbol{w}_I)]\!])-\boldsymbol{v}_I\cdot\boldsymbol{s}_I\big]\,\mathrm{d}\ell\,. \tag{136}$$

Since the integration domain is arbitrary the following local form is obtained

$$\boldsymbol{\tau}_I\cdot[\![\boldsymbol{\phi}+\boldsymbol{f}\times(\boldsymbol{v}-\boldsymbol{w}_I)]\!] = \boldsymbol{v}_I\cdot\boldsymbol{s}_I\,. \tag{137}$$

By noting that $\boldsymbol{v}_I = \boldsymbol{e}_I\times\boldsymbol{\tau}_I$ this equation can be rewritten

$$\boldsymbol{\tau}_I\cdot([\![\boldsymbol{\phi}+\boldsymbol{f}\times(\boldsymbol{v}-\boldsymbol{w}_I)]\!]-\boldsymbol{s}_I\times\boldsymbol{e}_I) = 0\,. \tag{138}$$

Since the bracket is parallel to the normal $\boldsymbol{e}_I$ one may multiply the bracket with $\boldsymbol{e}_I\times$ to find

$$\boldsymbol{e}_I\times[\![\boldsymbol{\phi}+\boldsymbol{f}\times(\boldsymbol{v}-\boldsymbol{w}_I)]\!] = \boldsymbol{s}_I\,, \tag{139}$$

where $\boldsymbol{e}_I\cdot\boldsymbol{s}_I$ is assumed.

References

1. Abraham, M.: Zur Elektrodynamik bewegter Körper. In: Rendiconti del Circolo Matematico di Palermo (1884-1940), **28**(1), 1–28 (1909)

2. Barnett, S.M., Loudon, R.: On the electromagnetic force on a dielectric medium. J. Phys. B: At. Mol. Opt. Phys. **39**(15), 671–684 (2006)
3. Bethune-Waddell, M., Chau, K.J.: Simulations of radiation pressure experiments narrow down the energy and momentum of light in matter. Rep. Prog. Phys. **78**(12) (2015)
4. Chang, J.S., Kelly, A.J., Crowley, J.M.: Handbook of Electrostatic Processes. Taylor & Francis (1995). ISBN: 9781420066166
5. Datsyuk, V.V., Pavlyniuk, O.R.: Maxwell stress on a small dielectric sphere in a dielectric. Phys. Rev. A **91**(2) (2015)
6. Feynman, R.P., Leighton, R.B., Sands, M.L.: The Feynman Lectures on Physics. **2.** The Electromagnetic Field. Addison-Wesley (1965)
7. Fitzpatrick, R.: Classical Electromagnetism (2006)
8. Fitzpatrick, R.: Maxwell's Equations and the Principles of Electromagnetism. Infinity Science Press (2008)
9. Fosdick, R., Tang, H.: Surface transport in continuum mechanics. Math. Mech. Solids **14**(6), 587–598 (2008)
10. Hiramatsu, Y., Oka, Y.: Determination of the tensile strength of rock by a compression test of an irregular test piece. Int. J. Rock Mech. Min. Sci. Geomech. Abst. **3**(2), 89–90 (1966). ISSN: 0148-9062
11. Hutter, K., Jöhnk, K.: Continuum Methods of Physical Modeling. Springer, Berlin (2004)
12. Hutter, K., Ven, A.A.F., Ursescu, A.: Electromagnetic Field Matter Interactions in Thermoelastic Solids and Viscous Fluids. Springer (2006)
13. Ivanova, E., Vilchevskaya, E. and Müller, W.H.: A Study of Objective Time Derivatives in Material and Spatial Description. In: Altenbach, H., Goldstein, R., Murashkin, E. (eds.) Mechanics for Materials and Technologies. Advanced Structured Materials, vol. 46, pp. 195–229. Springer, Cham (2017)
14. Jackson, J.D.: Classical electrodynamics, 3rd edn. Wiley (1999)
15. Kinsler, P., Favaro, A., McCall, M.W.: Four poynting theorems. Eur. J. Phys. **30**(5), 983–993 (2009)
16. Kovetz, A.: Electromagnetic Theory. Oxford University Press, Oxford (2000)
17. Liu, I.: Continuum Mechanics. Springer, Berlin (2002)
18. Lüders, K., Pohl, R.O. (eds.): Pohl's Introduction to Physics. Springer International Publishing (2018)
19. Mansuripur, M.: Resolution of the Abraham-Minkowski controversy. Opt. Commun. **283**(10), 1997–2005 (2010)
20. Minkowski, H.: Die Grundgleichungen für die elektromagnetischen Vorgänge in bewegten Körpern. Math. Ann. **68**(4), 472–525 (1910)
21. Müller, I.: Thermodynamics. Interaction of Mechanics and Mathematics Series. Pitman (1985). ISBN: 9780273085775
22. Reich, F.A.: Coupling of continuum mechanics and electrodynamics: an investigation of electromagnetic force models by means of experiments and selected problems. Doctoral Thesis. Berlin: Technische Universität Berlin (2017)
23. Reich, F.A. and x Müller, F.A.: Examination of electromagnetic powers with the example of a Faraday disc dynamo. Contin. Mech. Thermody. **30**(4), 861–877 (2018)
24. Reich, F.A., Stahn, O. and Müller, W.H.: A review of electrodynamics and its coupling with classical balance equations. In: APM–Proceedings of XLIII International Summer School, pp. 367–376 (2015)
25. Slattery, J.C., Sagis, L., Oh, E.S.: Interfacial Transport Phenomena. Springer, US (2007)
26. Truesdell, C.A. and Toupin, R.: The classical field theories. In: Handbuch der Physik, Bd. III/1. Berlin: Springer, 226–793; appendix, 794–858 (1960)

Printed in the United States
by Baker & Taylor Publisher Services